Air Conditioning: Home and Commercial

by Edwin P. Anderson
revised by Rex Miller

An Audel® Book

Macmillan Publishing Company
New York

Collier Macmillan Publishers
London

FOURTH EDITION

Copyright © 1986, 1990 by Macmillan Publishing Company, a division
of Macmillan, Inc.
Copyright © 1984 by The Bobbs-Merrill Co., Inc.
Copyright © 1978 by Howard W. Sams & Co., Inc.

All rights reserved. No part of this book may be reproduced or
transmitted in any form or by any means, electronic or mechanical,
including photocopying, recording or by any information
storage and retrieval system, without permission in writing
from the Publisher.

While every precaution has been taken in the preparation of
this book, the Publisher assumes no responsibility for errors or
omissions. Neither is any liability assumed for damages resulting
from the use of the information contained herein.

Macmillan Publishing Company
866 Third Avenue, New York, NY 10022
Collier Macmillan Canada, Inc.

Production services by the Walsh Group, Yarmouth, ME

Library of Congress Cataloging-in-Publication Data

Anderson, Edwin P., 1895-
  Air conditioning : home and commercial / by Edwin P. Anderson; revised by Rex
Miller. — 4th ed.
    p.   cm.
  At head of title: Audel.
  ISBN 0-02-584885-2
  1. Air conditioning.   I. Miller, Rex, 1929-   II. Title.
TH7687.A73   1990   697.9'3 — dc20   89-71316   CIP

Macmillan books are available at special discounts for bulk purchases
for sales promotions, premiums, fund-raising, or educational use.
For details, contact:

    Special Sales Director
    Macmillan Publishing Company
    866 Third Avenue
    New York, NY 10022

10 9 8 7 6 5 4 3 2 1

Printed in the United States of America

Foreword

This book is designed to fill the need for a concise, elementary text for those who are in need of acquiring practical information on the installation, operation, and servicing of air conditioning equipment in the home as well as in commercial and industrial applications. The subject is presented in a simple nontechnical language, yet the treatment is sufficiently complete to enable the reader to diagnose and correct troubles that may occur from time to time.

The material in this book has been carefully organized to provide a practical understanding of the construction, operation, and basic fundamentals so important in diagnosing operating faults in an air conditioning system. A complete description of the purpose and function of each operating component, along with a full coverage of refrigerants used in various applications, is included.

Each chapter is fully illustrated. Numerous troubleshooting charts will assist the operator or serviceman in the proper diagnosis and prompt repair procedure to use when a malfunction occurs.

REX MILLER
EDWIN P. ANDERSON

Contents

CHAPTER 1

AIR-CONDITIONING FUNDAMENTALS 9
Properties of air — air circulation — basic information — measurements and measuring devices — summary — review questions.

CHAPTER 2

PSYCHROMETRY 25
Air and water vapor mixtures — humidifying and dehumidifying — relative humidity — psychrometric charts — comfort chart — summary — review questions.

CHAPTER 3

HEAT LEAKAGE 41
Summary — review questions.

CHAPTER 4

VENTILATION REQUIREMENTS 65
Air leakage — natural ventilation — roof ventilators — fresh-air requirements — mechanical ventilation — air filtration — air filter classification — filter installation — humidity-control methods — humidifiers — dehumidifiers — air duct systems — heat gains in ducts — resistance losses in duct systems — fans and blowers — summary — review questions.

CHAPTER 5

ROOM AIR CONDITIONERS............................117
Operation — cooling capacity — capacity requirements — installation methods — double-hung window installation — casement window installation — outside support bracket — console-type installation — electrical system — electrical components — fan motors — thermostats — unit control switch — service operations — dismantling window air conditioners — electrical testing — compressor and motor — oiling of motors — leaks in system — filters — interior cleaning — winter care — refrigerant-system service — **troubleshooting guide** — summary — review questions.

CHAPTER 6

REFRIGERANTS..169
Desirable properties — classifications — Freon refrigerants — properties of Freons — operating pressures — pressure-temperature chart — refrigerant characteristics — simple refrigeration system — handling refrigerants — storing and handling refrigerant cylinders — cylinder capacity — first aid — summary — review questions.

CHAPTER 7

COMPRESSORS FOR AIR CONDITIONERS 197
Refrigeration cycle — refrigerants — compressors — rotary compressors — centrifugal compressors — installation and maintenance — service guide — **air conditioning troubleshooting guide** — summary — review questions.

CHAPTER 8

CONDENSING EQUIPMENT FOR AIR CONDITIONING 243
Classification — cooling-water circuits — double-tube-type condensers — shell-and-coil condensers — evaporative condensers — head-pressure control methods — condenser maintenance — summary — review questions.

CHAPTER 9

EVAPORATORS FOR AIR CONDITIONERS 269
Refrigerant control methods — evaporator calculations — evaporator maintenance — accessory equipment — summary — review questions.

CHAPTER 10

WATER-COOLING SYSTEMS 305
Water towers — cooling ponds — spray-cooling ponds — new developments — cooling-unit piping — liquid lines — suction lines — discharge lines — refrigerant-piping arrangement — summary — review questions.

CHAPTER 11

AIR-CONDITIONING CONTROL METHODS 325
Basic control types — limit controls — control valves, dampers, and relays — refrigerant control devices — summary — review questions.

CHAPTER 12

WEATHER DATA AND DESIGN CONDITIONS......................351
Indoor design conditions — outdoor design conditions — interpolation between stations — weather-oriented design factors — summary — review questions.

CHAPTER 13

YEAR-ROUND CENTRAL AIR CONDITIONING 371
Central system features — absorption-type air conditioning — reverse-cycle air conditioning (heat pumps) — **reversing-valve troubleshooting guide** — nonportable air conditioners — suspended-type air conditioners — **troubleshooting guide** — **self-contained air conditioner troubleshooting guide** — summary — review questions.

CHAPTER 14

AUTOMOBILE AIR CONDITIONING 411
Air-conditioning system operation — component description — service and maintenance — installation procedure — **troubleshooting guide** — **automobile air conditioning troubleshooting guide** — summary — review questions.

CHAPTER 15

MOTORS AND MOTOR CONTROLS 445
Motor voltage — polyphase motors — single-phase motors — ac motor control — single-phase motor control — compressor motor controls — belt drives — motor maintenance — **troubleshooting guide** — **electrical system troubleshooting guide** — summary — review questions.

CHAPTER 16

MAINTENANCE 481
General procedure — head pressures — suction pressure — back pressure — refrigerant charge — compressor service valves — gages — adding refrigerant — removing refrigerant — adding oil — removing oil — system pumpdown — purging system of noncondensables — leak test — evacuating system — checking compressor valves — **troubleshooting guide** — summary — review questions.

GLOSSARY 507

INDEX 539

CHAPTER 1

Air-Conditioning Fundamentals

Air conditioning may be defined as the simultaneous control of all, or at least the first three, of those factors affecting both the physical and chemical conditions of the atmosphere within any structure. These factors include *temperature, humidity, motion, distribution, dust, bacteria, odors,* and *toxic gases,* most of which affect human health and comfort to a greater or lesser degree.

Air that has been properly conditioned has had one or a combination of the foregoing processes performed on it. For example: it has been heated or cooled; it has had moisture removed from it (dehumidified); it has been placed in motion by means of fans or other apparatus; and it has been filtered and cleaned. These processes may be placed in the following order for ready reference:

Air Conditioning

1. Heating
2. Cooling
3. Humidifying
4. Dehumidifying
5. Circulating
6. Cleaning and filtering

The impression prevailing in many instances is that refrigerating or heating equipment cools or heats a room. This is only partly true since all the work performed by the equipment is on the *air* within the room, not on the room itself. In this connection, it is well to remember that air is only a vehicle or conveyance used to transport heat and moisture from one point to another. Air is a tangible item, and every cubic foot of air surrounding a person has a certain weight, depending on its temperature, the amount of moisture it is carrying, and its altitude above sea level.

PROPERTIES OF AIR

Air is a mixture made up primarily of two gases, being approximately 23 parts oxygen and 77 parts nitrogen by weight. There are still other gases in air, such as carbon dioxide, carbon monoxide, ozone, neon (in small quantities), and certain gases that are of no particular interest in the field of air conditioning.

Ozone is produced by sparks around electrical equipment and by the discharge of atmospheric electricity or lightning. *Neon* is a gas in its normal form and is used in signs for advertising. *Carbon monoxide* is not present in the atmosphere except in congested motor traffic. It is dangerous, being produced by the incomplete combustion of carbon. It is also given off by stoves, furnaces and cigarettes. Air containing carbon monoxide in excess of one-tenth of 1 percent is fatal to human beings.

Oxygen, the most important constituent of air, constitutes about one-fourth of the air by weight and one-fifth of the air by volume, and on it depends the existence of all animal life. *Nitrogen* is a relatively inert gas whose principal function is to dilute oxygen.

AIR CIRCULATION

Air-conditioning equipment must circulate a sufficient volume of air at all times for two main reasons:

1. The air must be constantly moving in order to carry away the moisture and heat immediately surrounding the body. If this is not done, the occupants soon become uncomfortable even though the relative humidity of the room as a whole is comparatively low.
2. The air must be constantly drawn into the conditioner and passed out over the cool evaporator in order that the moisture which it absorbs from the room may be condensed and eliminated through the drain.

Although the movement of a sufficient volume of air at all times is most essential, direct drafts must be prevented. The condensation of moisture on the evaporator surface during summer operation produces a measure of cleaning because this moisture absorbs a considerable amount of impurities from the air passing over the moist evaporator surfaces. As the condensation of moisture continues, it drips off the evaporator and is carried off to the drain, taking the impurities with it. For very dirty air, special provision must be made by installing air filters.

Cleaning and Filtering

There are numerous air cleaning and filtering devices on the market. Such devices serve to eliminate particles carried in the air that are detrimental to health and comfort and that cause property damage. These may be classified as *dust, fumes,* and *smoke*. Dust and fumes will settle in still air, whereas smoke is actuated by motion rather than by gravity and, if not removed, will remain in motion in the air.

Air washing is effective in removing dust and such fumes and smoke as are soluble in water, but carbons, soot, and similar substances are not removed by this method of cleaning. To make it possible to cleanse air of these substances, dry and viscous filters have been developed. To fulfill the essential requirements of clean air, an air cleaner should have the following qualifications:

Air Conditioning

1. It should be efficient in the removal of harmful and objectionable impurities in the air, such as dust, dirt, pollens, bacteria, etc.
2. It should be efficient over a considerable range of air velocities.
3. It should have a large dust-holding capacity without excessive increase of resistance.
4. It should be easy to clean and handle or be able to clean itself automatically.
5. It should leave the air passage through the filter or cleaner free from entrained moisture or charging liquids used in the cleaner.

BASIC INFORMATION

In order to obtain a clear concept in regard to the functioning of air-conditioning systems, it is necessary to understand the physical and thermal properties underlying the production of artificial cold. Since air conditioning deals largely with the problem of removal of heat from a room or space, the following definitions should be understood. First, it should be noted that heat is an active form of energy, much the same as mechanical and electrical energy. Heat may be transferred by three methods, namely, conduction, convection, and radiation.

Conduction is heat transfer that takes place chiefly in solids wherein the heat is passed from one molecule to another without any noticeable movement of the molecules.

Convection is heat transfer that takes place in liquids and gases where the molecules carry the heat from one point to another.

Radiation is heat transfer in wave motion, such as light and radio waves, that takes place through a transparent medium without affecting that medium's temperature. An illustration of this is the sun's rays passing through air. The air temperature is noticeably affected. Radiant heat is not apparent until it strikes an opaque surface where it is absorbed and manifests itself in a temperature rise.

Sensible Heat

Sensible heat is that form of heat which causes a change in the temperature of a substance and can be measured by a thermometer. Thus, for example, when the temperature of water is raised from 32 to 212°F, an increase in the sensible-heat content is taking place.

Specific Heat

The Btu required to raise the temperature of one pound of a substance one degree Fahrenheit is termed its *specific heat*. By definition, the specific heat of water is 1.00, but the amount of heat required to raise the temperature of various substances through a given temperature range will vary. Since water has a very large heat capacity, it has been taken as a standard; and since one pound of water requires one Btu to raise its temperature one degree Fahrenheit, its rating on the specific heat scale is 1.00. Iron has a lower specific heat — its average rating is 0.130; ice is 0.504, and air is 0.238. The more water an object contains, as in the case of fresh food or air, the higher the specific heat.

Latent Heat

Latent heat, meaning hidden heat, is that form of heat which causes a substance to change its physical state from a solid to liquid, a liquid to vapor, or vice versa. For example, when a liquid is evaporated to a gas, the change of physical state is always accompanied by the absorption of heat. Evaporation has a cooling effect on the surroundings of the liquid since the liquid obtains from its surroundings the necessary heat to change its molecular structure. This action takes place in the evaporator of an air-conditioning system. Any liquid tends to saturate the surrounding space with its vapor. This property of liquids is an important element in all air-conditioning work.

On the other hand, when a gas is condensed to a liquid, the change of physical state is always accompanied by the giving up of heat. This action takes place in the condensing unit of an air-conditioning system due to the mechanical work exerted on the gas by the compressor.

Latent Heat of Fusion

The change of a substance from a solid to a liquid or from a liquid to a solid involves the latent heat of fusion. One pound of water at a temperature of 32°F requires the extraction of 144 Btu to cause it to freeze into solid ice at 32°F. Every solid substance has a latent-heat value in varying degrees, and that amount required to convert it, or bring about a change of state, is termed the *latent heat of fusion.* This heat, assimilated or extracted, as the case may be, is not measurable with a thermometer because the heat units are absorbed or expanded in intermolecular work, separating the molecules from their attractive forces so that a change of state is effected.

Latent Heat of Evaporation

The change of a substance from a liquid to a vapor or from a vapor back to a liquid involves the *latent heat of evaporation.* Careful measurements have determined that the conversion of one pound of pure water at 212°F to steam at 212°F requires 970 Btu when carried out at the normal pressure of the atmosphere encountered at sea level. If heat is added and a count is kept of the Btu expended, it will be found that when all the water has been changed to steam, 970 heat units will have been used. The further addition of heat would serve only to heat the steam, such as would be possible if it had been trapped or the experiment performed in a closed vessel so that heat could be applied to it.

Superheat

Superheat is a term used frequently, especially for refrigerant control adjustment. Superheat is sensible heat absorbed by a vapor or gas not in contact with its liquid, and consequently it does not follow the temperature-pressure relationship. Therefore, superheat is sensible heat absorbed by the vapor raising the temperature of the vapor or gas without an appreciable change in pressure.

A gas is usually considered as a vapor in a highly superheated state or as a vapor not near its condensing point. Water in the air that is close to the condensing point is termed *water vapor*. Inasmuch as superheat is sensible heat, its effect can be measured with a thermometer and is merely the temperature rise in degrees Fahrenheit. Therefore, a 10°F superheat means a vapor that has absorbed sufficient heat to raise the vapor temperature 10°F above the temperature of the vaporizing liquid.

Pressure-Temperature Relationship

Extensive investigation of gases and their behavior has shown that a given weight expands or contracts uniformly $1/459$ of its original volume for each degree it is raised or lowered in temperature above or below 0°F, provided the pressure on the gas remains constant. This fact is known as the *law of Charles*. Following this same reasoning, assume −459°F as absolute zero.

Actually this temperature or condition has never been attained. The *law of conservation of matter* states that matter can be neither created nor destroyed, although it can be changed from one form into another. Temperature within a few degrees of absolute zero has been reached by liquefying oxygen, nitrogen, and hydrogen, but these, like other gases, change their physical state from gas into liquid and fail to disappear entirely at these low temperatures. The fact that absolute zero has never been reached is also explained by another law, known as the *law of conservation of energy*. It has already been explained that heat is a form of energy. This law states that energy can be neither created nor destroyed, although it can be changed from one form into another.

Having considered the effect of temperature on a gas, the next step is the effect of pressure on gases to aid the study of refrigeration. In 1662, Robert Boyle announced a simple relation existing between the volume of a gas and the pressure applied to it, which has since become known to scientists as *Boyle's law: At constant temperature, the volume of a given weight of gas varies inversely as the pressure to which it is subjected. The more pressure applied*

AIR CONDITIONING

to a gas, the smaller its volume becomes if the temperature remains the same; likewise, if the pressure is released or reduced, the volume of the gas increases. Mathematically, this might be expressed:

$$P \times V = p \times v$$

where P = pressure on the gas at volume V
p = pressure on the same weight of gas at volume v

Boyle's law has been found to be only approximately true, especially for the refrigerant gases, which are more easily liquefied. The variations from the law are greater as the point of liquefaction or condensing of any gas is reached, although the material movement of air is determined by this law. It will be found that if the temperature is held constant and sufficient pressure is applied to a given weight of gas, it will change from the gaseous state into the liquid state. The point at which this change of state takes place is known as the *point of liquefaction* or *condensing*.

It should now be evident that there is some kind of definite relationship existing between the pressure, temperature, and volume at which a given weight of gas may exist. There is such a relationship, and it is used extensively in scientific work. This relationship is known as the *combined law of Boyle and Charles* and may be expressed mathematically as follows:

$$\frac{P \times V}{T} = \frac{p \times v}{t}$$

where P and p are expressed in the absolute pressure scale in pounds per square foot; V and v are expressed in cubic feet; and T and t are expressed in degrees on the absolute-temperature scale.

When the pressure, temperature, or volume of a gas is varied, a new set of conditions is created under which a given weight of gas exists in accordance with the preceding mathematical equation. If a gas is raised to a certain temperature (which varies with each individual gas), no matter how much pressure is applied to it, it will be found impossible to condense. This temperature is known as the *critical temperature*. The pressure corresponding to the critical temperature is termed the *critical pressure*. Above the critical points, it is impossible to vaporize or condense a substance.

AIR-CONDITIONING FUNDAMENTALS

MEASUREMENTS AND MEASURING DEVICES

The instrument commonly used for measuring temperature is known as the *thermometer*, which operates on the principle of the expansion and contraction of liquids (and solids) under varying intensities of heat. The ordinary thermometer charged with mercury operates with a fair degree of accuracy over a wide range. It becomes useless, however, where temperatures below $-38°F$ (38 degrees below zero Fahrenheit) are to be indicated because mercury freezes at this point; and another liquid, such as alcohol (usually colored for easy observation), must be substituted. The upper range for mercurial thermometers is quite high, about $900°F$, so it is at once apparent that for ordinary service and general use the mercury thermometer is usually applicable.

Thermometer

In operation, the thermometer depends on the effect of heat on the main body of mercury or alcohol that expands or contracts in a bulb or reservoir. This action raises or lowers the height of the liquid in the capillary tube forming the thermometer stem. Several thermometer scales are in existence and are used in various countries. The *English*, or *Fahrenheit*, scale is commonly used in the United States, the *Celsius* in France, and the *Reamur* in Germany. Since the Celsius scale is so widely used in scientific work in all countries, an illustration of the comparison of thermometers is shown in Fig. 1-1 so that any one scale may be converted to another. The freezing point on the Fahrenheit scale is fixed at $32°$ and on both the Celsius and Reamur scales it is placed at $0°$. On the Fahrenheit scale, the boiling point of pure water under the normal pressure encountered at sea level is $212°$; on the Celsius scale, it is $100°$; and in the case of the Reamur, it is $60°$.

If it is desired to convert $50°$ Celsius to Fahrenheit, the method would be in accordance with the formula:

$$°F = {}^9/_5 \, °C + 32$$

Therefore

$$50 \times {}^9/_5 + 32 = 122°F$$

AIR CONDITIONING

or, using a calculator, it becomes

$$50 \times 1.8 + 32 = 122°F \text{ (for } 9/5 = 1.8)$$

To convert 50° F to Celsius:

$$°F = 5/9 \times (°C - 32), \text{ or } 0.5555555 \times °C - 32$$

Therefore,

$$0.5555555 \times 50 - 32 = 10°F \text{ (for } 5/9 = 0.5555555)$$

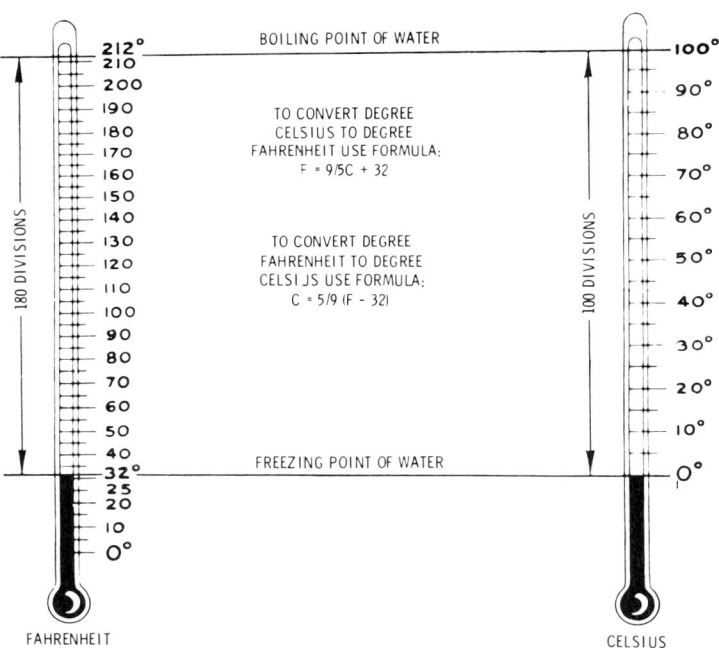

Fig. 1-1. Relationship between Fahrenheit and Celsius temperature scales.

Barometer

The *barometer* is an instrument for the measurement of atmospheric pressure. In its earliest form, it consisted simply of a glass tube somewhat in excess of 30 inches in length filled with mercury. This tube was inverted in a cup partially filled with mercury, as

18

AIR-CONDITIONING FUNDAMENTALS

shown in Fig. 1-2. The height of the mercury column in the tube is a measure of the existing atmospheric pressure.

Standard atmospheric pressure at sea level is 29.921 inches of mercury. Most of the pressure gages used in engineering calculations indicate gage pressure, or pounds per square inch (psi). Barometer readings may be converted into gage pressure by multiplying inches of mercury by 0.49116. Thus, if the barometer reads 29.921 inches, the corresponding gage pressure equals 0.49116 × x 29.921, or 14.696 psi. Table 1-1 is a convenient conversion table based on the standard atmosphere, which, by definition, equals 29.921 inches of mercury (in. Hg), or 14.696 psi.

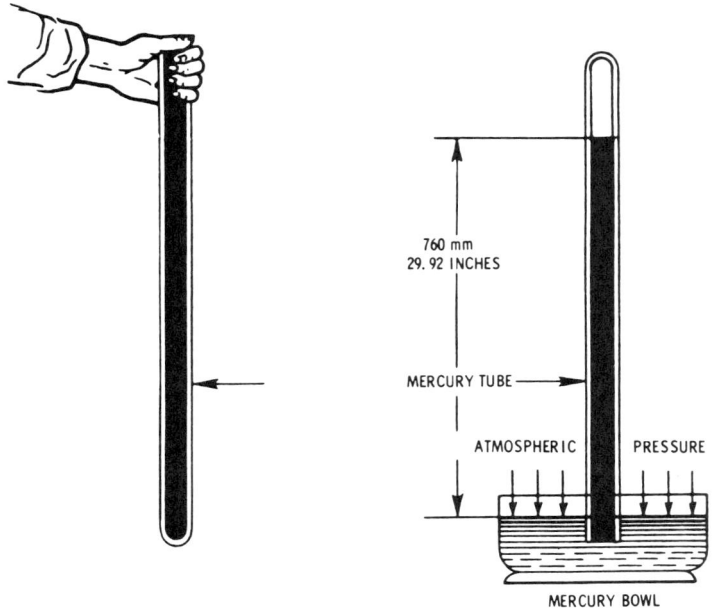

Fig. 1-2. Method of obtaining barometric pressure.

Sling Psychrometer

Relative humidity is measured by an instrument known as the *sling psychrometer*, which uses two mercury thermometers mounted side by side (Fig. 1-3). One of the thermometers is the

AIR CONDITIONING

same type used to measure the temperature of the air; it is called the *dry bulb thermometer*. The other, called the *wet-bulb thermometer*, has a small wick saturated with water attached to its bulb. If the air is dry (less humid), the evaporation from the saturated wick will be faster and consequently the temperature will be lower.

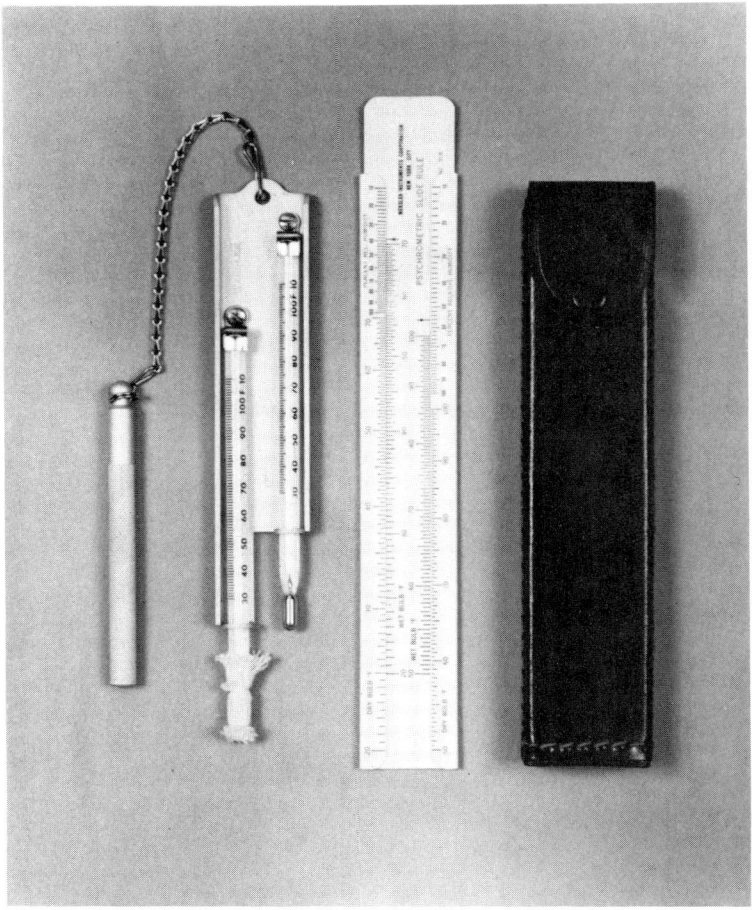

Fig. 1-3. Sling psychrometer. Note the thermometers plus a slide rule for obtaining relative humidity.

AIR-CONDITIONING FUNDAMENTALS

Pressure Gages

Pressure gages, as the name implies, are used for pressure measurements on refrigeration systems as a means of checking performance. Gages for the high-pressure side of the system have scales reading from 0 to 300 psig, or for higher pressures, from 0 to 500 psig (Fig. 1-4). Gages for the low-pressure part of the system are termed *compound gages* since their scales are graduated for pressures above atmospheric pressure in pounds per square inch gage (psig) and for pressures below atmospheric pressure (vacuum) in inches of mercury. The compound gage is calibrated from 30 inches of vacuum to pressures ranging from 60 to 150 psig, depending on gage design.

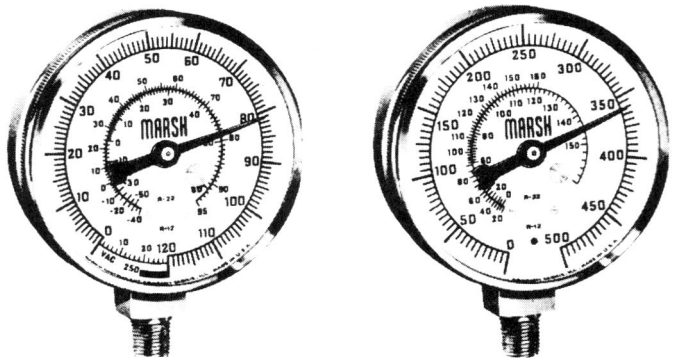

Courtesy Marsh Instrument Co., Inc.

Fig. 1-4. Compound and high-pressure gage, respectively.

British Thermal Unit

The unit of measuring the quantity of heat is the *British thermal unit* (Btu) and is the heat required to raise the temperature of *one pound* of pure water, at its greatest density, *one degree* Fahrenheit.

Refrigeration Capacity Measurements

The cooling effect is measured by a unit known as a *ton of refrigeration*. A ton of refrigeration is obtained when 1 ton (2000 lb)

AIR CONDITIONING

of ice at 32°F is melted to water at 32°F in 24 hr. If it is remembered that the latent heat of fusion is 144 Btu/lb, it follows that the ton represents a unit cooling effect of 144 × 2000, or 288,000 Btu/24 hr, 12,000 Btu/hr, or 200 Btu/min. Thus, for air conditioning calculation, the size of the required condensing unit, expressed in tons, can be obtained by dividing the heat gain of the structure, expressed in Btu/hr, by 12,000. The foregoing may be written:

$$\text{Refrigeration (in tons)} = \frac{\text{Btu/hr heat gain}}{12,000}$$

Table 1-1. Atmospheric Pressure for Various Barometer Readings

Barometer (in.Hg)	Pressure (psi)	Barometer (in. Hg)	Pressure (psi)
28.00	13.75	29.75	14.61
28.25	13.88	**29.921**	**14.696**
28.50	14.00	30.00	14.74
28.75	14.12	30.25	14.86
29.00	14.24	30.50	14.98
29.25	14.37	30.75	15.10
29.50	14.49	31.00	15.23

SUMMARY

Factors affecting both physical and chemical conditions of the atmosphere are temperature, humidity, motion, distribution, dust, bacteria, odor, and toxic gases. Air is a mixture of two gases: oxygen and nitrogen. There are still other gases in the air, such as carbon dioxide, carbon monoxide, ozone, and neon in small quantities. The air must be constantly moving in order to carry away the moisture and heat.

Air filters are used to eliminate particles of dirt or dust carried in the air that are detrimental to health and comfort. Water filters are effective in removing dust and some fumes and smoke, but carbon and soot are not removed by this method. Dry filters and electronic filters make it possible to clean the air of most harmful dust and pollen.

Heat is an active form of energy much the same as mechanical and electrical energy. Heat is transferred by three methods: conduction, convection, and radiation. Conduction means the flow of heat through a solid substance, such as iron. The transfer of heat by convection means the carrying of heat by air rising from a heated surface. Radiation takes place in the absence of matter, as in the passage of heat through the vacuum inside the bulb of an incandescent lamp.

Sensible heat is that form of heat which causes a change in the temperature of a substance. Specific heat is the Btu required to raise the temperature of one pound of substance one degree Fahrenheit. Latent heat is the form of heat that changes the physical state of a substance, such as solid to liquid or liquid to a vapor. Latent heat of fusion is changing a substance from a solid to a liquid or from a liquid to a solid.

Latent heat of evaporation is changing a substance from a liquid to a vapor or from a vapor back to a liquid. It takes 970 Btu to change 1 lb of pure water at 212°F to steam when atmospheric pressure is at sea level. Superheat is sensible heat absorbed by a vapor or gas not in contact with its liquid and consequently does not follow the temperature-pressure relationship.

A thermometer is an instrument used for determining the temperature of a body or space. The barometer is an instrument used for measuring the atmospheric pressure. Standard atmospheric pressure at sea level is 29.921 in. Hg, or 14.696 psi.

REVIEW QUESTIONS

1. Name the three most important factors that affect human health and comfort.
2. What factors besides heating and cooling are necessary in an air-conditioning system?
3. What are the chemical constituents of air?
4. Why is air movement a necessary part of air conditioning?
5. What are the devices used for cleaning and filtering air?
6. What are the three principal methods by which heat may be transferred through space?
7. If the specific heat of water is taken as 1.00, what is the specific heat of iron?

8. How does latent heat affect the change of state in various substances?
9. What is meant by the latent heat of evaporation?
10. What is superheat?
11. State the relations between pressure, temperature, and volume for a given weight of gas.
12. How many Btu are required to convert 1 lb of water at 32°F to 1 lb of ice at the same temperature?
13. Define the law of conservation of matter.
14. Define Boyle's law.
15. State the relations between the various temperature scales.
16. What instrument is commonly used for measurement of atmospheric pressure?
17. State the relations between absolute and gage pressure.
18. What is a British thermal unit?
19. What is a ton of refrigeration, and what is the Btu equivalent?

CHAPTER 2

Psychrometry

Psychrometry is that branch of physics relating to the measurement or determination of atmospheric conditions, particularly regarding the moisture mixed with air. In all air-conditioning calculations it should be understood that the dry air and water vapor composing the atmosphere are separate entities, each with its own characteristics. This water vapor is not dissolved in the air in the sense that it loses its own individuality, but merely serves to moisten the air.

AIR AND WATER-VAPOR MIXTURES

Water vapor is a gaseous form of water at a temperature below the boiling point of water. It is the most variable constituent of the atmosphere. At certain temperatures and barometric pressures it is extremely unstable in either gaseous or liquid form. This is evident by the formation and disappearance of clouds and fog. Water vapor constitutes about 3 percent of the total air by volume in hot, humid weather and about $1/5$ of 1 percent of total air by volume in dry, cold weather. Water vapor is actually steam at very low pressure; hence, its properties are those of steam at low temperatures, and its actions are comparable to steam.

HUMIDIFYING AND DEHUMIDIFYING

The air becomes humidified when moisture is added to it and is dehumidified when moisture is removed. Perhaps it may seem strange that the addition of moisture to air and the removal of moisture from air are two of the six important essentials of proper air conditioning. It does seem odd that the amount of moisture in the air of a room should have any effect on the personal comfort of the occupant. An excess or deficiency of moisture, however, does have a very noticeable effect that will immediately become apparent once the meaning of the somewhat mysterious term *relative humidity* is understood.

RELATIVE HUMIDITY

Relative humidity may be defined as the ratio of the quantity of vapor actually present in the air to the greatest amount possible at any given temperature. It follows that the relative humidity of air at any given temperature can be obtained merely by dividing the amount of moisture actually in the air by the amount of moisture that the air can hold at that temperature and multiplying the result by 100 in order to get the percentage factor.

Relative Humidity Measurements

Relative humidity is measured by an instrument known as the *sling psychrometer* (Fig. 2-1), which consists simply of two ordinary thermometers securely fastened in a frame to which a chain is attached. By means of this chain, the instrument can be rapidly whirled around so that it comes in contact with the maximum

PSYCHROMETRY

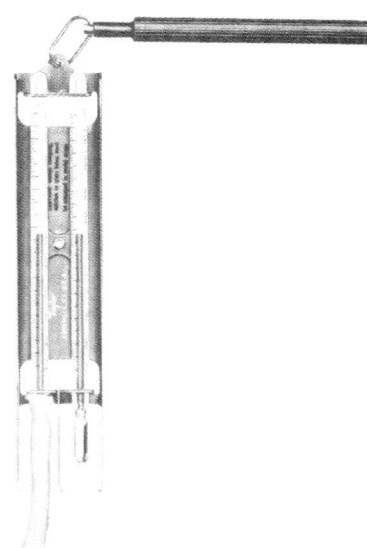

Fig. 2-1. Typical sling psychrometer.

Courtesy Taylor Instrument Company

amount of air. Around the bulb of one thermometer is a small piece of cloth, which is dampened with water before taking a reading. The theory of the instrument is simply that the evaporation of moisture from the bulb of the *wet thermometer* causes it to read lower than the one which is *dry*. The rate of evaporation depends directly on the amount of moisture in the air at the time the test is made. The difference between the readings of the two thermometers enables one to find the relative humidity. See the slide rule in Fig. 1-3 for how to arrive at the relative humidity by this method.

Grains of Moisture

Grains of moisture is the weight of water vapor present in one cubic foot of air. The grain is a unit of weight. It is the basic unit of the English weight system, derived from the weight of a grain of wheat. There are 7000 grains in a pound. Thus, one grain of moisture weighs $1/7000$ lb. or 0.000142 lb.

27

AIR CONDITIONING

Specific Volume

Specific volume is the number of cubic feet of moist air required to contain one pound of dry air molecules. Specific volume is the reciprocal of the specific gravity. It is equal to the number of cubic centimeters occupied by one gram of a substance [when the specific gravity is referred to water at 40°C (39.2°F) as a standard].

Dew-Point Temperature

The saturation temperature for any given quantity of water vapor in the atmosphere is known as the *dew point*. By definition, for a given atmospheric pressure, the dew point is the temperature of saturation at which moisture begins to change into the form of tiny water droplets or dew.

Effective Temperature

As applied to air conditioning, the *effective temperature* is an empirically determined index of the degree of warmth or cold as apparent to the human body and takes into account the temperature, moisture content, and motion of the surrounding air. Effective temperatures are not strictly a degree of heat in the same sense that dry-bulb temperatures are; for instance, the effective temperature can be lowered by increasing the rate of air flow even though wet- and dry-bulb temperatures remain constant. For space cooling and heating, however, the air-movement factor is considered a constant at approximately 20 ft/min; and under this condition, effective temperature is determined by the wet- and dry-bulb thermometer readings only.

Dry-Bulb Temperature

An ordinary thermometer is used to measure *dry-bulb temperature*. Two types of liquid are used in thermometers: colored alcohol and mercury. The alcohol thermometer is more common because it is less expensive and can measure the normal range of air temperature. Since mercury freezes at approximately −38°F, mercury thermometers are not practical for measuring extremely low temperatures.

Wet-Bulb Temperature

Wet-bulb temperature is the temperature at which the air becomes saturated if moisture is added to it without the addition or subtraction of heat. Thus, if the bulb of an ordinary thermometer is surrounded with a moistened wick, placed in a current of air, and superheated with water vapor, the reading obtained will be at some point below the dry-bulb temperature. The minimum reading thus obtained is the *wet-bulb temperature* of the air.

Wet-Bulb Depression

Since outdoor summer air is rarely fully saturated, there is normally a considerable difference between its dry- and wet-bulb temperatures. The difference between the two temperatures is called the *wet-bulb depression*. For example, if the dry-bulb temperature reading is 80 and the corresponding wet-bulb temperature is 65, the wet-bulb depression is 80 − 65, or 15.

Temperature-Humidity Index

The *temperature-humidity index* describes numerically the human discomfort resulting from temperature and moisture. It is computed by adding dry- and wet-bulb temperature readings, multiplying the sum by 0.4, and adding 15. Summer estimates indicate about 10 percent of the populace are uncomfortable before the index passes 70, more than 50 percent are uncomfortable after it passes 75, and almost all are uncomfortable at 80 or above.

Total Heat Content

Total heat content is the amount of heat energy stored in the gaseous air and water vapor, measured in British thermal units (Btu). All the different methods of treating air can be pictured on the psychrometric chart. The magnitude of the changes required can be determined by plotting the conditions of the air entering and leaving the apparatus and then connecting the two points by a straight line.

AIR CONDITIONING

PSYCHROMETRIC CHARTS

A psychrometric chart is a graphical representation of the fundamental mathematical relationship dealing with the thermodynamic properties of moist air. The various charts shown in the figures on the following pages will be found valuable in making air-conditioning calculations. The chart readings can be obtained for *dry-bulb*, *wet-bulb*, and *dew-point temperatures* and other heat-treatment properties.

Three basic formulas can be utilized to determine (1) the *total heat* to be removed from both the gases and water vapor; (2) the *sensible heat* removed from the air only; and (3) the *latent heat* removed in condensing the water vapor:

Total heat = cu ft/min × $4.5(h_1 - h_2)$
 = Btu/hr
Sensible heat = cu ft/min × $1.08(DB_1 - DB_2)$
 = Btu/hr
Latent heat = cu ft/min × $68(WB_1 - WB_2)$
 = Btu/hr

where, h_1 = total heat for entering air, Btu/lb
 h_2 = total heat for leaving air, Btu/lb
 DB_1 = entering air for dry-bulb temperature
 DB_2 = leaving air for dry-bulb temperature
 WB_1 = entering air for wet-bulb temperature
 WB_2 = leaving air for wet-bulb temperature

Most air-conditioning calculations are made assuming normal sea-level pressure and standard temperatures, but occasionally the altitude or high-temperature industrial applications will require converting from standard conditions to those actually present. Use Table 1-1 (Chapter 1) to obtain corrected atmospheric pressure.

Psychrometric Chart Instructions

In Fig. 2-2, note the point encircled in the center of the chart, which represents a condition of 80°F dry-bulb temperature and 50.9% relative humidity. The lines on the chart sloping upward to the left represent constant wet-bulb temperatures, which, in this

PSYCHROMETRY

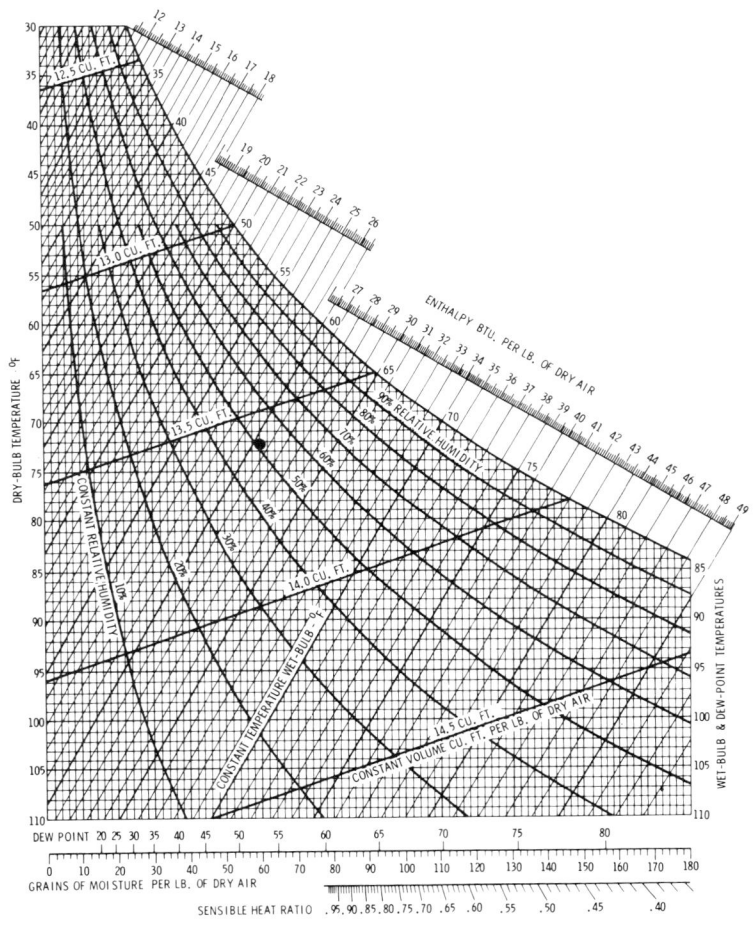

Fig. 2-2. Psychrometric charts.

case, is 67°F. The moisture content of the encircled point is 78 grains per pound of dry air. The total heat in Btu per pound, as noted, is scaled at the extreme right. The heat removed in the cooling of air can be obtained by finding the difference in the total heat corresponding to the wet-bulb temperature at the beginning and the final conditions. For example, if air is to be cooled

31

from a condition of 65°F dry-bulb temperature and 60% relative humidity, the total heat (from Fig. 2-2) is 24.5 Btu/lb of dry air.

When calculating the amount of refrigeration required for a particular job, it is generally necessary to make allowances for the various conditions, such as heat losses due to leakage (infiltration), the heating effect due to electrical apparatus, and body heat from persons, which exists in theaters and auditoriums where a large number of persons are gathered (Fig. 2-3).

It is now possible to proceed with illustrative examples of a general nature so that the solutions can be used on other problems if other details are known.

Example 1 — Given 80°F dry-bulb and 65°F wet-bulb temperature, find the dew point, relative humidity, and grains of moisture per pound of dry air.

Solution — Figure 2-4 shows the solution schematically. The intersection on the chart of the two temperature conditions shows a dew point of 57.0 and a relative humidity of 45%. There are 68 grains of moisture/lb of dry air.

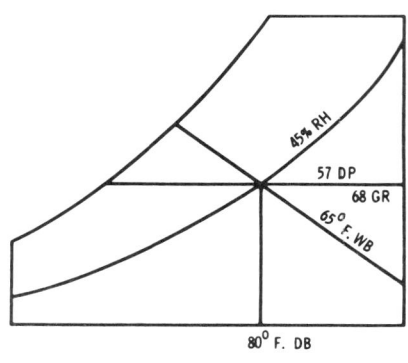

Fig. 2-3. Simplified psychrometric chart showing the method of obtaining readable values from chart in Fig. 2-2.

Example 2 — Find the total heat load for cooling 6000 cu ft/min of air entering at 83°F dry-bulb and 70°F wet-bulb, and leaving at 60°F dry-bulb and 58°F wet-bulb.

Solution — Figure 2-5 shows the schematic solution. Locate the entering and leaving conditions on chart. Read the total heat of 34.1 Btu/lb for entering air and 25.2 Btu/lb for leaving air. The total heat may be obtained by substitution of values:

PSYCHROMETRY

Total Heat = cu ft/min × 4.5$(h_1 - h_2)$
= 6000 × 4.5(34.1 − 25.2)
= 240,300 Btu/hr (approx.)

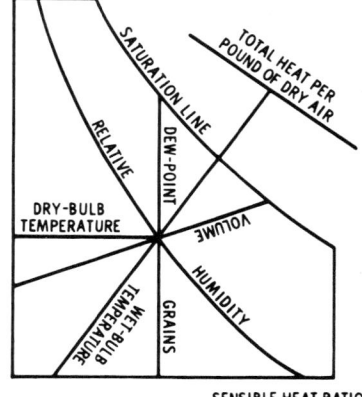

Fig. 2-4. Schematic solution for Example 1.

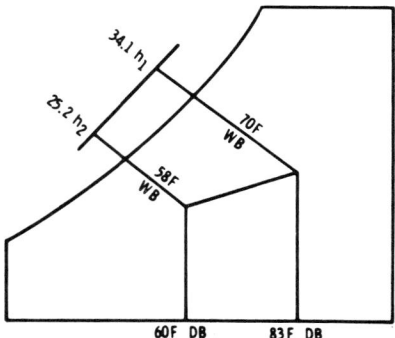

Fig. 2-5. Schematic solution for Example 2.

Example 3 — Find the apparatus dew point for air entering at 82°F dry-bulb and 68°F wet-bulb, and leaving at 59°F dry-bulb and 57°F wet-bulb.
Solution — The schematic solution is shown in Fig. 2-6. By extending the line to connect the entering and leaving conditions of air to the saturation curve, the apparatus dew point or average surface temperature of the coil can be found. Following this procedure, the apparatus dew point is 54.5.

33

AIR CONDITIONING

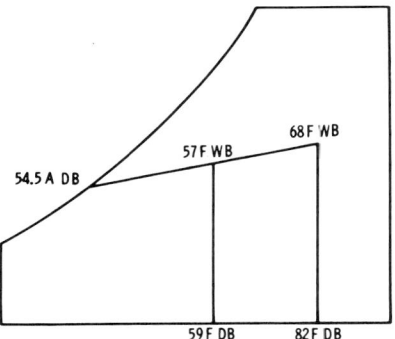

Fig. 2-6. Schematic solution for Example 3.

Example 4 — Assume that air is entering at 81°F dry-bulb and 69°F wet-bulb, and leaving at 56°F dry-bulb and 54°F wet-bulb. If the total heat load is 460,000 Btu/hr, what is the sensible heat factor?

Solution — Figure 2-7 shows the schematic solution. By using the sensible-heat factor, as noted on the chart, the ratio of the sensible heat to the total load can easily be found. First, locate the entering and leaving conditions and draw a straight line connecting the two points on the chart. Next, locate the sensible-heat ratio reference point on the chart for 81°F dry-bulb and 69°F wet-bulb. Draw a line through the reference point parallel with the line representing the entering and leaving conditions. Extend the line and read 0.54 sensible-heat factor. The sensible heat load equals 0.54 × 460,000, or 248,400 Btu/hr.

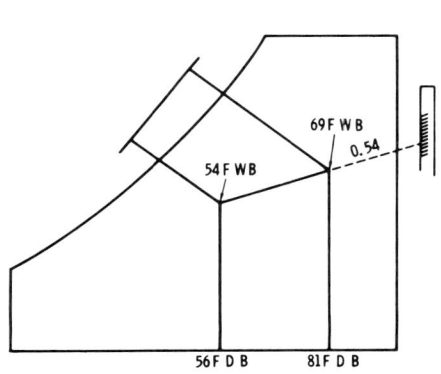

Fig. 2-7. Schematic solution for Example 4.

34

PSYCHROMETRY

Example 5 — Find the dry-bulb and wet-bulb temperatures for an air mixture of 7500 cfm of inside air at 76°F dry-bulb and 63°F wet-bulb, mixed with 2500 cfm of outside air entering at 95°F dry-bulb and 75°F wet-bulb.

Solution — With reference to the schematic solution in Fig. 2-8, the following chart readings will indicate the procedure. Locate the conditions for inside and outside air on the chart. Draw a line connecting the two points. The ratio of inside to outside air is 2:1. Divide the line into four equal parts. Count up one equal part of the line from the inside air condition. This point represents the conditions of the air mixture, which are 80.9°F dry-bulb and 66.4° wet-bulb.

Fig. 2-8. Schematic solution for Example 5.

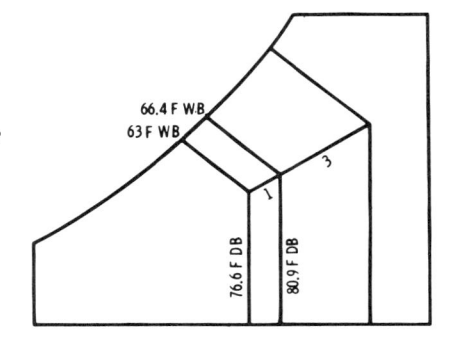

Example 6 — It is desired to install a central air-conditioning system in a 3000-seat theater for outside conditions of 90°F dry-bulb and 60 percent relative humidity, and inside conditions of 80°F dry-bulb and 63°F dew-point temperature. The heating effect due to the electric lights is 1500 Btu/min, and the heat leakage through the walls, ceiling, etc., is 3000 Btu/min. The sensible heat is assumed to be 7.0 Btu/min per person, and the amount of water vapor emitted is 18.9 grains. To summarize, the heating values are as follows:

Sensible heating effect of audience	= 7 × 3000
	= 21,000 Btu
Heat leakage	= 3000 Btu
Heating effect due to illumination	= 1500 Btu

Heat equivalent due to fan (est. 45 hp) = 1900 Btu
Total heating effect per minute = 27,400 Btu

Solution — If the specific heat of air at 64.7°F and saturated with moisture is taken as 0.2472 and the temperature rise of the air is 15.3°F, the heat capacity of the air will be 15.3 × 0.2472 = 3.78 Btu. The total amount of air to be circulated by the fan to absorb 27,400 Btu/min is 27,400 ÷ 3.78, or 7250 lb, approximately. The volume of 1 lb of air saturated with moisture at 64.7°F is 13.5 cu ft. Thus the total capacity of the fan per minute is 7250 × 13.5, or 97,875 cu ft, and the amount supplied per person is 32.6 cu ft/min.

From the preceding discussion, it will be noted that the amount of water vapor emitted is 18.9 × 3000 ÷ 7000, or 8.1 lb/min. The refrigeration required to condense this amount will be 8.1 × 1061, or 8600 Btu, making a total of 36,000 Btu/min to be removed from the air if all of it is to be recirculated. If only one-half is to be recirculated and the other half is to be exhausted into the atmosphere, there will be 18,000 Btu/min of refrigeration required for the air to be recirculated. The fresh air will enter the spray chamber at 90°F and 60% relative humidity, which corresponds to a dew-point temperature of 74.2 and a moisture content of 127 grains/lb of dry air. The refrigeration required to cool this air to the temperature of the spray chamber, taken from the total heat curve, is 41.3 − 29.3, or 12.0 Btu/lb, and the total refrigeration is 7250 ÷ 2 × 12.0, or 43,500 Btu. The total demand on the refrigeration equipment becomes 43,500 + 18,000, or 61,500 Btu. Adding 5 percent for safety, the amount becomes 64,600 Btu/min, or 323.0 tons of refrigeration.

COMFORT CHART

Because of the possibility of shock to the human system, it is not desirable to have too great a difference between indoor and outdoor temperatures, particularly during the summer season. Some of the indoor temperatures given in Table 2-1 may seem high at

Table 2-1. Recommended Air Velocities in Ducts

Outside Temperature	Desirable Indoor Temperature	Desirable Temperature Reduction
100°	82°	18°
95°	80°	15°
90°	78°	12°
85°	76.5°	8.5°
80°	75°	5°
75°	73.5°	1.5°
70°	72°	

first, but they have been selected as the result of many tests and can be considered reasonably correct. Deviations from this table must be made for climatic characteristics of the various sections of the country and the length of time spent in the conditioned area. Where several hours are spent in the conditioned area, the temperatures shown in the table are in order.

It should be remembered that, from the standpoint of physical comfort, low humidity with high temperatures is comparable to high humidity with lower temperatures. From this, it is apparent that to obtain the greatest physical comfort in cold weather without high temperatures, it is necessary to increase the humidity of the room conditioned. It also explains why persons in some parts of the country do not suffer unduly even though the outside air temperatures are 120°F or more but with low relative humidity.

With reference to the comfort chart in Fig. 2-9, it should be noted that both summer and winter comfort zones apply to inhabitants of the United States only. Application of the winter comfort line is further limited to rooms heated by central systems of the convection type. The line does not apply to rooms heated by radiant methods. Application of the summer comfort line is limited to homes, offices, etc., where the occupants become fully adapted to the artificial air conditions. The line does not apply to theaters, department stores, etc., where the exposure is less than three hours. The optimum comfort line shown pertains to Pittsburgh and other cities in the northern portion of the United States and Southern Canada at elevations not in excess of 1000 feet above sea level. An increase of approximately 1°F should be made for every 5° reduction in north latitude.

AIR CONDITIONING

Fig. 2-9. Comfort charts.

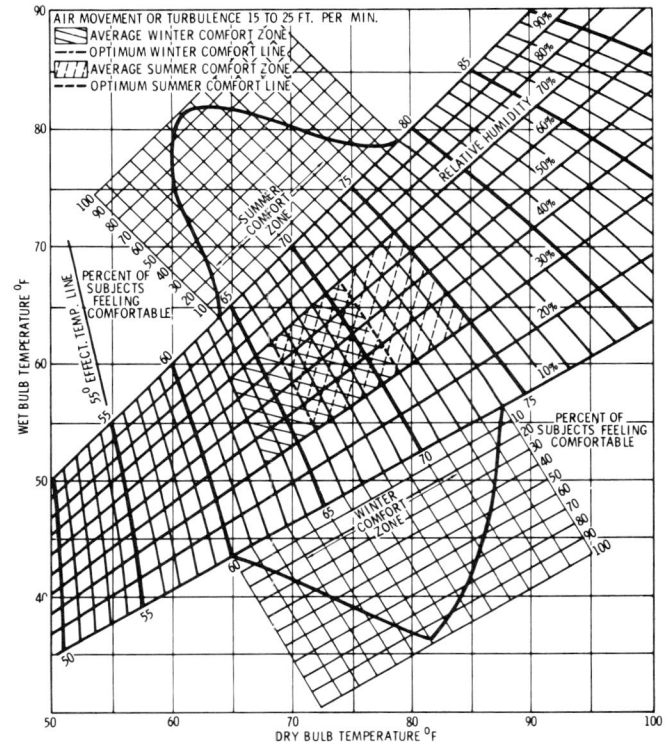

Courtesy American Society of Heating & Ventilating Engineers

SUMMARY

Water vapor is a gaseous form of water at a temperature below its boiling point. At certain temperatures and barometric pressures, water is extremely unstable in either gaseous or liquid form. The air becomes humidified when moisture is added and is dehumidified when moisture is removed. An excess or deficiency of moisture has a very noticeable effect on comfort.

Relative humidity may be defined as the ratio of the quantity of

PSYCHROMETRY

moisture actually present in the air to the greatest amount possible at any given temperature. The relative humidity of the air at any given temperature can be obtained merely by dividing the amount of moisture actually in the air by the amount of moisture the air can hold at that temperature. Relative humidity is measured by an instrument called the sling psychrometer.

Grains of moisture is the weight of water vapor present in one cubic foot of air. A grain is the basic unit of the English weight system, and 7000 grains = 1 lb. One grain of moisture weighs $1/7000$ lb or 0.000142 lb.

The saturation temperature for any given quantity of water vapor in the atmosphere is known as the dew point. For a given atmospheric pressure, it is the temperature at which moisture begins to change into the form of tiny water droplets or dew. Total heat content is the amount of heat energy stored in the gaseous air and water vapor, and is measured in British thermal units (Btu).

When calculating the amount of refrigeration required for a particular job, it is generally necessary to make allowances for various conditions, such as heat loss from infiltration, heat from electrical apparatus, and body heat. Most air-conditioning calculations are made assuming sea-level pressure and standard temperature.

REVIEW QUESTIONS

1. What is meant by the term *psychrometry?*
2. State the amount of water vapor present in the air during hot humid weather.
3. What is the purpose of humidification and dehumidification of air?
4. What is meant by the term *relative humidity?*
5. How is relative humidity measured?
6. Define *dew point, effective temperature,* and *dry-* and *wet-bulb temperature.*
7. What is meant by *wet-bulb depression?*
8. What is the function of the psychrometric chart in making air-conditioning calculations?

AIR CONDITIONING

9. Describe the construction and use of the sling psychrometer.
10. If it is assumed that the dry- and wet-bulb temperatures register 80°F and 65°F, respectively, what is the temperature humidity index?

CHAPTER 3

Heat Leakage

The thermal properties of building materials affect the design of the air-conditioning and heating systems. The rate of heat flow through walls, floors, and ceilings is usually the basis for calculating the heat and/or cooling load required for a particular building or space.

In addition, consideration must be given to air leakage that takes place through various apertures, which must be properly evaluated. This air leakage takes place through cracks around windows and doors, through walls, fireplaces, and chimneys. Although the leakage or air infiltration through fireplaces and chimneys may be considerable, it is usually neglected since the dampers should be closed during periods of extremely hot weather.

In this connection it should be noted that, because of the numerous variables affecting air-conditioning loads, it is often difficult to calculate precisely the required size of an air conditioning unit for a particular building or space. This will easily be realized when consideration is given to the fact that most of the compo-

AIR CONDITIONING

nents of the cooling load vary greatly during a 24-hour period. Economic consideration should thus be the determining factor in the selection of equipment for cooling-season operation in comfort air conditioning since available weather and other necessary data vary for different locations.

Heat leakage is always given in Btu per hour per degree Fahrenheit temperature difference per square foot of exposed surface (Btu/sq ft/hr/°F). Prior to designing an air-conditioning system, an estimate must be made of the maximum probable heat loss for each room or space to be cooled. Therefore, before attempting to place even a small room cooler into service, check the walls and determine just what size unit will do a good job of cooling and dehumidifying.

Heat losses may be divided into two groups: (1) losses through confining walls, floors, ceilings, glass, or other surfaces, and (2) infiltration losses due to leaks through cracks and crevices around doors and windows. The heat leakage through walls, floors, and ceilings can be determined by means of a formula and depends on the type and thickness of the insulating material used. The formula for heat leakage is:

$$H = KA(t_1 - t_2)$$

where H = the heat required
K = heat-transfer coefficient, Btu/sq ft/hr/°F
A = area, sq ft
$t_1 - t_2$ = temperature gradient through wall, °F

Example — Calculate the heat leakage through an 8-inch brick wall having an area of 200 sq ft if the inside temperature is 70°F and the outside temperature is 10°F.
Solution — If it is assumed that the heat-transfer coefficient of a plain brick wall is 0.50, substitution of values in the foregoing equation will be:

$$H = 0.50 \times 200(70 - 10) = 6000 \text{ Btu/hr}$$

HEAT LEAKAGE

The heat leakage through floors, ceilings, and roofs may be estimated in the same manner. The K (heat-transfer coefficient) depends on the construction and particular insulating materials used. Tables 3-1 to 3-20 cover almost every type and combination of walls, floors, and ceilings encountered in the field. Partition walls are also included so that these may be estimated for heat leakage. Although the K values given in the tables do not agree entirely with the data given in various handbooks, they will give sufficiently close values for adaptation since in most instances only approximate values can be obtained. In making use of the data, it must be remembered that the outside area is used as a basis for estimating. Ceiling and floor construction must also be determined, and in many instances the four walls of the room may not be of the same construction since one or more of the walls may be partitions.

Table 3-1. Concrete Wall (No Exterior Finish)

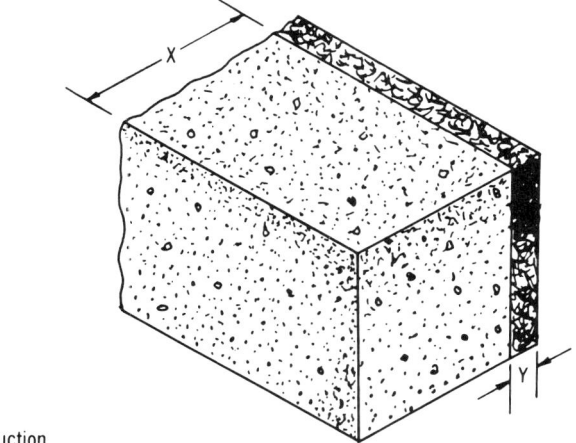

K = X + Y

X = Wall

Y = Interior Construction

Concrete Wall
(No Exterior Finish)
Values of K in Btu/hr/1°F/sq ft

Wall Construction (Y)	Thickness (X)				
	6"	8"	10"	12"	16"
Plain wall—no interior finish	0.58	0.51	0.46	0.41	0.34
½" Plaster—direct on concrete	0.52	0.46	0.42	0.38	0.32
½" Plaster on wood lath, furred	0.31	0.29	0.26	0.24	0.22
¾" Plaster on metal lath, furred	0.34	0.32	0.29	0.27	0.24
½" Plaster on ⅜" plasterboard, furred	0.32	0.30	0.27	0.25	0.22
½" Plaster on ½" board insulation, furred	0.21	0.20	0.19	0.18	0.17
½" Plaster on 1" corkboard, set in ½" cement	0.16	0.15	0.14	0.14	0.13
½" Plaster on 1½" corkboard, set in ½" cement	0.15	0.14	0.14	0.13	0.12
½" Plaster on 2" corkboard, set in ½" cement	0.13	0.12	0.11	0.10	0.09
½" Plaster on wood lath on 2" fur., 1⅛" gypsum fill	0.20	0.19	0.18	0.17	0.16

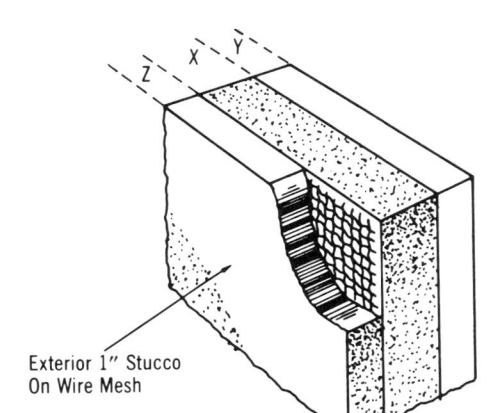

Table 3-2. Concrete Wall (Exterior Stucco Finish)

Exterior 1" Stucco
On Wire Mesh

X = Concrete
Y = Insulation
Z = Exterior Stucco 1"
K = X + Z + Y

Concrete Wall
(Exterior Stucco Finish)
Values of K in Btu/hr/1°F/sq ft

Wall Construction (Y)	Thickness of Concrete (X)						
	6"	8"	10"	12"	16"	18"
Plain walls—no interior finish	0.54	0.48	0.43	0.39	0.33	0.28
½" Plaster direct on concrete	0.49	0.44	0.40	0.36	0.31	0.27
½" Plaster on wood lath, furred	0.31	0.29	0.27	0.25	0.23	0.22
¾" Plaster on metal lath, furred	0.32	0.30	0.28	0.26	0.24	0.23
½" Plaster on ⅜" plasterboard, furred	0.31	0.29	0.27	0.25	0.23	0.22
½" Plaster on wood lath on 2" furring strips with 1⅝" cellular gypsum fill	0.20	0.19	0.18	0.17	0.16	0.15
½" Plaster on ½" board insulation, furred	0.22	0.21	0.20	0.19	0.18	0.17
1"	0.17	0.16	0.15	0.145	0.140	0.13
½" Plaster on 1½" sheet cork set in cement	0.145	0.140	0.135	0.130	0.12	0.11
2"	0.12	0.118	0.115	0.110	0.105	0.100

Table 3-3. Concrete Wall (Brick Veneer)

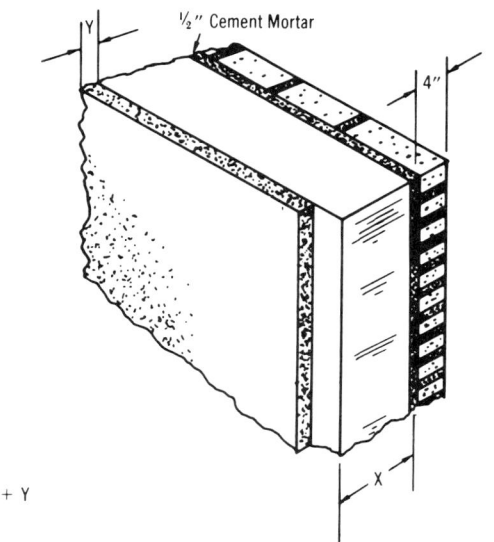

X = Concrete
Y = Interior Finish

K = X + ½" Cement + 4" Brick + Y

Concrete Wall (Brick Veneer)
Values of K in Btu/hr/1°F/sq ft

Wall Construction (Y)	Concrete (X)			
	6"	8"	10"	12"
Plain wall—no interior finish	0.39	0.36	0.33	0.30
½" Plaster direct on concrete	0.37	0.34	0.30	0.29
½" Plaster on wood lath, furred	0.24	0.23	0.21	0.20
¾" Plaster on metal lath, furred	0.27	0.25	0.23	0.22
½" Plaster on wood lath, on 2" furring strips, with 1⅝" cellular gypsum fill	0.18	0.17	0.16	0.15
½" Plaster on ⅜" plasterboard, furred	0.25	0.24	0.22	0.21
½" Plaster on ½" board insulation, furred	0.17	0.16	0.14	0.12
½" Plaster on 1" board insulation, furred	0.15	0.14	0.12	0.11
½" Plaster on 1½" board insulation, furred	0.13	0.12	0.11	0.10
½" Plaster on 2" board insulation, furred	0.11	0.10	0.09	0.08

HEAT LEAKAGE

Table 3-4. Concrete Wall (4-in. Cut Stone)

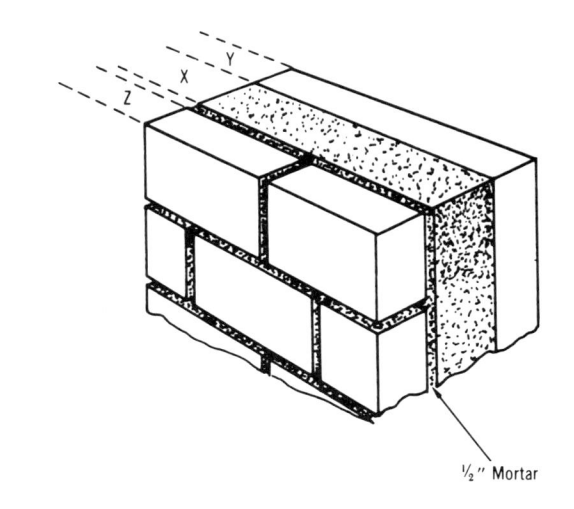

K = X + Y + Z
X = Concrete
Y = Insulation
Z = 4" Cut Stone

½" Mortar

Concrete Wall
(4-in. Cut Stone)
Values of K in Btu/hr/1°F/sq ft

Wall Construction (Y)	Thickness of Concrete (X)				
	6"	8"	10"	12"	16"
Plain walls—no interior finish	0.46	0.42	0.38	0.34	0.30
½" Plaster direct on concrete	0.42	0.38	0.34	0.32	0.28
¾" Plaster on metal lath, furred	0.29	0.27	0.26	0.24	0.22
½" Plaster on wood lath, furred	0.28	0.26	0.25	0.23	0.21
½" Plaster on ⅜" plasterboard, furred	0.28	0.26	0.25	0.23	0.21
½" Plaster on wood lath on 2" furring strips, filled with 1⅝" gypsum	0.18	0.175	0.17	0.16	0.15
½" Plaster on ½" board insulation, furred	0.21	0.20	0.19	0.18	0.17
½" Plaster on 1" board insulation, furred	0.16	0.155	0.15	0.14	0.13
½" Plaster on 1" sheet cork, set in ½" cement or mortar	0.15	0.145	0.140	0.135	0.130
1½"	0.138	0.135	0.130	0.125	0.12
2"	0.115	0.110	0.108	0.105	0.100

Table 3-5. Cinder and Concrete Block Wall

K = X + Y
X = Wall
Y = Interior Finish

Cinder and Concrete Block Wall
Values of K in Btu/hr/1 °F/sq ft

Wall construction (Y)	Concrete 8"	Concrete 12"	Cinder 8"	Cinder 12"
Plain wall—no interior finish	0.46	0.34	0.31	0.23
½" Plaster—direct on blocks	0.42	0.32	0.29	0.21
½" Plaster on wood lath, furred	0.28	0.23	0.22	0.17
¾" Plaster on metal lath, furred	0.29	0.24	0.23	0.17
½" Plaster on ⅜" plasterboard, furred	0.28	0.23	0.22	0.17
½" Plaster on ½" board insulation, furred	0.21	0.18	0.17	0.14
½" Plaster on 1" board insulation, furred	0.16	0.14	0.14	0.11
½" Plaster on 1½" sheet cork set in ½" cement	0.13	0.12	0.12	0.10
½" Plaster on 2" sheet cork set in ½" cement	0.11	0.10	0.10	0.09
½" Plaster on wood lath, on 2" furring strips, 1⅝" Gypsum fill	0.18	0.16	0.16	0.13

Table 3-6. Cinder and Concrete Block Wall (Brick Veneer)

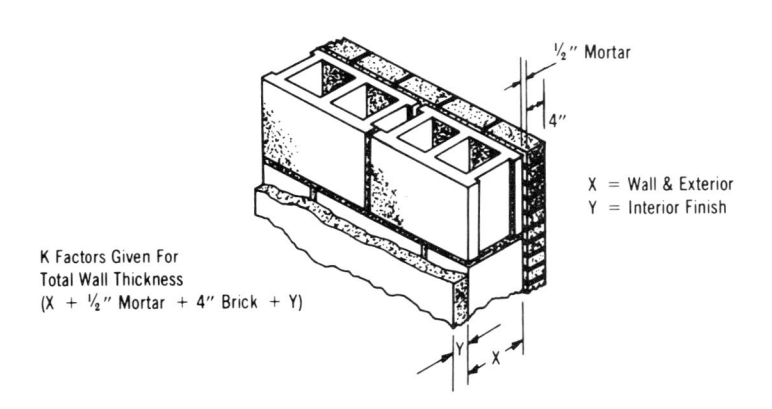

X = Wall & Exterior
Y = Interior Finish

K Factors Given For
Total Wall Thickness
(X + ½" Mortar + 4" Brick + Y)

Cinder and Concrete Block Wall
(Brick Veneer)
Values of K in Btu/hr/1°F/sq ft

Wall construction (Y)	Concrete 8"	Concrete 12"	Cinder 8"	Cinder 12"
Plain wall—no interior finish	0.33	0.26	0.25	0.18
½" Plaster—direct on blocks	0.31	0.25	0.23	0.18
½" Plaster on wood lath, furred	0.23	0.19	0.18	0.15
¾" Plaster on metal lath, furred	0.24	0.20	0.19	0.15
½" Plaster on wood lath, on 2" furring strips, 1⅝" gypsum fill	0.16	0.14	0.14	0.12
½" Plaster on ½" board insulation, furred	0.18	0.16	0.15	0.13
½" Plaster on 1" board insulation, furred	0.14	0.13	0.12	0.10
½" Plaster on ⅜" plasterboard, furred	0.23	0.19	0.18	0.15
½" Plaster on 1½" sheet cork, set in ½" cement mortar	0.12	0.11	0.11	0.09
½" Plaster on 2" sheet cork, set in ½" cement mortar	0.10	0.09	0.09	0.08

Table 3-7. Brick Veneer on Hollow Tile Wall

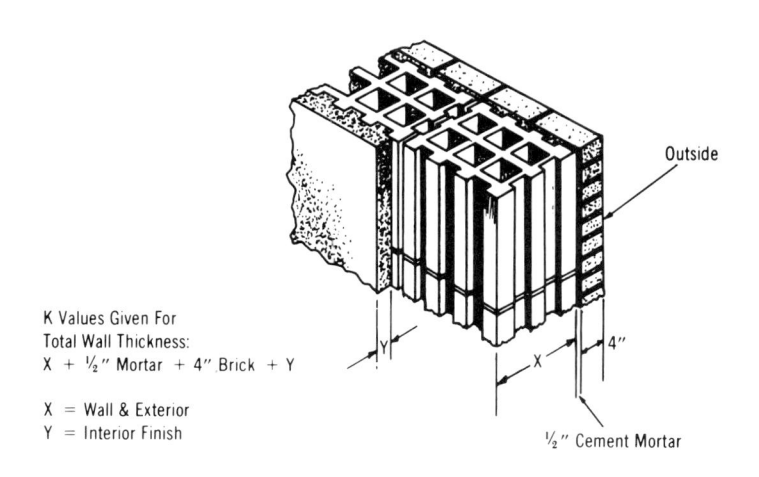

K Values Given For
Total Wall Thickness:
X + ½" Mortar + 4" Brick + Y

X = Wall & Exterior
Y = Interior Finish

Brick Veneer on Hollow Tile Wall
(Brick Veneer)
Values of K in Btu/hr/1 °F/sq ft

Wall Construction (Y)	Tile Thickness (X)				
	4"	6"	8"	10"	12"
Plain wall—no interior finish	0.29	0.27	0.25	0.22	0.18
½" Plaster on hollow tile	0.28	0.25	0.24	0.21	0.17
½" Plaster on wood lath, furred	0.20	0.19	0.18	0.17	0.15
¾" Plaster on metal lath, furred	0.21	0.20	0.19	0.18	0.16
½" Plaster on ⅜" plasterboard, furred	0.20	0.19	0.18	0.17	0.15
½" Plaster on wood lath, on 2" furring strips, with 1⅝" cellular gypsum fill	0.14	0.13	0.12	0.11	0.10
½" Plaster on ½" board insulation, furred	0.16	0.15	0.14	0.13	0.12
½" Plaster on 1" board insulation, furred	0.14	0.13	0.12	0.11	0.10
½" Plaster on 1½" board insulation, furred	0.13	0.12	0.11	0.10	0.09
½" Plaster on 2" board insulation, furred	0.12	0.11	0.10	0.09	0.08

HEAT LEAKAGE

Table 3-8. Hollow Tile Wall (4-in. Cut-Stone Veneer)

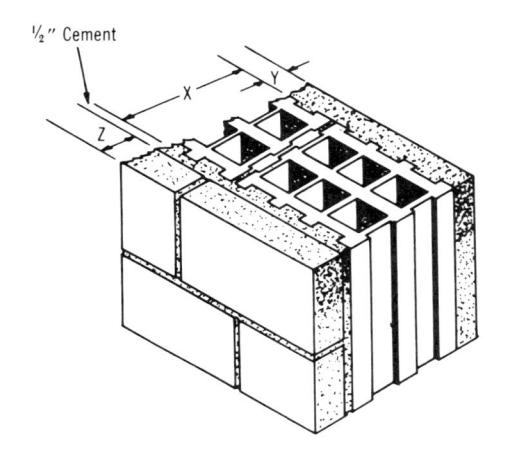

$\frac{1}{2}''$ Cement

X = Tile
Y = Insulation
Z = 4" Cut Stone

$K = X + Y + Z$

Hollow Tile Wall
(4-in. Cut-Stone Veneer)
Values of K in Btu/hr/1 °F/sq ft

Wall Construction (Y)	Thickness of Tile (X)			
	6"	8"	10"	12"
Plain walls—no interior finish	0.30	0.29	0.28	0.22
$\frac{1}{2}''$ Plaster—direct on hollow tile	0.28	0.27	0.26	0.21
$\frac{3}{4}''$ Plaster on metal lath, furred	0.22	0.21	0.20	0.18
$\frac{1}{2}''$ Plaster on wood lath, furred	0.21	0.20	0.19	0.17
$\frac{1}{2}''$ Plaster on $\frac{3}{8}''$ plasterboard, furred	0.21	0.20	0.19	0.17
$\frac{1}{2}''$ Plaster on $\frac{1}{2}''$ cork board set in $\frac{1}{2}''$ cement	0.17	0.16	0.15	0.14
$\frac{1}{2}''$ Plaster on 1" corkboard set in $\frac{1}{2}''$ cement	0.14	0.13	0.13	0.12
$\frac{1}{2}''$ Plaster on $1\frac{1}{2}''$ corkboard set in $\frac{1}{2}''$ cement	0.13	0.12	0.11	0.10
$\frac{1}{2}''$ Plaster on 2" corkboard set in $\frac{1}{2}''$ cement	0.11	0.10	0.09	0.08
$\frac{1}{2}''$ Plaster on wood lath, 2" furring & $1\frac{5}{8}$ gypsum fill	0.15	0.14	0.13	0.12
$\frac{1}{2}''$ Plaster on $\frac{1}{2}''$ board insulation	0.18	0.17	0.16	0.15

Table 3-9. Hollow Tile Wall (Stucco Exterior)

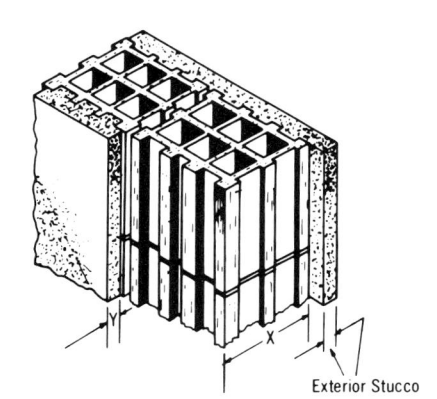

X = Tile Thickness
Y = Interior Finish

K Values Given For
Total Wall Thickness
X + Y + Stucco Finish

Exterior Stucco

Hollow Tile Wall
(Stucco Exterior)
Values of K in Btu/hr/1°F/sq ft

Wall Construction (Y)	Tile Thickness (X)				
	6"	8"	10"	12"	16"
Plain wall with stucco—no interior finish	0.32	0.30	0.28	0.22	0.18
½" Plaster—direct on hollow tile	0.31	0.29	0.27	0.21	0.17
½" Plaster on wood lath, furred	0.22	0.20	0.18	0.16	0.12
¾" Plaster on metal lath, furred	0.23	0.21	0.19	0.17	0.13
½" Plaster on ⅜" plasterboard, furred	0.21	0.20	0.18	0.16	0.12
½" Plaster on wood lath, on 2" furring strips, with 1⅝" cellular gypsum fill	0.16	0.15	0.14	0.13	0.12
½" Plaster on ½" board insulation, furred	0.15	0.14	0.13	0.12	0.11
½" Plaster on 1" board insulation, furred	0.13	0.12	0.11	0.10	0.09
½" Plaster on 1½" board insulation, furred	0.12	0.11	0.10	0.09	0.08
½" Plaster on 2" board insulation, furred	0.11	0.10	0.09	0.08	0.07

HEAT LEAKAGE

Table 3-10. Limestone or Sandstone Wall

K = X + Y
X = Wall
Y = Interior Construction

Limestone or Sandstone Wall
Values of K in Btu/hr/1°F/sq ft

Wall Construction (Y)	Thickness (X)			
	8"	10"	12"	16"
Plain wall—no interior finish	0.56	0.50	0.48	0.39
½" Plaster—direct on stone	0.50	0.45	0.42	0.36
¾" Plaster on metal lath, furred	0.33	0.31	0.29	0.26
½" Plaster on wood lath, furred	0.31	0.29	0.28	0.25
½" Plaster on ⅜" plasterboard, furred	0.31	0.30	0.28	0.25
½" Plaster on wood lath, on 2" furring strips, with 1⅝" cellular gypsum fill	0.20	0.19	0.18	0.17
½" Plaster on ½" board insulation, furred	0.23	0.22	0.21	0.19
½" Plaster on 1" board insulation, furred	0.17	0.16	0.15	0.14
½" Plaster on 1½" corkboard, set in ½" cement	0.14	0.14	0.13	0.12
½" Plaster on 2" corkboard, set in ½" cement	0.11	0.11	0.10	0.10
½" Plaster on wood lath, on 2" furring strips, with 2" cellular gypsum fill	0.19	0.18	0.17	0.16

53

Table 3-11. Brick Wall (4-in. Cut Stone)

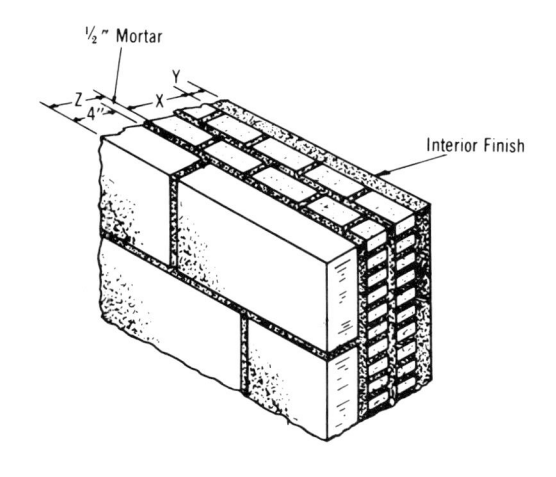

K = X + Y
X = Wall
Y = Interior
Z = 4" Stone 8½" Mortar

Brick Wall
(4-in. Cut Stone)
Values of *K* in Btu/hr/1°F/sq ft

Wall Construction (*Y*)	Thickness (*X*)						
	8"	9"	12"	13"	16"	18"	24"
Plain wall—no interior finish	0.33	0.29	0.26	0.25	0.22	0.20	0.16
½" Plaster—direct on brick	0.31	0.28	0.25	0.24	0.21	0.19	0.15
¾" Plaster on metal lath, furred	0.23	0.22	0.20	0.19	0.18	0.17	0.15
½" Plaster on wood lath, furred	0.22	0.21	0.19	0.18	0.17	0.15	0.12
½" Plaster on ⅜" plasterboard, furred	0.23	0.22	0.20	0.19	0.18	0.17	0.14
½" Plaster, lath, 2" furring strips, with 1⅝" gypsum fill	0.16	0.15	0.14	0.13	0.12	0.11	0.10
½" Plaster on ½" sheet cork	0.18	0.16	0.15	0.14	0.13	0.12	0.11
½" Plaster on 1" sheet cork	0.14	0.13	0.12	0.11	0.10	0.10	0.09
½" Plaster on 1½" sheet cork	0.12	0.12	0.11	0.11	0.10	0.10	0.09
½" Plaster on 2" sheet cork	0.11	0.10	0.10	0.09	0.09	0.08	0.08

Table 3-12. Solid Brick Wall (No Exterior Finish)

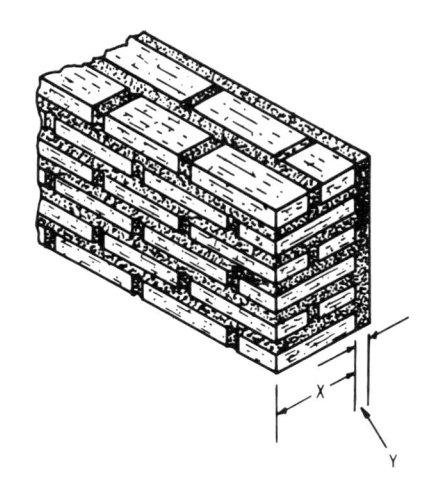

K = X + Y
X = Wall
Y = Interior Construction

Solid Brick Wall
(No Exterior Finish)
Values of K in Btu/1 °F/sq ft

Wall construction (Y)	Thickness (X)						
	8"	9"	12"	13"	16"	18"	24"
Plain wall—no interior finish	0.39	0.37	0.30	0.29	0.24	0.23	0.18
½" Plaster—direct on brick	0.36	0.35	0.28	0.27	0.23	0.20	0.16
¾" Plaster on metal lath, furred	0.26	0.25	0.23	0.21	0.19	0.18	0.15
½" Plaster on wood lath, furred	0.25	0.24	0.21	0.19	0.18	0.17	0.14
½" Plaster on ⅜" plasterboard, furred	0.24	0.23	0.20	0.18	0.17	0.16	0.14
½" Plaster on ½" board insulation, furred	0.20	0.19	0.18	0.17	0.16	0.15	0.13
½" Plaster on ½" sheet cork, set in ½" cement	0.18	0.17	0.16	0.15	0.14	0.13	0.12
½" Plaster on 1" sheet cork, set in ½" cement	0.15	0.14	0.13	0.12	0.11	0.10	0.09
½" Plaster on 1½" sheet cork, set in ½" cement	0.14	0.13	0.12	0.11	0.10	0.09	0.08
½" Plaster on 2" sheet cork, set in ½" cement	0.13	0.12	0.11	0.10	0.09	0.08	0.07
½" Plaster on 1½" split furring, tiled	0.28	0.26	0.23	0.22	0.20	0.19	0.17

AIR CONDITIONING

Table 3-13. Wood Siding Clapboard Frame or Shingle Walls

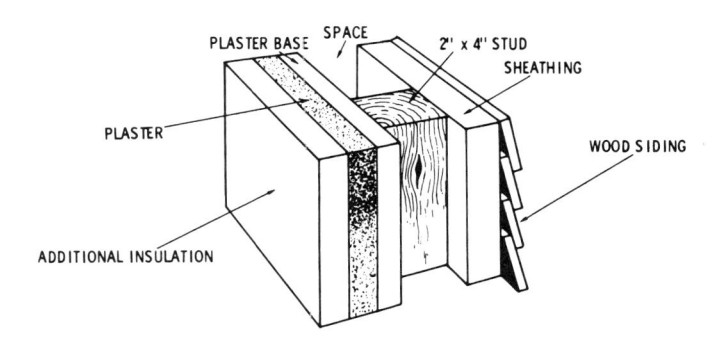

Wood Siding Clapboard Frame or Shingle Walls
Values of K in Btu/hr/1°F/sq ft

Type of Sheathing	Insulation between Studding	Plaster Base						
		Wood lath	Metal lath	$3/8"$ Plaster board	$1/2"$ Rigid insulation	$1"$ Rigid insulation	$1 1/2"$ Sheet cork	$2"$ Sheet cork
$1/2"$ Plaster-board	None	0.30	0.31	0.30	0.22	0.16	0.12	0.10
	Cellular gypsum, 18#	0.12	0.125	0.12	0.10	0.09	0.08	0.07
	Flaked gypsum, 24#	0.10	0.108	0.10	0.09	0.08	0.07	0.06
	$1/2"$ flexible insulation	0.16	0.17	0.16	0.14	0.11	0.09	0.08
$1"$ Wood	None	0.26	0.28	0.26	0.20	0.15	0.12	0.10
	Cellular gypsum, 18#	0.11	0.118	0.11	0.10	0.09	0.08	0.07
	Flaked gypsum, 24#	0.10	0.108	0.10	0.09	0.08	0.07	0.06
	$1/2"$ flexible insulation	0.15	0.16	0.15	0.13	0.11	0.09	0.08
$1/2"$ Rigid insulation (board form)	None	0.22	0.23	0.22	0.17	0.14	0.11	0.09
	Cellular gypsum, 18#	0.10	0.11	0.10	0.09	0.08	0.07	0.06
	Flaked gypsum, 24#	0.09	0.10	0.09	0.08	0.07	0.06	0.05
	$1/2"$ flexible insulation	0.14	0.148	0.14	0.12	0.10	0.08	0.07

HEAT LEAKAGE

Table 3-14. Interior Walls and Plastered Partitions (No Fill)

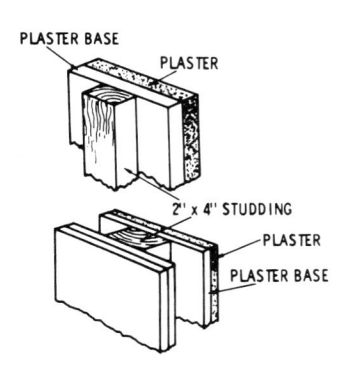

Interior Walls and Plastered Partitions (No Fill)
Values of K in Btu/hr/1°F/sq ft

Plaster base	Single partition, one side plastered	Double partition, two sides plastered
Wood lath	0.60	0.33
Metal lath	0.67	0.37
3/8" Plasterboard	0.56	0.32
1/2" Board (rigid) insulation	0.33	0.18
1" Board (rigid) insulation	0.22	0.12
1½" Corkboard	0.16	0.08
2" Corkboard	0.12	0.06

AIR CONDITIONING

Table 3-15. Frame Floors and Ceilings (No Fill)

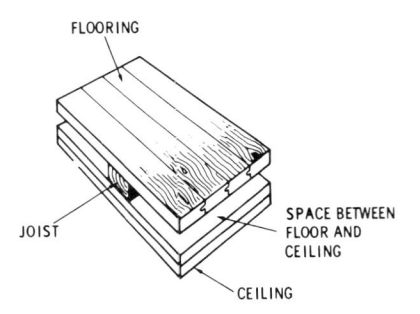

Frame Floors and Ceilings (No Fill)
Values of K in Btu/hr/1°F/sq ft

Type of ceiling with ½" plaster	No flooring	1" Yellow pine flooring	1" Yellow pine flooring and ½" Fiberboard	1" Sub oak or maple on 1" yellow pine flooring
No ceiling		0.45	0.27	0.32
Metal lath	0.66	0.24	0.20	0.24
Wood lath	0.60	0.27	0.19	0.23
⅜" Plasterboard	0.59	0.28	0.20	0.24
½" Rigid insulation	0.34	0.20	0.15	0.17

HEAT LEAKAGE

Table 3-16. Masonry Partitions

BARE NO PLASTER

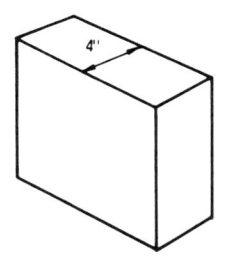

4"

PLASTERED ONE SIDE

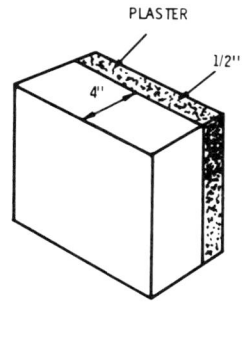

PLASTER

1/2"
4"

PLASTERED BOTH SIDES

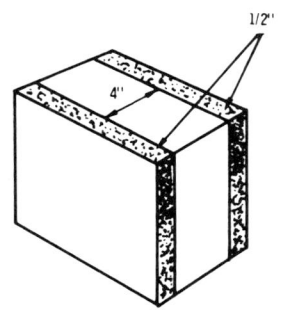

1/2"
4"

Masonry Partitions
Values of K in Btu/hr/1°F/sq ft

Wall	Bare, no plaster	One side plastered	Two sides plastered
4" Brick	0.48	0.46	0.41
4" Hollow gypsum tile	0.29	0.27	0.26
4" Hollow clay tile	0.44	0.41	0.39

Table 3-17. Concrete Floors (on Ground)

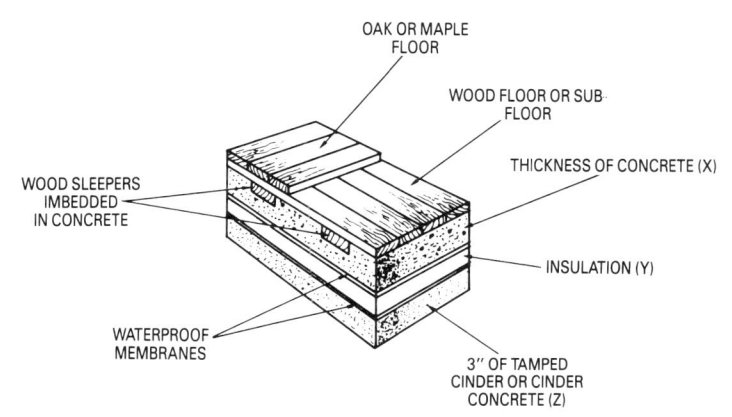

Diagram labels: OAK OR MAPLE FLOOR; WOOD FLOOR OR SUB-FLOOR; THICKNESS OF CONCRETE (X); WOOD SLEEPERS IMBEDDED IN CONCRETE; INSULATION (Y); WATERPROOF MEMBRANES; 3" OF TAMPED CINDER OR CINDER CONCRETE (Z)

Concrete Floors (on Ground)
Values of K in Btu/hr/1°F/sq ft

Type of Insulation	Thickness of Y	Concrete thickness X	No flooring	Tile or terrazzo floor	1" yellow pine on wood sleepers	Oak or maple on yellow pine sub floor on sleepers
None, and no cinder base (Z)	0	4	1.020	0.930	0.510	0.370
	0	5	0.940	0.830	0.480	0.360
	0	6	0.860	0.790	0.460	0.340
	0	8	0.760	0.700	0.430	0.320
None, with 3" cinder base (Z)	0	4	0.580	0.550	0.510	0.310
	0	5	0.550	0.520	0.470	0.300
	0	6	0.520	0.490	0.440	0.290
	0	8	0.500	0.470	0.400	0.280
1" Rigid insulation (Y) with 3" cinder base (Z)	1"	4	0.220	0.210	0.180	0.165
	1"	5	0.210	0.200	0.175	0.160
	1"	6	0.200	0.190	0.170	0.155
	1"	8	0.190	0.180	0.160	0.150
2" Corkboard insulation (Y) with 3" cinder base (Z)	2"	4	0.125	0.120	0.115	0.105
	2"	5	0.120	0.115	0.110	0.100
	2"	6	0.115	0.110	0.105	0.095
	2"	8	0.110	0.105	0.100	0.090

HEAT LEAKAGE

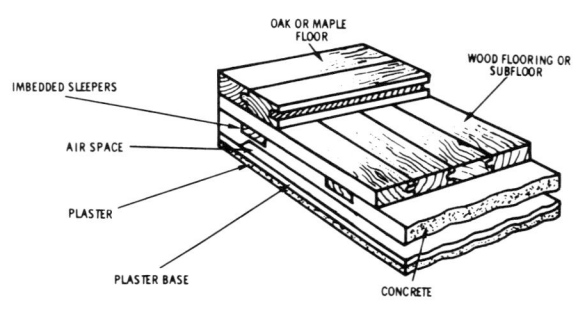

Table 3-18. Concrete Floors and Ceilings

Concrete Floors and Ceilings
Values of K in Btu/hr/1°F/sq ft

Type of ceiling and base	Thickness	No flooring	Tile or Terrazzo floor	1" Yellow pine on wood sleepers	Oak or maple on yellow pine, sub floor on sleeper
No ceiling	4	0.60	0.57	0.38	0.29
	6	0.54	0.52	0.35	0.28
	8	0.48	0.46	0.33	0.26
	10	0.46	0.44	0.31	0.25
½" Plaster direct on concrete	4	0.56	0.53	0.38	0.30
	6	0.51	0.49	0.35	0.29
	8	0.47	0.44	0.33	0.27
	10	0.42	0.41	0.31	0.26
Suspended or furred ceiling on ⅜" plasterboard, ½" plaster	4	0.33	0.32	0.24	0.20
	6	0.31	0.30	0.23	0.19
	8	0.29	0.28	0.22	0.18
	10	0.28	0.27	0.21	0.17
Suspended or furred metal lath, ¾" plaster	4	0.36	0.34	0.26	0.21
	6	0.34	0.32	0.24	0.20
	8	0.32	0.30	0.23	0.19
	10	0.31	0.29	0.22	0.18
Suspended or furred on ½" rigid insulation, ½" plaster	4	0.22	0.21	0.20	0.18
	6	0.21	0.20	0.19	0.17
	8	0.20	0.19	0.18	0.16
	10	0.19	0.18	0.17	0.15
1½" corkboard in ½" mortar, ½" plaster	4	0.16	0.15	0.14	0.13
	6	0.15	0.14	0.13	0.12
	8	0.14	0.13	0.12	0.11
	10	0.13	0.12	0.11	0.10

Table 3-19. Frame Floors and Ceilings (with Fill)

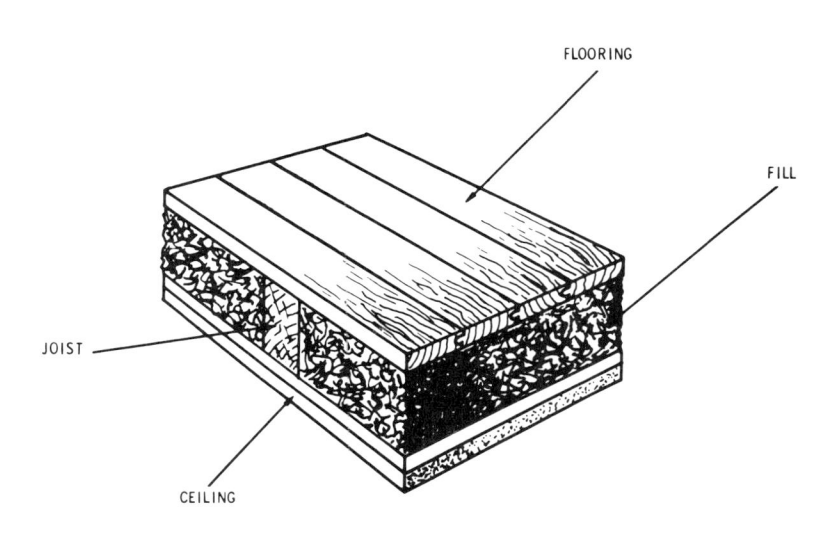

Frame Floors and Ceilings (with Fill)
Values of K in Btu/hr/1°F/sq ft

Type of ceiling with ½" plaster	Insulation or fill between joists	No flooring	1" yellow pine flooring	1" yellow pine flooring and ½" fiberboard	1" suboak, maple on 1" yellow pine flooring
Wood lath	½" flexible insulation	0.220	0.150	0.120	0.140
Wood lath	½" rigid insulation	0.240	0.160	0.130	0.150
Wood lath	Flaked 2" gypsum	0.150	0.120	0.105	0.120
Wood lath	Rock wool, 2"	0.120	0.095	0.080	0.090
Corkboard ½"	None	0.155	0.115	0.100	0.105
Corkboard ½"	None	0.120	0.100	0.085	0.095

HEAT LEAKAGE

Table 3-20. Brick Veneer (Frame Walls)

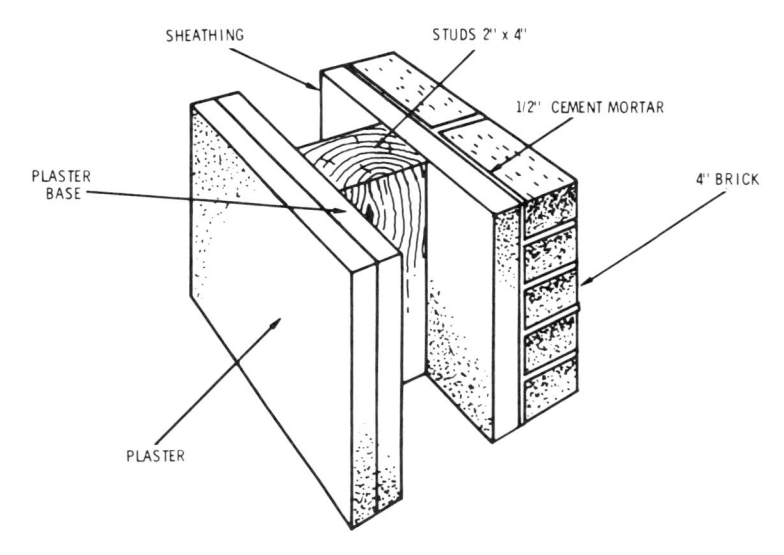

Brick Veneer (Frame Walls)

Values of K in Btu/hr/1°F/sq ft

Type of sheathing	Insulation between studding	Plaster Base							
		Wood lath	Metal lath	$\frac{1}{2}''$ Fiber board	$\frac{3}{8}''$ Plaster-board	$\frac{1}{2}''$ Cork-board	1'' Cork-board	$1\frac{1}{2}''$ Cork-board	2'' Cork-board
1'' Wood sheathing	None	0.260	0.270	0.20	0.27	0.18	0.15	0.12	0.095
	$\frac{1}{2}''$ fiberboard	0.170	0.180	0.14	0.16	0.14	0.11	0.09	0.080
	Flaked gypsum 24#	0.095	0.100	0.09	0.10		0.08	0.07	0.060
	Cellular gypsum 18#	0.110	0.150	0.10	0.11		0.09	0.08	0.070
$\frac{1}{2}''$ Plaster-board sheathing	None	0.320	0.340	0.24	0.30		0.16	0.13	0.110
	$\frac{1}{2}''$ fiberboard	0.180	0.190	0.14	0.17		0.11	0.09	0.080
	Flaked gypsum 24#	0.100	0.110	0.09	0.10		0.08	0.07	0.060
	Cellular gypsum 18#	0.120	0.120	0.10	0.12		0.09	0.08	0.070
$\frac{1}{2}''$ Fiber-board (rigid) insulation sheathing	None	0.250	0.260	0.19	0.26		0.15	0.12	0.100
	$\frac{1}{2}''$ fiberboard	0.160	0.170	0.13	0.15		0.10	0.09	0.080
	Flaked gypsum 24#	0.090	0.090	0.08	0.09		0.07	0.07	0.060
	Cellular gypsum 18#	0.100	0.105	0.09	0.10		0.08	0.07	0.060

63

AIR CONDITIONING

SUMMARY

Building materials affect the design of air-conditioning and heating systems. Heat loss through walls, floors, and ceilings is a prime consideration when calculating and installing an air conditioner.

Heat leakage is always given in Btu per hour per degree Fahrenheit temperature difference per square foot of exposed surface (Btu/sq ft/hr/°F). Prior to designing an air-conditioning system, an estimate must be made of the maximum heat loss of each room or space to be cooled. One ton of refrigeration equals 12,000 Btu/hr. It is an easy matter to calculate the required size of equipment since all that is necessary is to divide the combined sum of leakage and cooling loads (in Btu/hr) by 12,000 to obtain the condensing unit capacity.

REVIEW QUESTIONS

1. How do the thermal properties of building materials affect the design of air-conditioning systems?
2. What is meant by *air leakage* in buildings, and what is the method of air-leakage reduction?
3. State the various factors affecting the size of an air-conditioning system in residential projects.
4. Define the method of heat-leakage calculation.
5. Upon what factors is the heat-leakage formula based?
6. Calculate the heat leakage in Btu/hr for a plain 10-inch concrete wall (no exterior finish) having an area of 325 sq ft for a temperature difference of 40°F.
7. Why is the heat leakage through walls, ceilings, and floors an important consideration when installing air-conditioning plants?

CHAPTER 4

Ventilation Requirements

By definition, ventilation is the process of supplying or removing air to or from any building or space. Such air may or may not have been conditioned. Methods of supplying or removing the air are accomplished by natural ventilation and mechanical methods.

Natural ventilation is obtained by opening or closing window sashes and by using vents or ventilators above the roof, chimney drafts, etc.; mechanical ventilation is produced by the use of fans or other means to force the air through the space to be ventilated.

AIR LEAKAGE

Air leakage due to cold air outside and warm air inside takes place when the building contains cracks or openings at different levels. This results in the cold and heavy air entering at low levels and pushing the warm and light air out at high levels; the same draft takes place in a chimney. When storm sashes are applied to well-fitted windows, very little redirection in infiltration is secured, but the application of the sash does give an air space that reduces heat transmission and helps prevent window frosting. By applying storm sashes to poorly fitted windows, a reduction in leakage of up to 50 percent may be obtained; the effect, insofar as air leakage is concerned, will be roughly equivalent to that obtained by the installation of weather stripping.

NATURAL VENTILATION

There are two natural forces available for moving air through and out of buildings: wind forces and temperature difference between the inside and the outside. Air movement may be caused by either of these forces acting alone or by a combination of the two, depending on atmospheric conditions, building design, and location.

Wind Forces

In considering the use of natural forces for producing ventilation, the conditions which must be considered are as follows:

1. Average wind velocity
2. Prevailing wind direction
3. Seasonal and daily variations in wind velocity and direction
4. Local wind interference by buildings and other obstructions

When the wind blows without encountering any obstructions to change its direction, the movements of the airstream, as well as the pressure, remain constant. If, on the other hand, the airstream meets an obstruction of any kind, such as a house or ventilator, the airstream will be pushed aside as illustrated in Fig. 4-1. In the case of a simple ventilator (Fig. 4-2), the closed end forms an obstruction that changes the direction of the wind, expanding at the closed end and converging at the open end, thus producing a

VENTILATION REQUIREMENTS

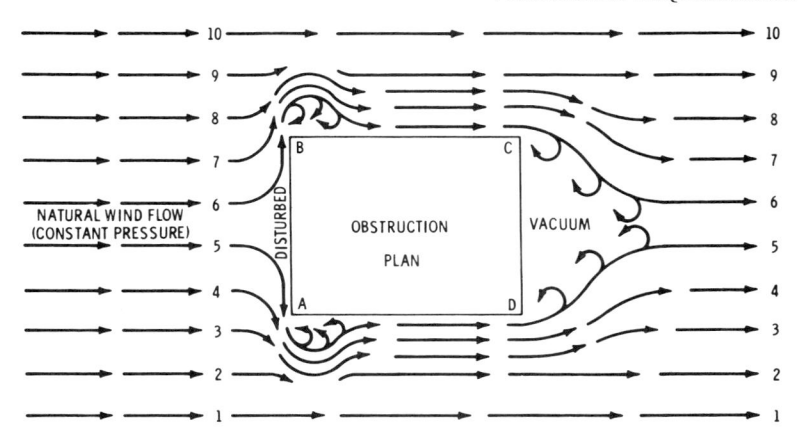

Fig. 4-1. Diagram of wind action where the airstream meets an obstruction such as a building or ventilator.

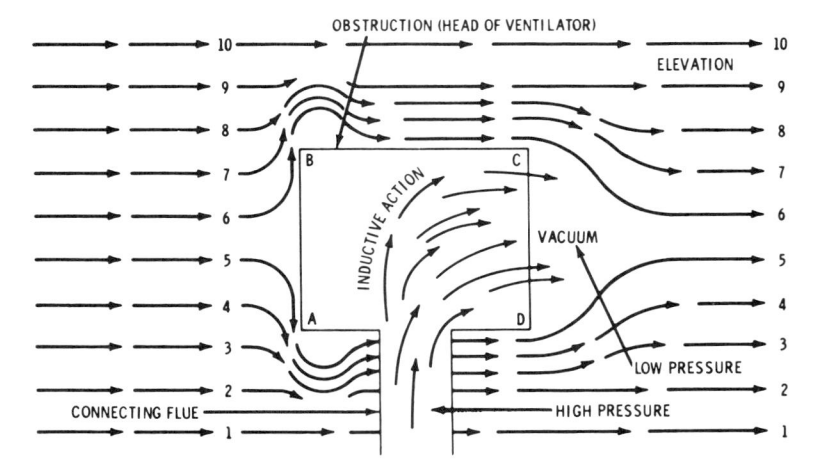

Fig. 4-2. Airflow in a roof ventilator.

vacuum inside the head, which induces an upward flow of air through the flue and through the head.

Temperature-Difference Forces

Perhaps the best example of the thermal effect is the draft in a stack or chimney known as the *flue effect*. The flue effect of a

67

stack is produced within a building when the outdoor temperature is lower than the indoor temperature and is due to the difference in weight of the warm column of air within the building and the cooler air outside. The flow due to flue effect is proportional to the square root of the draft head, or approximately:

$$Q = 9.4A \sqrt{h(t - t_0)}$$

where Q = air flow, cu ft/min
A = free area of inlets or outlets (assumed equal), sq ft
h = height from inlets to outlets, feet
t = average temperature of indoor air, °F
t_0 = temperature of outdoor air, °F
9.4 = constant of proportionality (including a value of 65% for effectiveness of openings)

The constant of proportionality should be reduced by 50% (constant = 7.2) if conditions are not favorable.

Combined Wind and Temperature Forces

It should be noted that when both forces are acting together, even without interference, the resulting air flow is not equal to the sum of the two estimated quantities, but the flow through any opening is proportional to the square root of the sum of the heads acting on that opening. When the two heads are equal in value, and the ventilating openings are operated so as to coordinate them, the total airflow through the building is about 10 percent greater than that produced by either head acting independently under ideal conditions. This percentage decreases rapidly as one head increases over the other. The larger head will predominate.

ROOF VENTILATORS

The function of a roof ventilator is to provide a storm and weatherproof air outlet. For maximum flow by induction, the ventilator should be located on that part of the roof where it will receive the full wind without interference. Roof ventilators are made in a variety of shapes and styles (Fig. 4-3). Depending on

VENTILATION REQUIREMENTS

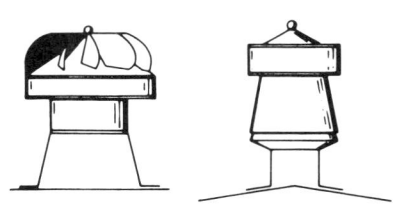

Fig. 4-3. Essential parts of a turbine and syphonage roof ventilator.

their construction and application, they may be termed as stationary, revolving, turbine, ridge, and syphonage.

Ventilator Capacity

Several factors must be taken into consideration in selecting the proper ventilator for any specific problem:

1. Mean temperature difference
2. Stack height (chimney effect)
3. Induction effect of the wind
4. Area of ventilator opening

Of the several other minor factors, only one requires close attention. This is the area of the inlet air openings (Fig. 4-4). The action of a roof ventilator is to let air escape from the top of a building, which naturally means that a like amount of air must be admitted to the building to take the place of the air exhausted. The nature, size, and location of these inlet openings is of importance in determining the effectiveness of the ventilating system.

FRESH-AIR REQUIREMENTS

Table 4-1 is a guide to the amount of air required for efficient ventilation. This information should be used in connection with the ventilator-capacity tables provided by fan manufacturers to assist in the proper selection of the number and size of the units required.

AIR CONDITIONING

Table 4-1. Fresh-Air Requirements

Type of Building or Room	Minimum Air Changes per Hour	Cubic Feet of Air per Minute per Occupant
Attic spaces (for cooling)	12-15	
Boiler room	15-20	
Churches, auditoriums	8	20-30
College classrooms		25-30
Dining rooms (hotel)	5	
Engine rooms	4- 6	
Factory buildings		
Ordinary manufacturing	2- 4	
Extreme fumes or moisture	10-15	
Foundries	15-20	
Galvanizing plants	20-30*	
Garages		
Repair	20-30	
Storage	4- 6	
Homes (night cooling)	9-17	
Hospitals		
General		40-50
Children's		35-40
Contagious diseases		89-90
Kitchens		
Hotel	10-20	
Restaurant	10-20	
Libraries (public)	4	
Laundries	10-15	
Mills		
Paper	15-20*	
Textile-general buildings	4	
Textile-dyehouses	15-20*	
Offices		
Public	3	
Private	4	
Pickling plants	10-15†	
Pump rooms	5	
Restaurants	8-12	
Schools		
Grade		15-25
High		30-35
Shops		
Machine	5	
Paint	15-20*	
Railroad	5	
Woodworking	5	
Substations (electric)	5-10	
Theaters		10-15
Turbine rooms (electric)	5-10	
Warehouses	2	
Waiting rooms (public)	4	

*Hoods should be installed over vats or machines.
†Unit heaters should be directed on vats to keep fumes superheated.

Fig. 4-4. Typical ventilator arrangement.

MECHANICAL VENTILATION

Mechanical ventilation differs from natural ventilation mainly in that the air circulation is performed by mechanical means, such as fans or blowers; in natural ventilation, the air is caused to move by natural forces. In mechanical ventilation, the required air changes are effected partly by diffusion but chiefly by positive currents put in motion by electrically operated fans or blowers, as shown in Fig. 4-5. Fresh air is usually circulated through registers connected with the outside and warmed as it passes over and through the intervening radiators.

Volume of Air Required

The volume of air required is determined by the size of the space to be ventilated and the number of times per hour that the air in the space is to be changed. In many cases, existing local regulations or codes will govern the ventilating requirements. Some of these codes are based on a specified amount of air per person, and others on the air required per square foot of floor area.

Duct-System Resistance

Air ducts may be designed with either a round or rectangular cross section. The radius of elbows should preferably be at least

AIR CONDITIONING

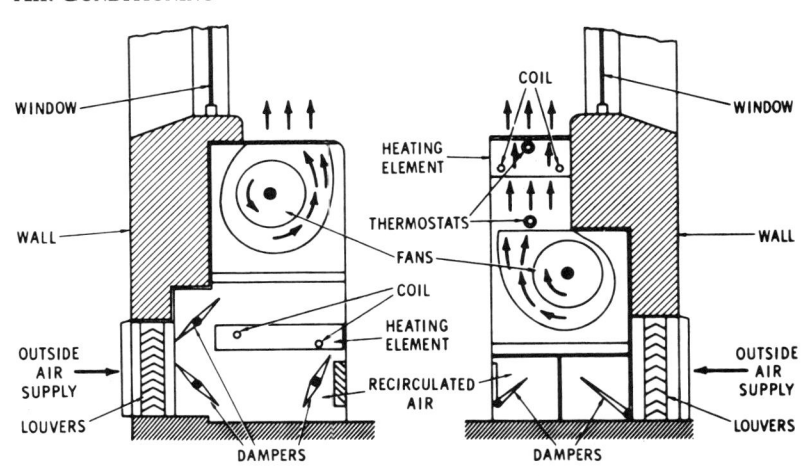

Fig. 4-5. Typical mechanical ventilators for residential use showing placement of fans and other details.

one and one-half times the pipe diameter for round pipes or the equivalent round pipe size in the case of rectangular ducts. Accuracy in estimating the resistance to the flow of air through the duct system is important in the selection of blowers for application in duct systems (Fig. 4-6).

Resistance should be kept as low as possible in the interest of economy since underestimating the resistance will result in failure of the blower to deliver the required volume of air. Careful study

Fig. 4-6. Central-station air-handling unit.

Courtesy Buffalo Forge Company

72

VENTILATION REQUIREMENTS

should be made of the building drawings with consideration being given to duct locations and clearances. Keep all duct runs as short as possible, bearing in mind that the airflow should be conducted as directly as possible from the source to the delivery points. Select the locations of the duct outlets so as to ensure proper air distribution. The ducts should be provided with cross-sectional areas that will permit air to flow at suitable velocities. Moderate velocities should be used in all ventilating work to avoid waste of power and reduce noise. Lower velocities are more frequently used in schools, churches, theaters, etc., instead of factories and other places where noise due to airflow is not objectionable.

AIR FILTRATION

The function of air filters as installed in air-conditioning systems is to remove the airborne dust that tends to settle in the air of the ventilated space and may become a menace to human health if not properly removed.

Effect of Dust on Health

The effect of dust on health has been properly emphasized by competent medical authorities. The normal human breathes about seventeen times a minute. The air taken into the lungs may contain large quantities of dust, soot, germs, bacteria, and other deleterious matter. Most of this solid matter is removed in the nose and air passages of the normal person; but if these passages are dry and permit the passage of these materials, colds and respiratory diseases result. Air-conditioning apparatus removes these contaminants from the air and further provides the correct amount of moisture so that the respiratory tracts are not dehydrated but are kept properly moist. Among the airborne diseases are mumps, measles, scarlet fever, pneumonia, colds, tuberculosis, hay fever, grippe, influenza, and diphtheria.

Dust is more than just dry dirt. It is exceedingly complex, with variable mixtures of materials, and, as a whole, it is rather uninviting, especially the type found in and around human habitation. Dust contains fine particles of sand, soot, earth, rust, fiber, animal and vegetable refuse, hair, chemicals, and compounds, all of which are either abrasive, irritating, or both. The U.S. Weather Bureau estimates that there are 115,000 particles of dust per cubic inch of ordinary city air, and that each grain of this dust at breathing level contains from 85,000 to 125,000 germs. The proximity of factories using oil or coal-burning equipment and the presence of considerable street traffic will aggravate this condition and increase the dust and germ content.

The Mellon Institute of Industrial Research conducted a series of experiments to determine dust precipitation in three large cities. The measurements published were the average amounts measured in more than ten stations in each of the cities and were for a period of one year. It was observed that precipitation was considerably greater in the industrial districts. It is certainly plain that filtered air is almost a necessity, especially in large buildings where a number of individuals are gathered. Hotels, theaters, schools, stores, hospitals, factories, and museums require the removal of dust from the atmosphere admitted to their interiors, not only because a purer product is made available for human health and consumption, but also so that the fittings, clothing, furniture, and equipment will not be damaged by the dust particles borne by the fresh airstream.

The air filter is one part of the air-conditioning system that should be operated the year around. There are times when the air washer (the cooling and humidifying apparatus) is not needed, but the filter is one part of the system that should be kept in continuous operation, purifying the air.

Various Dust Sources

The dust removed in the filters may be classified as poisonous, infective, obstructive, or irritating and may originate from an animal, vegetable, mineral, or metallic source. Industry produces many deleterious fumes and dusts, for example, those emanating from grinding, polishing, dyeing, gilding, painting, spraying, cleaning, pulverizing, baking, mixing of poisons, sawing, and sand blasting. This dust must be filtered out of the air because of its effect on health and efficiency, to recover valuable materials, to save heat by reusing filtered air, and to prevent damage to equipment and products in the course of manufacturing.

VENTILATION REQUIREMENTS

AIR-FILTER CLASSIFICATION

Air-cleaning equipment may be classified according to the following methods used: (1) filtering, (2) washing, and (3) combined filtering and washing.

There are several types of filters divided into two main classes: dry and wet. The wet filters may again be subdivided into what is known as the manual and automatic self-cleaning types.

Dry Filters

Dry filters are usually made in standard units of definite rated capacity and resistance to the passage of air. At the time the installation is designed, ample filter area must be included so that the air velocity passing through the filters is not excessive; otherwise proper filtration will not be secured. The proper area is secured by mounting the proper number of filters on an iron frame. A dry filter (Fig. 4-7) may be provided with permanent filtering material from which the dirt and dust must be removed pneumatically or by vibration; or it may be provided with an inexpensive filter medium, which can be dispensed with when loaded with dirt and replaced with a new slab.

Fig. 4-7. Typical air-filter panel absorber containing activated charcoal for heavy-duty control of odors and vapors.

Courtesy Barneby Cheney Corporation

AIR CONDITIONING

A dry-type filter is known as a *dry plate*. The cells are composed of a number of perforated aluminum plates, arranged in series and coated with a fireproof filament. Air drawn through the cell is split into thin streams by the perforations so that the dust particles are thrown against the filament by the change in the direction of the air between the plates. The filament tentacles hold the dust particles and serve to purify the air. Dust particles are deposited on the intervening flat surfaces. When the plates are loaded with dirt and require cleaning, this may be accomplished in two ways.

In moderate-sized installations, the operator removes the cells from the sectional frame, rests one edge on the table or floor, and gently raps the opposite side with the open hand, causing the collected dirt to drop out and restoring the efficiency of the filter. Larger installations (Fig. 4-8), where a great number of cells are

Courtesy Barneby Cheney Corporation

Fig. 4-8. Activated-charcoal room-purification filter panel absorber. Filter absorbers of this type are recommended for central air-conditioning-system purification of recirculated air in occupied spaces, purification of exhaust to prevent atmospheric pollution, etc. They are suitable for light-, medium-, and heavy-duty requirements.

76

VENTILATION REQUIREMENTS

used, or where an abnormally high dust load is encountered, use a mechanical vibrator in which the cell is placed and rapped for about a minute to free it of dust. Under average dust conditions, cells must be cleaned or replaced about twice a year.

Packed-type filters contain a fibrous mat so that the slow movement and irregular path taken by the air permits it to leave behind all particles of dirt in the mazes of the fibrous mat (Fig. 4-8). This type of unit must be removed every 12 to 15 weeks under average conditions.

Wet, or Viscous, Filters

The wet, or viscous, filter makes use of an adhesive in which the dust particles are caught and held upon impingement. It also makes use of densely packed layers of viscous-coated metal baffles, screens, sinuous passages, crimped wire, and glass wool inserted in the path of the air. The air divides into small streams that constantly change their direction and force the heavier dirt particles against the viscous-coated surfaces, where they are held.

There are three major types of wet, or viscous, air filters: (1) replaceable, (2) manually cleaned, and (3) automatically cleaned.

Wet Replacement Type — The glass-wool air filter is of the viscous-type replacement cell. Glass wool, being noncorrosive and non-absorbing, maintains its density and leaves all the viscous adhesive free to collect and hold dirt and dust. This type of filter lends itself to both small and large installations. A carton of 12 filters weighs about 37 lb. After the filter replacement has been completed, the dirty filters can be put into the original shipping container and removed from the premises.

Manually Cleaned Type — The manually cleaned type is usually made in the form of a standard cell with a steel frame containing the filter media. The cell is fitted into the cell frame (Fig. 4-9), with the use of a felt gasket to prevent leakage of air past the cell. As a usual practice, cleaning is necessary about once every eight weeks. This is made a simple matter by the use of automatic latches so that the filter sections can be pulled out of the frame, immersed in a solution of water and washing soda or cleaning compound, allowed to dry, then dipped into charging oil, drained to remove excess oil, and then replaced in the filter.

AIR CONDITIONING

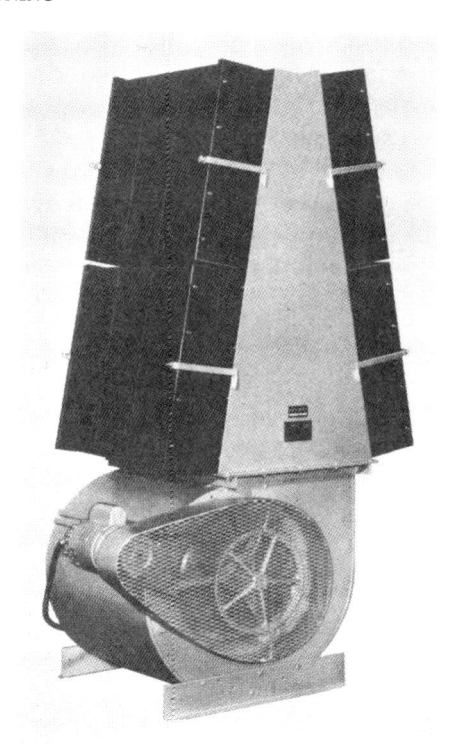

Courtesy Barneby Cheney Corporation

Fig. 4-9. Typical open-frame industrial air purifier. Each filter unit contains one or more activated charcoal absorbers, including blower and motor.

Automatic Viscous Filters — The automatic filter is of the self-cleaning type and utilizes the same principle of adhesive dust impingement as the manual and replacement types, but the removal of the accumulated dirt from the filter medium is entirely automatic. The self-cleaning type may be divided into two distinct systems: (1) immersion variety and (2) flushing type.

The immersion-type air filter (Fig. 4-10) makes use of an endless belt that rotates the filter medium and passes it through an oil bath, washing the dirt off. The dirt, being heavier than the oil, settles to the bottom into special containers, which are removed

VENTILATION REQUIREMENTS

Courtesy Rockwell-Standard Corporation

Fig. 4-10. Typical automatic viscous filter. In a self-cleaning filter of this type, brushes on an endless carrier chain periodically sweep collected materials from the upstream side of the panels. Used in textile mills, laundries, and other ventilation systems involving a high volume of coarse, bulky contaminants.

and cleaned from time to time. The usual speed of the filter is from 1 1/2 to 3 inches every 12 minutes. Such apparatus is particularly adapted for continuous operation. The sediment can be removed and the oil changed or added without stopping the apparatus.

The flushing type is constructed of cells in the form of a unit laid on shelves and connected by metal aprons. The dirt caught by the cells is flushed down into a sediment tank by flooding pipes, which travel back and forth over the clean-air side of the cells. The aprons catch the heavier dust particles before they get to the cells so that the cells are more efficient. This type does not flush the cells while the system is in operation. In many cases, duplicate units are provided so that one unit can be flushed while the other bears the full load. Some of the flushing mechanisms are interlocked with the fan circuit, especially where a single unit is used, so that flushing cannot take place unless the fan is shut down. If

AIR CONDITIONING

air is permitted to pass through the filter while it is being flushed, oil will be carried along with the air.

A cylindrical absorber, shown in Fig. 4-11, can be used for the control of odor problems by forcing air through two perforated metal walls enclosing an annular bed of activated charcoal. A motor-operated blower draws contaminated air through the charcoal-filled container and recirculates pure odor-free air through the room or space to be purified.

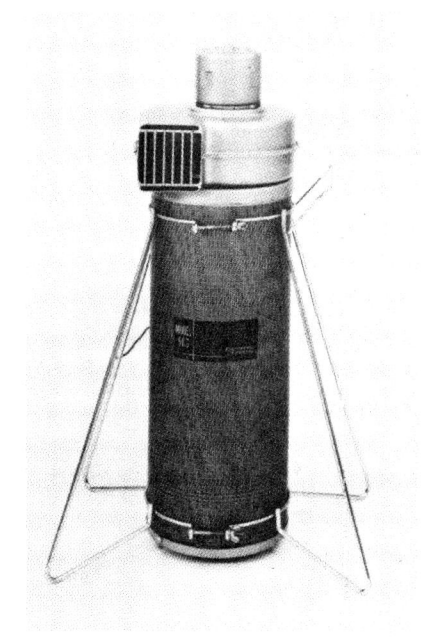

Fig. 4-11. Cylindrical-type canister air purifier.

Courtesy Barneby Cheney Corporation

The wall-mounted air purifier shown in Fig. 4-12 is equipped with a washable dust filter and centrifugal blower. Air purifiers of this type are ideal for toilets, utility rooms, public rest rooms, and similar odor-problem areas.

In residential-type air conditioners, however, unit filters provide the necessary dust protection, particularly since they are manufactured in a large variety of types and sizes. Where lint in a dry state predominates, a dry filter is preferable because of its lint-

VENTILATION REQUIREMENTS

Courtesy Barneby Cheney Corporation

Fig. 4-12. Wall-mounted air purifier.

holding capacity. Throwaway filters are used increasingly where the cleaning process needs to be eliminated.

FILTER INSTALLATION

Air filters are commonly installed in the outdoor air intake ducts of the air-conditioning system and often in the recirculating air ducts as well. Air filters are logically mounted ahead of heating or cooling coils and other air-conditioning equipment in the system to prevent dust from entering. Filters should be installed so that the face area is at right angles to the airflow wherever possible (Fig. 4-13). In most cases, failure of an air-filter installation can be traced to faulty installation, improper maintenance, or both.

AIR CONDITIONING

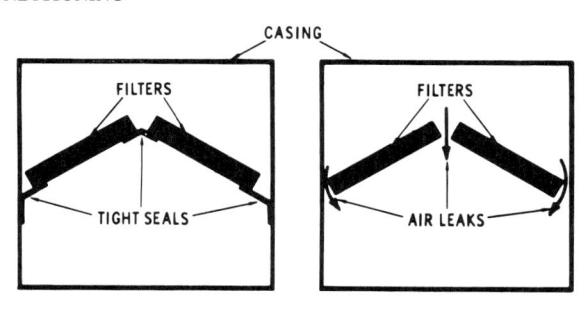

Fig. 4-13. Good (left) and poor (right) air-filter installations. They should fit tightly to prevent air from bypassing the filter.

The American Society of Heating and Ventilating Engineers gives the most important requirements for a satisfactory and efficiently functioning air-filter installation as follows:

1. The filter must be of ample size for the amount of air it is expected to handle. An overload of 10 to 15 percent is regarded as the maximum allowable. When air volume is subject to increase, a larger filter should be installed.
2. The filter must be suited to the operating conditions, such as degree of air cleanliness required, amount of dust in the entering air, type of duty, allowable pressure drop, operating temperatures, and maintenance facilities.
3. The filter type should be the most economical for the specific application. The first cost of the installation should be balanced against depreciation as well as expense and convenience of maintenance.

The following recommendations apply to filters and washers installed with central fan systems:

1. Duct connections to and from the filter should change their size or shape gradually to ensure even air distribution over the entire filter area.
2. Sufficient space should be provided in front as well as behind the filter to make it accessible for inspection and service. A distance of 2 feet may be regarded as the minimum.

VENTILATION REQUIREMENTS

3. Access doors of convenient size should be provided in the sheet-metal connections leading to and from the filters.
4. All doors on the clean-air side should be lined with felt to prevent infiltration of unclean air. All connections and seams of the sheet-metal ducts on the clean-air side should be air tight.
5. Electric lights should be installed in the chamber in front of and behind the air filter.
6. Air washers should be installed between the tempering and heating coils to protect them from extreme cold in winter.
7. Filters installed close to air inlets should be protected from the weather by suitable louvers, in front of which a large-mesh wire screen should be provided.
8. Filters should have permanent indicators to give a warning when the filter resistance reaches too high a value.

HUMIDITY-CONTROL METHODS

Humidifiers, by definition, are devices for adding moisture to the air. Thus, to humidify is to increase the density of water vapor within a given space or room. Air humidification is effected by vaporization of water and always requires heat for its proper functioning. Thus, devices that function to add moisture to the air are termed *humidifiers*, whereas devices that function to remove moisture from the air are termed *dehumidifiers*.

HUMIDIFIERS

As previously noted, air humidification consists of adding moisture. The types of humidifiers used in air-conditioning systems are:

1. Spray-type air washers
2. Pan evaporative humidifiers
3. Electrically operated humidifiers
4. Air-operated humidifiers

Air-Washer Method

An air washer consists essentially of a row of spray nozzles inside a chamber or casing. A tank at the bottom of the chamber provides for collection of water as it falls through the air and comes into intimate contact with the wet surface of the chamber baffles. The water is generally circulated by means of a pump, the warm water being passed over refrigerating coils or blocks of ice to cool it before being passed to the spray chamber. The water lost in evaporation is usually replaced automatically by the use of a float arrangement, which admits water from the main as required. In many locations, the water is sufficiently cool to use as it is drawn from the source. In other places, the water is not cool enough and must be cooled by means of ice or through the use of a refrigerating machine.

The principal functions of the air washer are to cool the air passed through the spray chamber and to control humidity. In many cases, the cooling coils are located in the bottom of the spray chamber so that as the warm spray descends, it is cooled and ready to be again sprayed by the pump. In some cases, the water is passed through a double-pipe arrangement and is cooled on the counter-current principle.

A sketch of an air washer is shown in Fig. 4-14. In this case, the spray pipes are mounted vertically. In some instances, the spray pipes are horizontal so that the sprays are directed downward. As some of the finer water particles tend to be carried along with the air current, a series of curved plates or baffles is generally used, which force the cooled and humidified air to change the direction of flow, throwing out or eliminating the water particles in the process.

Pan Humidifiers

The essential parts of the pan-type humidifier are shown in Fig. 4-15. The main part is a tank of water heated by low-pressure steam or forced hot water where a water temperature of 200°F or higher is maintained. The evaporative-type humidifier is fully automatic, the water level being controlled by means of a float control. In operation, when the relative humidity drops below the

VENTILATION REQUIREMENTS

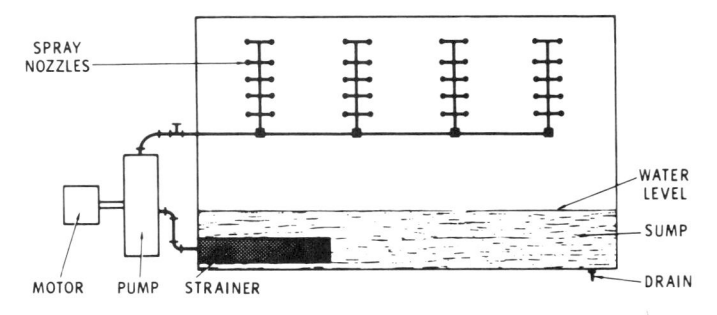

Fig. 4-14. Elementary diagrams showing essential parts of air-washer unit with identification of parts.

humidity-control setting, the humidifier fan blows air over the surface of the heated water in the tank. The air picks up moisture and is blown to the space to be humidified. When the humidity control is satisfied, the humidifier fan stops.

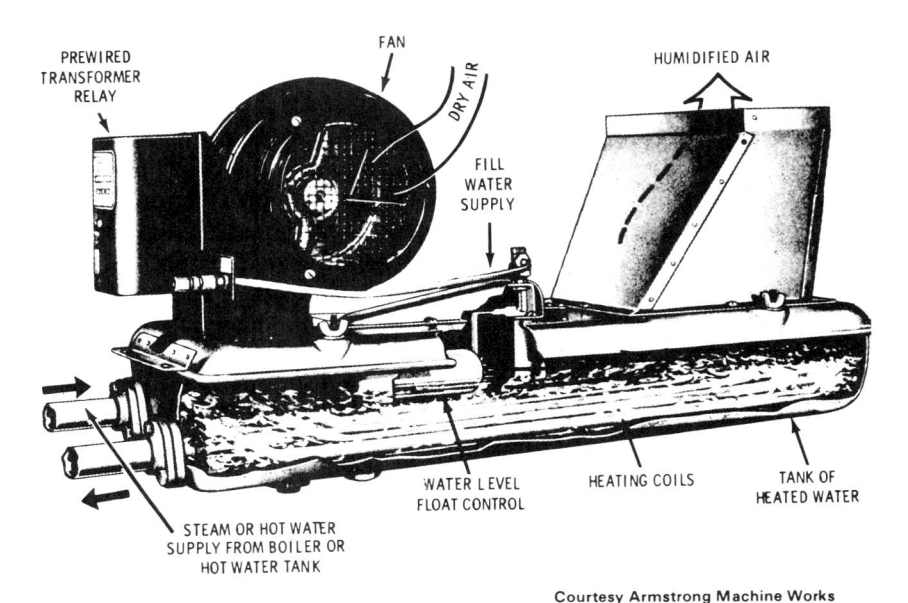

Courtesy Armstrong Machine Works

Fig. 4-15. Typical evaporative pan humidifier showing operative components.

85

AIR CONDITIONING

Electrically-Operated Humidifiers

Dry-steam electrically-operated humidifiers operate by means of a solenoid valve, which is energized by a humidistat. When the relative humidity drops slightly below the desired level set by the humidistat (Fig. 4-16), a solenoid valve actuated by the humidistat admits steam from the separating chamber to the reevaporating chamber. Steam passes from this chamber through the muffler directly to the atmosphere. The fan, which is energized when the

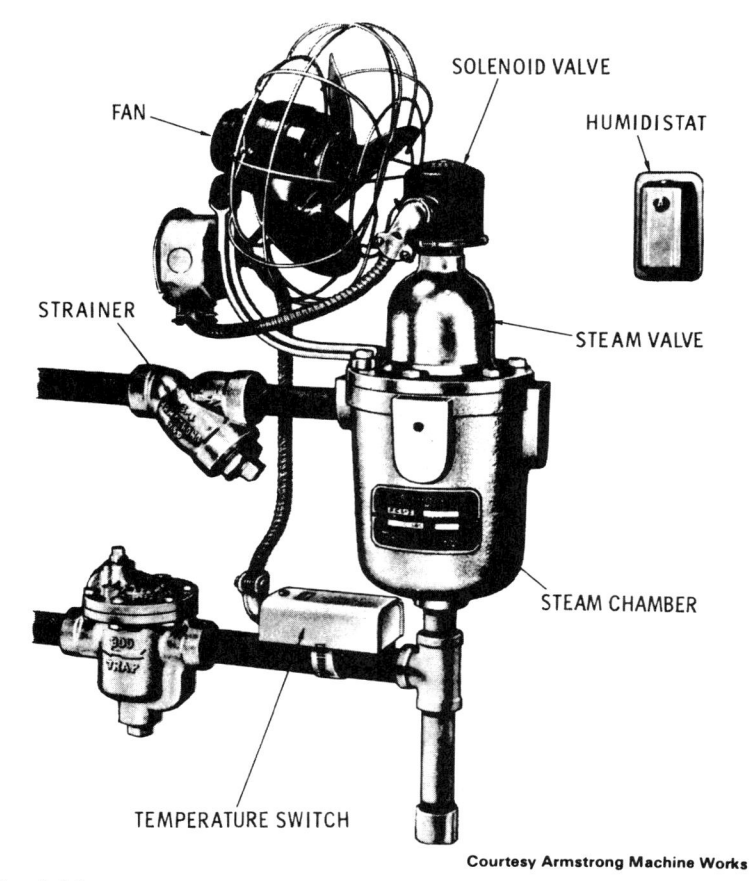

Courtesy Armstrong Machine Works

Fig. 4-16. Armstrong dry-steam humidifier for direct discharge of steam into the atmosphere of the area to be humidified.

VENTILATION REQUIREMENTS

solenoid valve opens, assists in dispersing the steam into the area to be humidified. When the relative humidity reaches the desired level, the humidistat closes the solenoid valve and stops the fan.

Air-Operated Humidifiers

Air-operated humidifying units operate in the same manner as electrical units, except that they utilize a pneumatic hygrostat as a humidity controller and an air operator to open or close the steam valve (Fig. 4-17). A decrease in relative humidity increases the air pressure under a spring-loaded diaphragm to open the steam valve wider. An increase in relative humidity reduces the pressure under the diaphragm and allows the valve to restrict the steam flow. In a humidifier operation of this type, the steam supply is taken off the top of the header (Fig. 4-18). Any condensate formed in the supply line is knocked down to the humidifier drain by a baffle inside the inlet of the humidifier separating chamber.

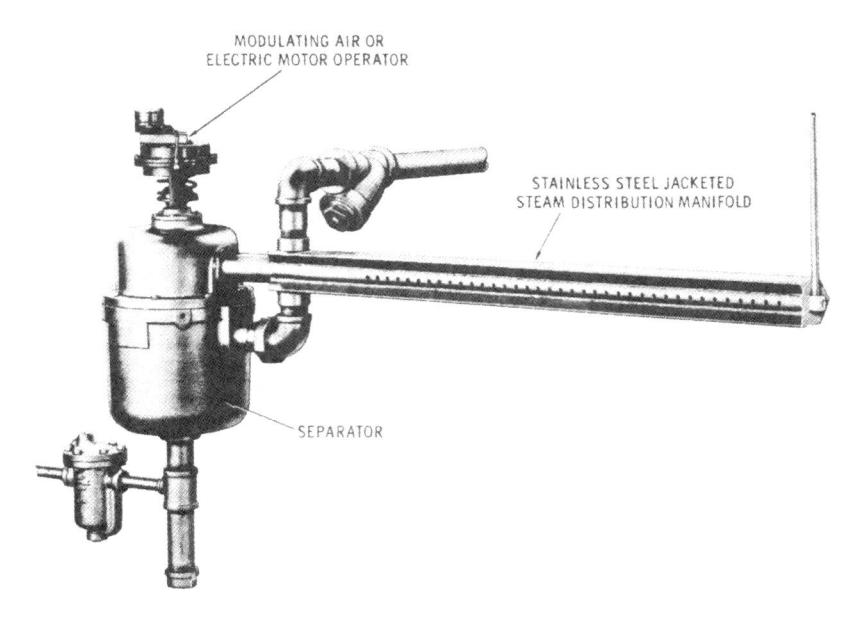

Fig. 4-17. Armstrong dry-steam humidifier with steam-jacket distribution.

87

AIR CONDITIONING

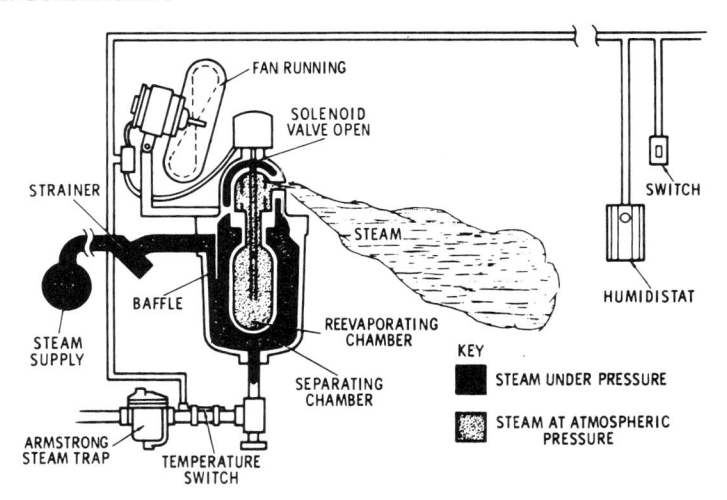

Fig. 4-18. Operation of Armstrong dry-steam humidifier for area humidification.

Any droplets of condensation picked up by the stream as it flows through the humidifier cap when the steam valve opens will be thrown to the bottom of the reevaporating chamber. Pressure in this chamber is essentially atmospheric. Since it is surrounded by steam at supply pressure and temperature, any water is reevaporated to provide dry steam at the outlet. The humidifier outlet is also surrounded by steam at supply pressure to ensure that there will be no condensation or drip at this point. A clamp-on temperature switch is attached to the condensate drain line to prevent the electric or pneumatic operator from opening the steam valve until the humidifier is up to steam temperature.

DEHUMIDIFIERS

The removal of moisture from the air is termed *dehumidification*. Air dehumidification is accomplished by one of two methods: cooling and adsorption. Dehumidification can be accomplished by an air washer, providing the temperature of the spray is lower than the dew point of the air passing through the unit. If the temperature of the spray is higher than the dew point, condensation will not take place. Air washers having refrigerated sprays usually have their own recirculating pump.

Electric Dehumidification

An electric dehumidifier operates on the refrigeration principle. It removes moisture from the air by passing the air over a cooling coil; the moisture in the air condenses to form water, which then runs off the coil into a collecting tray or bucket. The amount of water removed from the air varies, depending on the relative humidity and volume of the area to be dehumidified. In locations with high temperature and humidity conditions, 3 to 4 gallons of water per day can usually be extracted from the air in an average-size home.

When the dehumidifier is first put into operation, it will remove relatively large amounts of moisture until the relative humidity in the area to be dried is reduced to the value where moisture damage will not occur. After this point has been reached, the amount of moisture removed from the air will be considerably less. This reduction in the amount of moisture removal indicates that the dehumidifier is operating normally and that it has reduced the relative humidity in the room or area to a safe value.

The performance of the dehumidifier should be judged by the elimination of dampness and accompanying odors rather than by the amount of moisture that is removed and deposited in the bucket. A dehumidifier cannot act as an air conditioner to cool the room or area to be dehumidified. In operation, the air that is dried when passed over the coil is slightly compressed, raising the temperature of the surrounding air, which further reduces the relative humidity of the air.

Controls

As mentioned previously, the dehumidifier (Fig. 4-19) operates on the principles of the conventional household refrigerator. It contains a motor-operated compressor, a condenser, and a receiver. In a dehumidifier, the cooling coil takes the place of the

AIR CONDITIONING

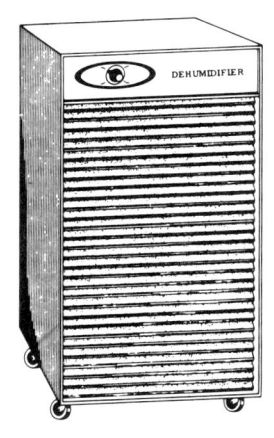

Fig. 4-19. Automatic electric dehumidifier.

Courtesy Westinghouse Electric Corporation

evaporator, or chilling unit, in a refrigerator. The refrigerant (usually Freon) is circulated through the dehumidifier in the same manner as in a refrigerator. The refrigerant flow is controlled by a capillary tube. The moisture-laden air is drawn over the refrigerated coil by means of a motor-operated fan or blower.

The dehumidifier operates by means of a humidistat (Fig. 4-20), which starts and stops the unit to maintain a selected humid-

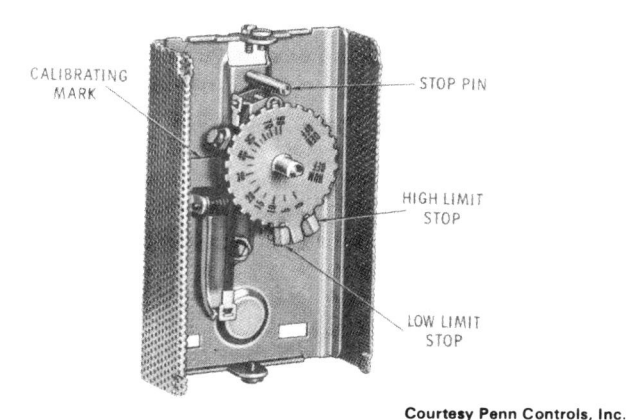

Courtesy Penn Controls, Inc.

Fig. 4-20. Typical humidistat designed to control humidifying or dehumidifying equipment or both with one instrument.

VENTILATION REQUIREMENTS

ity level. In a typical dehumidifier, the control settings range from DRY to EXTRA DRY to CONTINUOUS to OFF. For best operation, the humidistat control knob is normally set at EXTRA DRY for initial operation over a period of 3 to 4 weeks. After this period of time, careful consideration should be given to the dampness in the area being dried. If sweating on cold surfaces has discontinued and the damp odors are gone, the humidistat control should be reset to DRY At this setting, more economical operation is obtained, but the relative humidity probably will be higher than at the EXTRA DRY setting.

After 3 or 4 weeks of operation at the DRY setting, if the moisture condition in the area being dried is still satisfactory, the operation of the dehumidifier should be continued with the control set at this position. However, if at this setting the dampness condition is not completely corrected, the control should be returned to the EXTRA DRY setting. Minor adjustments will usually be required from time to time, but it should be remembered that the control must be set near EXTRA DRY to correct the dampness conditions but as close to DRY as possible to obtain the most economical operation.

Adsorption-Type Dehumidifiers

Adsorption-type dehumidifiers operate on the use of sorbent materials for adsorption of moisture from the air. The sorbents are substances that contain a vast amount of microscopic pores, which afford a great internal surface to which water adheres or is adsorbed. A typical dehumidifier based on the HoneyCombe desiccant wheel principle is shown schematically in Fig. 4-21. The wheel is formed from thin corrugated and laminated asbestos sheets rolled to form wheels of various desired diameters and thicknesses. The wheels are impregnated with a desiccant cured and reinforced with a heat-resistant binder. The corrugations in the HoneyCombe wheel form narrow flutes perpendicular to the wheel diameter. Approximately 75 percent of the wheel face area is available for the adsorption or dehumidifying flow circuit, and 25 percent is available for the reactivation circuit. In the smaller units, the reactivated air is heated electrically; in the larger units, it is heated by electric, steam, or gas heaters.

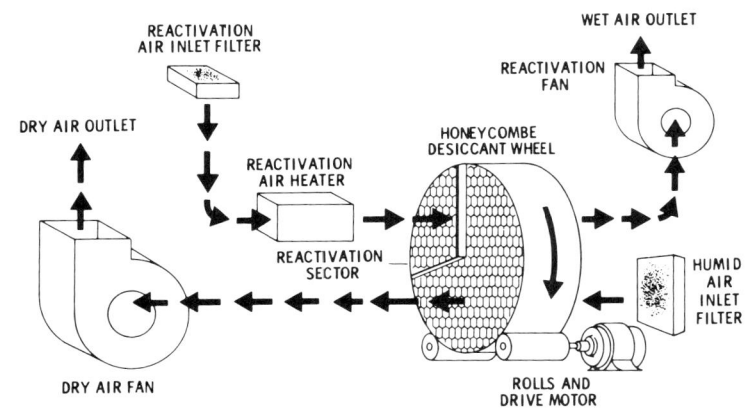

AIR CONDITIONING

Fig. 4-21. Assembly of components operating on the honeycomb method of dehumidification.

Another industrial adsorbent dehumidifier of the stationary-bed type is shown in Fig. 4-22. It has two sets of stationary adsorbing beds arranged so that one set is dehumidifying the air while the other set is drying. With the dampers in the position shown, air to be dried flows through one set of beds and is dehumidified while the drying air is heated and circulated through the other set. After completion of drying, the beds are cooled by shutting off the drying air heaters and allowing unheated air to circulate through them. An automatic timer controller is provided to allow the dampers to rotate to the opposite side when the beds have adsorbed moisture to a degree that begins to impair performance.

AIR-DUCT SYSTEMS

Air ducts for transmission of air in a forced-air heating, ventilation, or air-conditioner system must be carefully designed from the standpoint of economy as well as for proper functioning. When designing air ducts, the following methods may be used:

1. Compute the total amount of air to be handled per minute by the fan as well as the fractional volumes comprising the

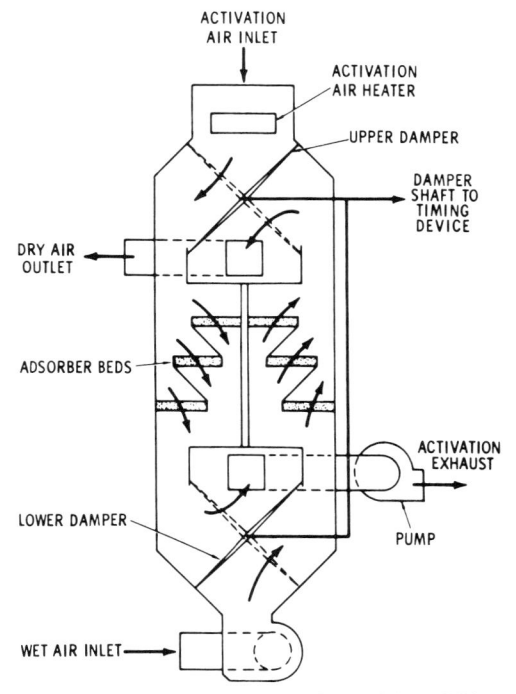

Fig. 4-22. Stationary-bed-type solid-absorbent dehumidifier.

total, which are to be supplied to or withdrawn from different parts of the building.
2. Locate the supply unit in the most convenient place and as close as possible to the center of distribution.
3. Divide the building into zones, and proportion the air volumes per minute in accordance with the requirements of the different zones.
4. Locate the air inlets or outlets for supply and recirculation, respectively; and at the positions so located on the building plans, indicate the air volumes to be dealt with. The position of the outlets and inlets should be such as to produce a thorough diffusion of the conditioned air throughout the space supplied.
5. Determine the size of each outlet or inlet on the basis of

AIR CONDITIONING

passing the required amount of air per minute at a suitable velocity.
6. Calculate the areas, and select suitable dimensions for all branch and main ducts on the basis of creating equal frictional losses per foot of length. This involves reducing the velocities in smaller ducts.
7. Ascertain the resistance of the ducts that sets up the greatest friction. In most cases this will be the longest run, although not invariably so. This will be the resistance offered by the duct system as a whole to the flow of the required amount of air.
8. Revise the dimensions and areas of the shorter runs so that the ducts themselves will create resistances equal to the longest run. This will cut down the cost of the sheet metal, and the result will be just the same as if dampers were used. Too high a velocity, however, must be avoided.
9. In order to compensate for unforeseen contingencies, volume dampers should be provided for each branch.

HEAT GAINS IN DUCTS

In any air-conditioning installation involving a duct system, invariably there is an accession of heat by the moving air in the ducts between the coils and supply grilles when air is supplied below room temperature. If the ducts are located through much of their length in the conditioned space, then, of course, this heat absorption has no effect on the total load and frequently may be disregarded.

More frequently, however, the supply ducts must pass through spaces that are not air conditioned. Under these conditions, the heat absorbed by the air in the ducts can be regarded as an additional load on the cooling equipment. The temperature rise in a duct system of a cooling installation depends on the following factors:

1. Temperature of the space through which the duct passes
2. Air velocity through the duct
3. Type and thickness of insulation, if any

VENTILATION REQUIREMENTS

The first factor establishes the temperature differential between the air on either side of the duct walls. The dew point of the air surrounding the duct may also have some effect on the heat pickup, as condensation on the duct surface gives up the heat of vaporization to the air passing through the system. The highest air velocities consistent with the acoustic requirements of the installation should be used, not only for economy in the sheet metal material used, but also to reduce the heat pickup in the ducts.

The amount of heat absorbed by a unit area of sheet-metal duct conveying chilled air is almost directly proportional to the temperature difference between the atmosphere surrounding the duct and the chilled air, irrespective of the velocity of the latter. The heat pickup rate will be influenced somewhat by the outside finish of the duct and also by the air motion, if any, in the space through which the duct passes.

Heat leakage in Btu per hour per square foot per degree difference in temperature for uncovered galvanized iron duct work will be between 0.5 and 1.0, with an average value of 0.73. The rate of leakage, of course, will be greatest at the start of the duct run and will gradually diminish as the air temperature rises. Covering the duct with the equivalent of $1/2$-inch rigid insulation board and sealing cracks with tape will reduce the average rate of heat pickup per square foot of surface per hour to 0.23 Btu per degree difference.

To summarize, the designer of a central unit system should observe the following:

1. Locate the equipment as close to the conditioned space as possible.
2. Use duct velocities as high as practical, considering the acoustic level of the space and operating characteristics of the fans.
3. Insulate all supply ducts with covering equivalent to at least $1/2$-inch of rigid insulation, and seal cracks with tape.

Suitable Duct Velocities

Table 4-2 gives the recommended air velocities in feet per minute for different requirements. These air velocities are in accord-

AIR CONDITIONING

Table 4-2. Air Velocities

Designation	Recommended Air Velocities		
	Residences, Broadcasting Studios, etc.	Schools, Theaters, Public Buildings	Industrial Applications
Initial air intake	750	800	1000
Air washers	500	500	500
Extended surface heaters or coolers (face velocity)	450	500	500
Suction connections	750	800	1000
Through fan outlet			
For 1.5" static pressure	—	2200	2400
For 1.25" static pressure	—	2000	2200
For 1" static pressure	1700	1800	2000
For 0.75" static pressure	1400	1550	1800
For 0.5" static pressure	1200	1300	1600
Horizontal ducts	700	900	1000-2000
Branch ducts and risers	550	600	1000-1600
Supply grilles and openings	300	300 grille	400 opening
Exhaust grilles and openings	350	400 grille	500 opening
Duct outlets at high elevation	—	1000	—

ance with good practice. It should be understood that the fan-outlet velocities depend somewhat on the static pressure, as static pressures and fan-housing-velocity pressures are interdependent for good operating conditions. Fan-outlet velocities are also affected by the particular type of fan installed.

Air-Duct Calculations

For the sake of convenience, a simplified air-duct sizing procedure is outlined, which eliminates the usual complicated engineering calculation required when designing a duct system. Refer to Figs. 4-23 to 4-25, which show the plans of a typical family residence having a total cubic content of approximately 19,000 cu ft. It is desired to provide humidification, ventilation, filtration, and air movement to all rooms on both the first and second floor.

The air conditioner located as shown in Fig. 4-23 had a total air-handling capacity of 1000 cu ft/min (cubic feet per minute).

VENTILATION REQUIREMENTS

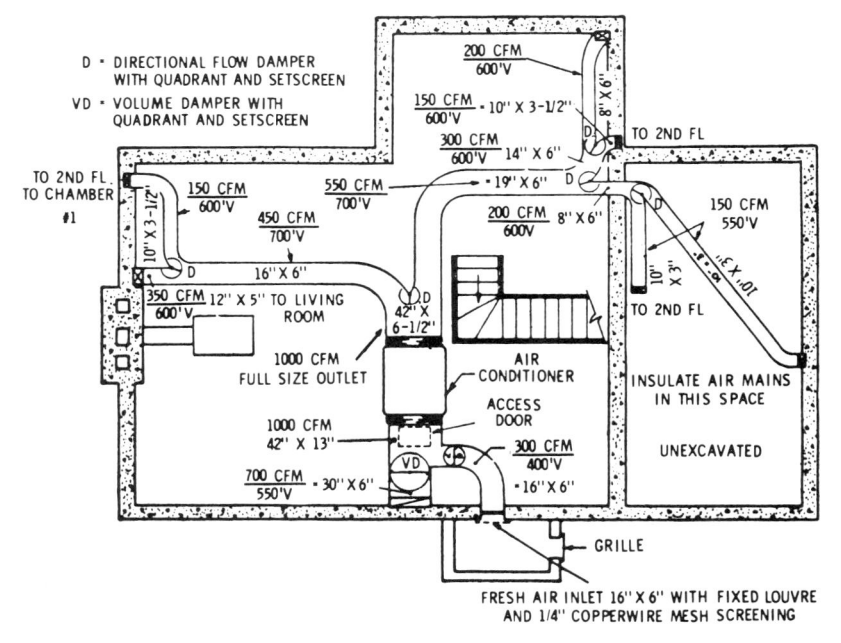

Fig. 4-23. Basement floor of a typical family residence showing air-conditioner and heat-duct layout.

Assuming that the rooms are to have individual air mains, Table 4-3 shows the methods of computing the amount of air to be supplied each room.

Column 2 in the table expresses the capacity of the individual room as a percentage of the total volume. For example, 3000 cu ft is 30 percent of the total space (10,000 cu ft) to which air mains will lead. Column 3 indicates the cubic feet of air per minute to be supplied to the individual rooms. These figures are attained as follows: The air conditioner will handle 1000 cu ft of air per minute; 30 percent of this is 300 cu ft/min. Similarly, 10 percent is 100 cu ft/min, which would indicate the quantity of air to be supplied to the living room and chamber No. 3, respectively. Having thus established the air quantity to be delivered to each of the rooms, the design of the ducts can now be considered.

Duct Sizes — consider the branch duct to both the living room

97

AIR CONDITIONING

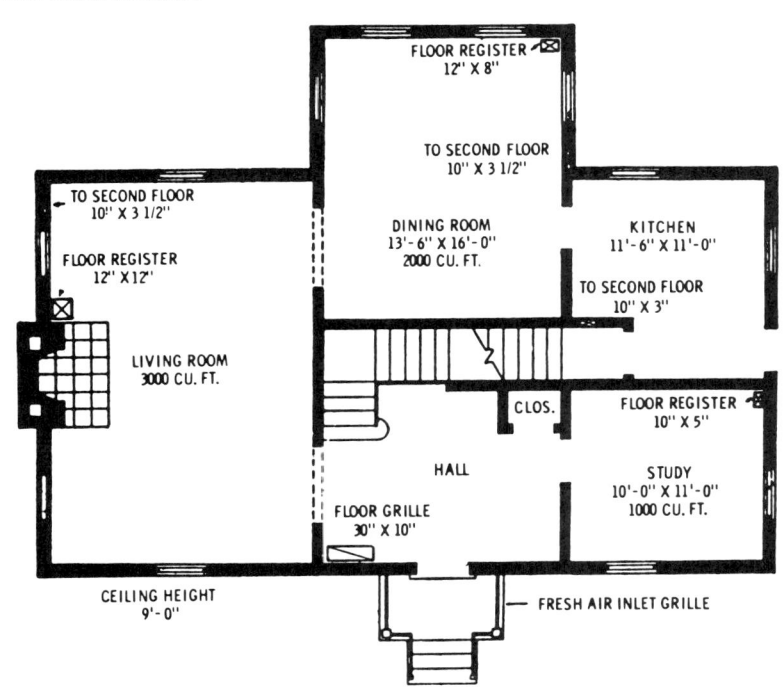

Fig. 4-24. First-floor layout of residence of Fig. 4-23. Vertical air discharges are located in the floor of the living room, dining room, and study. Air is returned to the unit through the floor grilles in the entrance hall. Sizes for air mains are shown on the basement plan.

Table 4-3. Air-Duct Calculation

Rooms		Col. 1 Room Volume, cu ft	Col. 2 % of Total Volume	Col. 3 Supply, cu ft/min
First Floor	Living Room	3000	30	300
	Dining Room	2000	20	200
	Study	1000	10	100
Second Floor	Chamber No. 1	1500	15	150
	Chamber No. 2	1500	15	150
	Chamber No. 3	1000	10	100
	Total Volume	10,000	100	1000

VENTILATION REQUIREMENTS

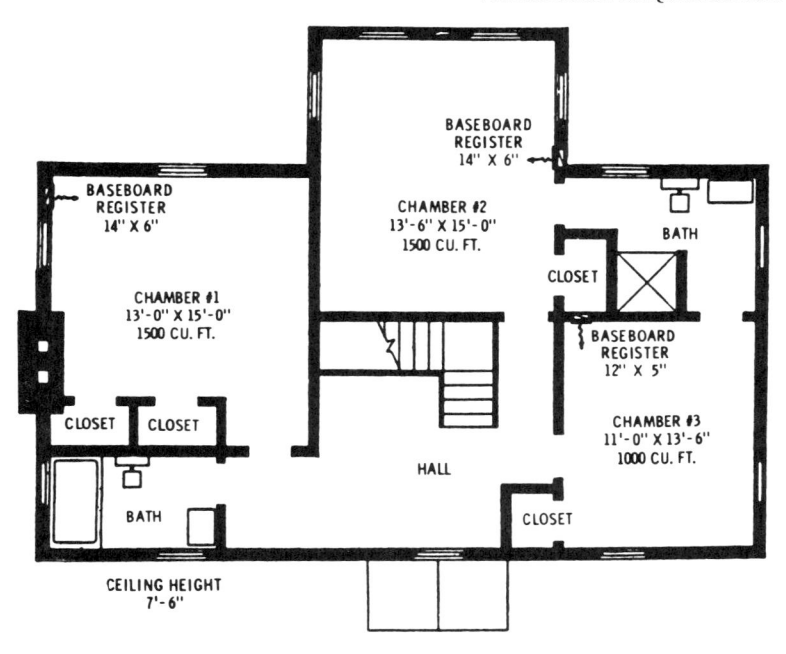

Fig. 4-25. Second-floor layout of residence of Figs. 4-23 and 4-24. Air discharges are located in the basement of the three chambers.

and chamber No. 1. From Table 4-3 it will be noted that the branch leading into chamber No. 1 must handle 150 cu ft/min while the duct to the living room handles 300 cu ft/min. Obviously, the connecting air main will handle 300 + 150, or 450 cu ft/min, in accordance with the foregoing recommendation, allowing a velocity of 600 ft/min for the branches and 700 ft/min for the supply air main. Therefore, the necessary duct areas will be as follows, using the formula:

$$\frac{a \times 144 \text{ (sq in/sq ft)}}{b}$$

where a = necessary air supply to room, cu ft/min
b = recommended velocity for a main or branch

Living-room (300 cu ft/min) branch:

$$\frac{300 \times 144}{600} = 72 \text{ sq in.} = 12 \times 6 \text{ in. branch}$$

Chamber No. 1 (150 cu ft/min) branch:

$$\frac{150 \times 144}{600} = 36 \text{ sq in} = 10 \times 3^{1/2} \text{ in. branch}$$

Total (450 cu ft/min) main:

$$\frac{450 \times 144}{700} = 93 \text{ sq in} = 16 \times 6 \text{ in. main}$$

The remaining duct work may be calculated similarly. It is recommended that the main air supply leaving the unit be the same size as that of the outlet of the unit up to the first branch takeoff. The return main to the unit should run the same size as the inlet of the unit for a distance of approximately 24 inches and should be provided with a large access door in the bottom of this length of the full-size duct. Figure 4-26 will be helpful as a further simplification in sizing the air ducts.

Example — It is desired to size a main duct for 250 cu ft/min at 500 ft/min velocity. What cross-sectional area is required?
Solution — Locate 250 cu ft/min on the left-hand side of Fig. 4-26. With a ruler or straightedge, carry a line across horizontally to the 500 velocity line and read off on the base line 72 sq in., or $1/2$ sq ft, which is the area required. All branches, risers, or grilles may be sized in the same manner.

RESISTANCE LOSSES IN DUCT SYSTEMS

In general, it can be said that duct sizes and the depth of a duct in particular are affected by the available space in the building.

VENTILATION REQUIREMENTS

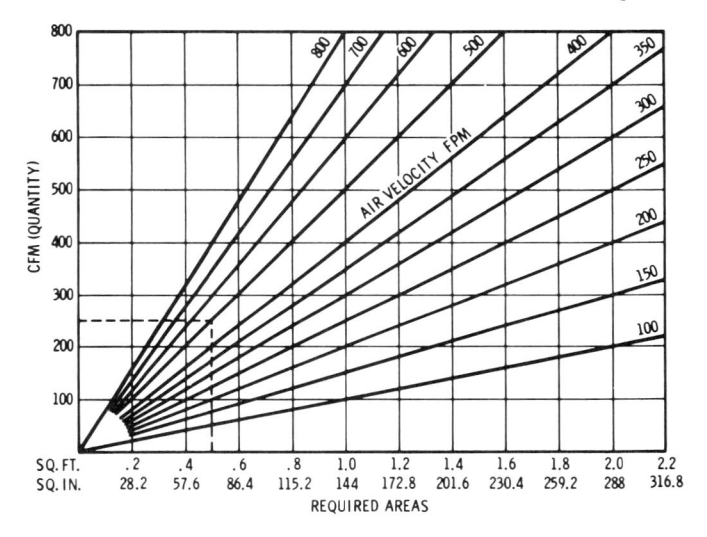

Fig. 4-26. Air-duct sizing method.

For this reason, although the round duct is the most economical shape from the standpoint of friction per unit area and also from the standpoint of metal required for construction per unit of area, it is rarely possible, except in industrial buildings, to use round ducts to any great extent. Square duct is the preferable shape among those of rectangular cross section. Headroom limitations usually require that the duct be flattened.

To illustrate the use of charts when sizing a duct system, the following example is submitted:

Example — Assume a system requiring the delivery of 5000 cu ft/min. The distribution requirement is the movement of the entire volume a distance of approximately 80 ft, with the longest branch beyond that point conveying 1000 cu ft/min for an additional 70 feet. Assume, further, that the operating characteristics of the fan and the resistance of the coils, filters, etc., allow a total supply-duct resistance of 0.10-inch water-gage resistance pressure. The supply duct is not to be more than 12 inches deep.
Solution — The total length of the longest run is 80 + 70 = 150 feet:

101

AIR CONDITIONING

$$\frac{100}{150} \times 0.10 = 0.067\text{-inch water gage}$$

Starting at this resistance at the bottom of Fig. 4-27, follow upward to the horizontal line representing 5000 cu ft/min. At this point, read the equivalent size of round duct required that is approximately 28 inches in diameter. Move diagonally upward to the right on the 28-inch-diameter line and then across horizontally on this line in Fig. 4-27 to the vertical line representing the 12-inch side of a rectangular duct. At this point, read 60 inches as the width of the rectangular duct required on the intersecting curved line.

Thus, for the main duct run, the duct size will be 60 × 12 inches. For the branch conveying 1000 cu ft/min, from the point where the 0.067-inch resistance line intersects the 1000 cu ft/min line, read 16 inches as the equivalent round duct required. Following through on Fig. 4-27 (page 104) as for a larger duct, read 12 × 18 inches as the size of the branch duct.

Duct runs should take into account the number of bends and offsets. Obstructions of this kind are usually represented in terms of the equivalent length of straight duct necessary to produce the same resistance value. Where conditions require sharp or right-angle bends, vane elbows composed of a number of curved deflectors across the airstream should be used.

FANS AND BLOWERS

The various devices used to supply air circulation in air-conditioning applications are known as fans, blowers, exhausts, or propellers. The different types of fans may be classified with respect to their construction as follows:

1. Propeller
2. Tubeaxial
3. Vaneaxial
4. Centrifugal

A *propeller fan* consists essentially of a propeller or disk-type wheel within a mounting ring or plate and includes the driving-

VENTILATION REQUIREMENTS

mechanism supports for either belt or direct drive. A *tubeaxial fan* consists of a propeller or disk-type wheel within a cylinder and includes the driving-mechanism supports for either belt drive or direct connection. A *vaneaxial fan* consists of a disk-type wheel within a cylinder and a set of air guide vanes located either before or after the wheel; it includes the driving-mechanism supports for either belt drive or direct connection. A *centrifugal fan* consists of a fan rotor or wheel within a scroll-type housing and includes the driving-mechanism supports for either belt drive or direct connection. See Fig. 4-28 for mounting arrangements.

Fan performance may be stated in various ways, the air volume per unit time, total pressure, static pressure, speed, and power input being the most important. The terms, as defined by the National Association of Fan Manufacturers, are as follows:

1. *Volume* handled by a fan is the number of cubic feet of air per minute expressed as fan-outlet conditions.
2. *Total pressure* of a fan is the rise of pressure from fan inlet to fan outlet.
3. *Velocity pressure* of a fan is the pressure corresponding to the average velocity determination from the volume of airflow at the fan outlet area.
4. *Static pressure* of a fan is the total pressure diminished by the fan-velocity pressure.
5. *Power output* of a fan is expressed in horsepower and is based on fan volume and the fan total pressure.
6. *Power input* to a fan is expressed in horsepower and is measured as horsepower delivered to the fan shaft.
7. *Mechanical efficiency* of a fan is the ratio of power output to power input.
8. *Static efficiency* of a fan is the mechanical efficiency multiplied by the ratio of static pressure to the total pressure.
9. *Fan-outlet area* is the inside area of the fan outlet.
10. *Fan-inlet area* is the inside area of the inlet collar.

Air Volume

The volume of air required is determined by the size of the space to be ventilated and the number of times per hour that the

AIR CONDITIONING

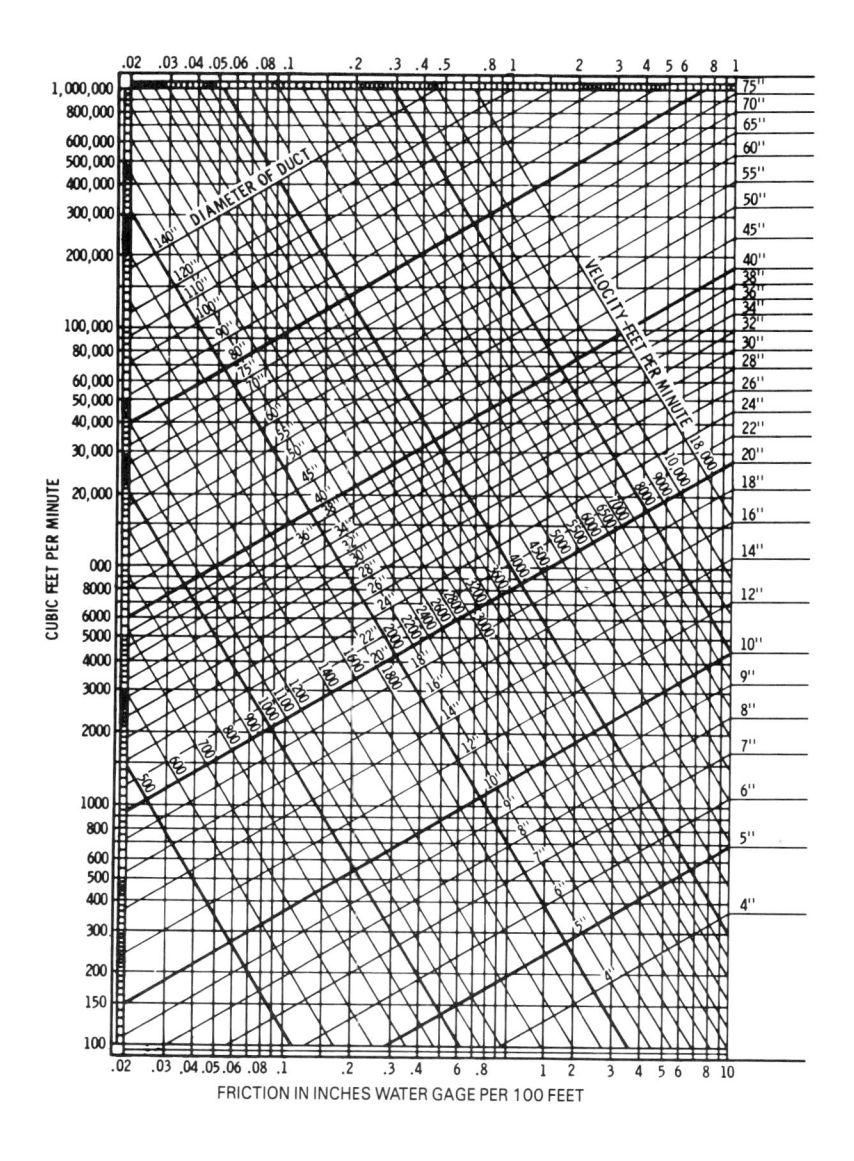

Fig. 4-27. Graphic representation

Ventilation Requirements

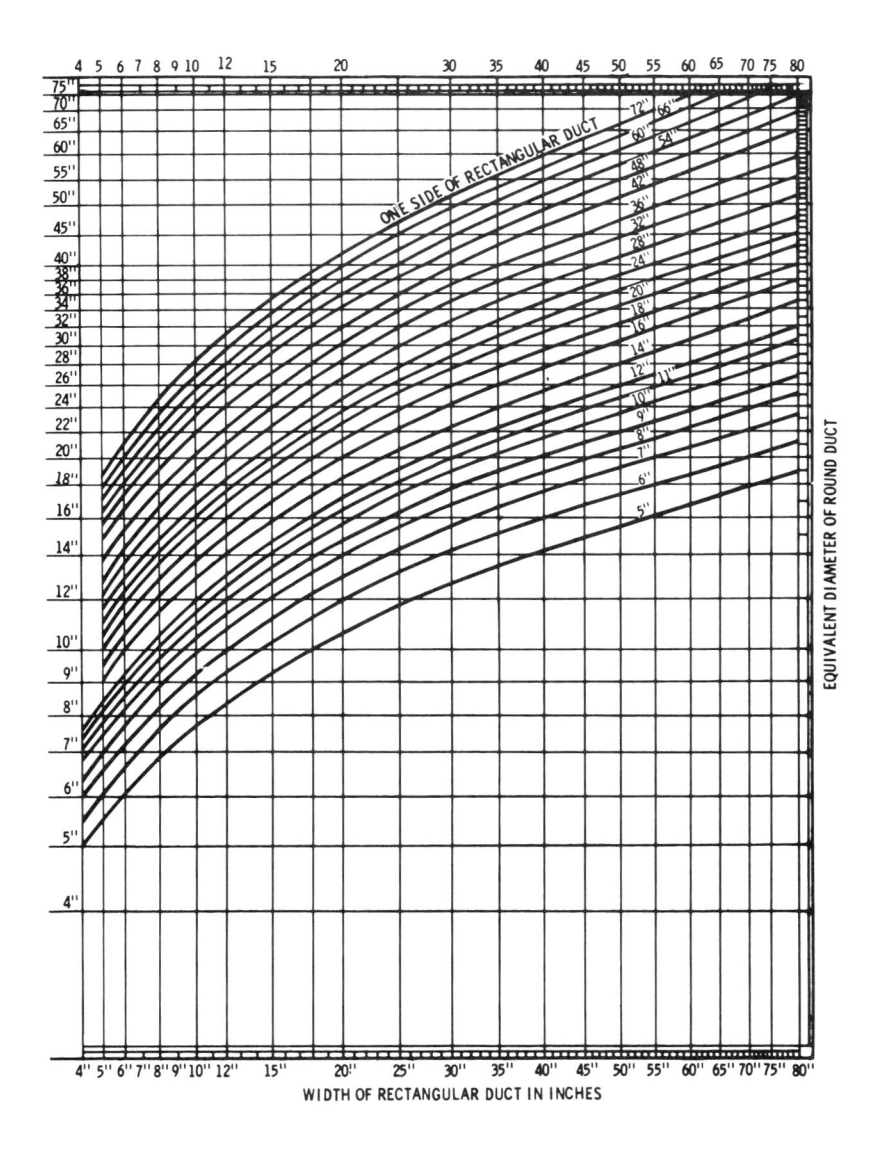

of air-duct areas.

AIR CONDITIONING

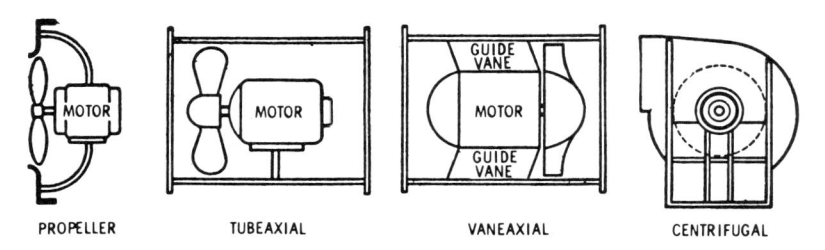

Fig. 4-28. Various fan classifications showing mounting arrangement.

air in the space is to be changed. Table 4-4 shows the recommended rate of air change for various types of spaces.

In many cases, existing local regulations or codes will govern the ventilating requirements. Some of these codes are based on a specified amount of air per person and on the air required per square foot of floor area. Table 4-4 should serve as a guide to average conditions. Where local codes or regulations are involved, they should be taken into consideration. If the number of persons occupying the space is larger than would be normal for such a space, the air should be changed more often than shown.

Table 4-4. Volume of Air Required

Space to be Ventilated	Air Changes per Hour	Minutes per Change
Auditoriums	6	10
Bakeries	20	3
Bowling alleys	12	5
Club rooms	12	5
Churches	6	10
Dining rooms (restaurants)	12	5
Factories	10	6
Foundries	20	3
Garages	12	5
Kitchens (restaurants)	30	2
Laundries	20	3
Machine shops	10	6
Offices	10	6
Projection booths	60	1
Recreation rooms	10	6
Sheet-metal shops	10	6
Ship holds	6	10
Stores	10	6
Toilets	20	3
Tunnels	6	10

Horsepower Requirements

The horsepower required for any fan or blower varies directly as the cube of the speed provided that the area of the discharge orifice remains unchanged. The horsepower requirements of a centrifugal fan generally decrease with a decrease in the area of the discharge orifice if the speed remains unchanged. The horsepower requirements of a propeller fan increase as the area of the discharge orifice decreases if the speed remains unchanged.

Drive Methods

Whenever possible, the fan wheel should be directly connected to the motor shaft. This can usually be accomplished with small centrifugal fans and with propeller fans up to about 60 inches in diameter. The deflection and the critical speed of the shaft, however, should be investigated to determine whether or not it is safe.

When selecting a motor for fan operation, it is advisable to select a standard motor one size larger than the fan requirements. It should be kept in mind, however, that direct-connected fans do not require as great a safety factor as that of belt-driven units. It is desirable to employ a belt drive when the required fan speed or horsepower is in doubt since a change in pulley size is relatively inexpensive if an error is made (Fig. 4-29).

Directly connected small fans for various applications are usually driven by single-phase ac motors of the split-phase, capacitor, or shaded-pole type. The capacitor motor is more efficient electrically and is used in districts where there are current limitations. Such motors, however, are usually arranged to operate at one speed. With such a motor, if it is necessary to vary the air volume or pressure of the fan or blower, the throttling of air by a damper installation is usually made.

In large installations, such as when mechanical draft fans are required, various drive methods are used:

1. A slip-ring motor to vary the speed
2. A constant-speed, directly connected motor, which, by means of moveable guide vanes in the fan inlet, serves to regulate the pressure and air volume.

AIR CONDITIONING

Fig. 4-29. Drive and mounting arrangement for various types of propeller fans.

Fan Selection

Most often the service determines the type of fan to use. When operation occurs with little or no resistance, and particularly when no duct system is required, the propeller fan is commonly used because of its simplicity and economy in operation. When a duct system is involved, a centrifugal or axial type of fan is usually employed. In general, centrifugal and axial fans are comparable with respect to sound effect, but the axial fans are somewhat lighter and require considerably less space. The following information is usually required for proper fan selection:

1. Capacity requirement in cubic feet per minute
2. Static pressure or system resistance
3. Type of application or service
4. Mounting arrangement of system

VENTILATION REQUIREMENTS

5. Sound level or use of space to be served
6. Nature of load and available drive

The various fan manufacturers generally supply tables or characteristic curves that ordinarily show a wide range of operating particulars for each fan size. The tabulated data usually include static pressure, outlet velocity, revolutions per minute, brake horsepower, tip or peripheral speed, etc.

Fan Applications

The numerous applications of fans in the field of air conditioning and ventilation are well known, particularly to engineers and air-conditioning repair and maintenance personnel. The various fan applications are as follows:

1. Exhaust fans
2. Circulating fans
3. Cooling-tower fans
4. Kitchen fans
5. Attic fans

Exhaust fans are found in all types of applications, according to the American Society of Heating and Ventilating Engineers. Wall fans are predominantly of the propeller type since they operate against little or no resistance. They are listed in capacities from 1000 to 75,000 cu ft/min. They are sometimes incorporated in factory-built penthouses and roof caps or provided with matching automatic louvers. Hood exhaust fans involving duct work are predominantly centrifugal, especially in handling hot or corrosive fumes.

Spray-booth exhaust fans are frequently centrifugal, especially if built into self-contained booths. Tubeaxial fans lend themselves particularly well to this application where the case of cleaning and of suspension in a section of ductwork is advantageous. For such application, built-in cleanout doors are desirable.

Circulating fans are invariably propeller or disk-type units and are made in a vast variety of blade shapes and arrangements. They are designed for appearance as well as utility. *Cooling-tower fans* are predominantly the propeller type; but axial types are also used for packed towers, and occasionally a centrifugal fan is used to supply draft. *Kitchen fans* for domestic use are small propeller fans arranged for window or wall mounting and with various useful fixtures. They are listed in capacity ranges of from 300 to 800 cu ft/min.

Attic fans are used during the summer to draw large volumes of outside air through the house or building whenever the outside temperature is lower than that of the inside. It is in this manner that the relatively cool evening or night air is utilized to cool the interior in one or several rooms, depending on the location of the air-cooling unit. It should be clearly understood, however, that the attic fan is not strictly a piece of air-conditioning equipment since it only moves air and does not cool, clean, or dehumidify. Attic fans have found use mainly due to their low cost and economy of operation, combined with their ability to produce comfort cooling by circulating air rather than conditioning it.

Fan Operation

Fans may be centrally located in an attic or other suitable space, such as a hallway, and arranged to move air proportionately from several rooms. A local unit may be installed in a window to provide comfort cooling for one room only when desired. Attic fans are usually propeller types and should be selected for low velocities to prevent excessive noise. The fans should have sufficient capacity to provide at least 30 air changes per hour.

In order to decrease the noise associated with air-exchange equipment, the following rules should be observed:

1. The equipment should be properly located to prevent noise from affecting the living area.
2. The fans should be of the proper size and capacity to obtain reasonable operating speed.
3. Equipment should be mounted on rubber or other resilient material to assist in preventing transmission of noise to the building.

If it is unavoidable to locate the attic air-exchange equipment above the bedrooms, it is essential that every precaution be taken to reduce the equipment noise to the lowest possible level. As high-speed ac motors are usually quieter than low-speed ones, it is often preferable to use a high-speed motor connected to the fan by means of an endless V-belt, if the floorspace available permits such an arrangement.

Attic-Fan Installation

Due to the low static pressures involved, which are usually less than $1/8$ inch of water, disk or propeller fans are generally used instead of the blower or housed types. It is important that the fans should have quiet operating characteristics and sufficient capacity to give at least 30 air changes per hour. For example, a house with 10,000 cu ft content would require a fan with a capacity of 300,000 cu ft/hr or 5000 cu ft/min to provide 30 air changes per hour.

The two general types of attic fans in common use are boxed-in fans and centrifugal fans. The boxed-in fan is installed within the attic in a box or suitable housing located directly over a central ceiling grille or in a bulkhead enclosing an attic stair. This type of fan may also be connected by means of a direct system to individual room grilles. Outside cool air entering through the windows in the downstairs room is discharged into the attic space and escapes to the outside through louvers, dormer windows, or screened openings under the eaves.

Although an air-exchange installation of this type is rather simple, the actual decision about where to install the fan and where to provide the grilles for the passage of air up through the house should be left to a ventilating engineer. The installation of a multi-blade centrifugal fan is shown in Fig. 4-30. At the suction side, the fan is connected to exhaust ducts leading to grilles, which are placed in the ceiling of the two bedrooms. The air exchange is accomplished by admitting fresh air through open windows and up through the suction side of the fan; the air is finally discharged through louvers as shown.

Another installation is shown in Fig. 4-31. This fan is a centrifugal curved-blade type, mounted on a light angle-iron frame, which supports the fan wheel, shaft, and bearings. The air inlet in this installation is placed close to a circular opening, which is cut in an air-tight board partition that serves to divide the attic space

AIR CONDITIONING

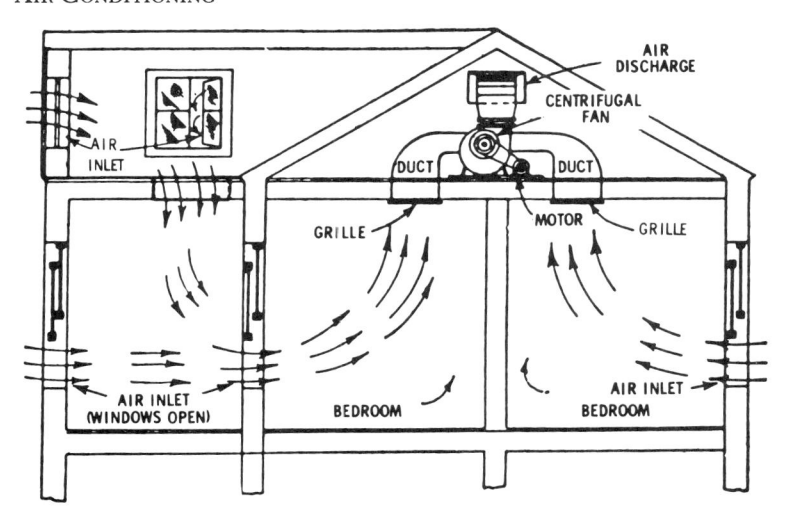

Fig. 4-30. Installation of a centrifugal fan in a one-family dwelling.

into a suction and discharge chamber. The air is admitted through open windows and doors and is then drawn up the attic stairway through the fan into the discharge chamber.

Fan Operation

The routine of operation in order to secure the best and most efficient results with an attic fan is important. A typical operating routine might require that, in the late afternoon when the outdoor temperature begins to fall, the windows on the first floor and the grilles in the ceiling or the attic floor be opened and the second-floor windows kept closed. This will place the principal cooling effect in the living rooms. Shortly before bedtime, the first-floor windows may be closed and those on the second floor opened to transfer the cooling effect to the bedrooms. A suitable time clock may be used to shut the motor off before arising time.

VENTILATION REQUIREMENTS

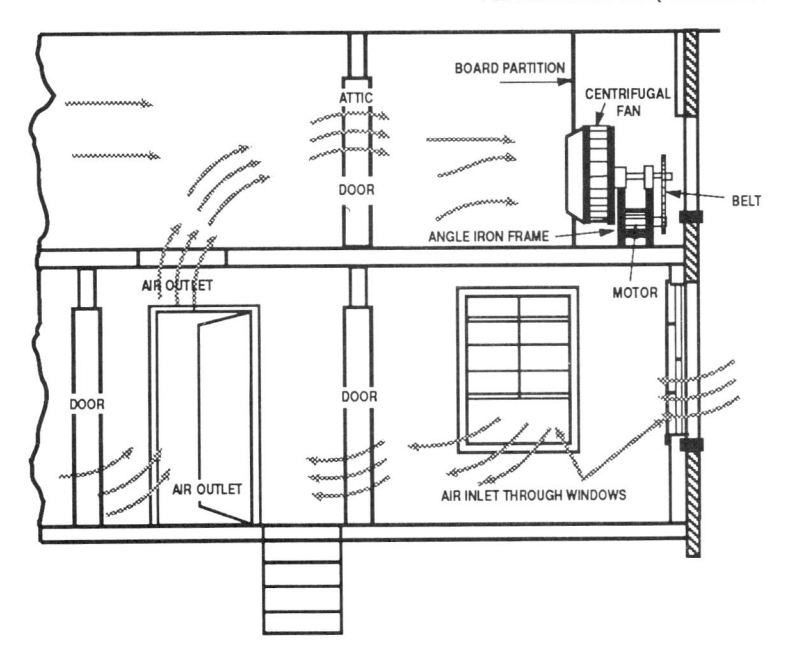

Fig. 4-31. Belt-driven fan in a typical attic installation.

SUMMARY

Ventilation is produced by two basic methods: natural and mechanical. Natural ventilation is obtained by open windows, vents, or drafts, whereas mechanical ventilation is produced by the use of fans.

Thermal effect is possibly better know as flue effect. Flue effect is the draft in a stack or chimney that is produced within a building when the outdoor temperature is lower than the indoor temperature. This is due to the difference in weight of the warm column of air within the building and the cooler air outside.

Air may be filtered two ways: dry filtering and wet filtering. Various air-cleaning equipment, such as filtering, washing, or combined filtering and washing devices are used to purify the air. When designing the duct network, ample filter area must be included so that the air velocity passing through the filters is sufficient. Accuracy in estimating the resistance to the flow of air through the duct system is important in the selection of blower motors. Resistance should be kept as low as possible in the interest of economy. Ducts should be installed as short as possible.

The effect of dust on health has been properly emphasized by competent medical authorities. Air-conditioning apparatus removes these contaminants from the air and further provides the correct amount of moisture so that the respiratory tracts are not dehydrated but are kept properly moist. Dust is more than just dry dirt; it is a complex, variable mixture of materials and, as a whole, is rather uninviting, especially the type found in and around human habitation. Dust contains fine particles of sand, soot, earth, rust, fiber, animal and vegetable refuse, hair, and chemicals.

Humidifiers add moisture to dry air. The types of humidifiers used in air-conditioning systems are spray-type air washers, pan evaporative, electrically operated, and air operated.

The function of an air washer is to cool the air and to control humidity. An air washer usually consists of a row of spray nozzles and a chamber or tank at the bottom that collects the water as it falls through the air coming in contact with many baffles. Air passing over the baffles picks up the required amount of humidity.

Dehumidification is the removal of moisture from the air and is accomplished by two methods: cooling, and adsorption. Cooling-type dehumidification operates on the refrigeration principle. It removes moisture from the air by passing the air over a cooling coil. The moisture in the air condenses to form water, which then runs off of the coil into a collecting tray or bucket.

Various types of fans are used in air-conditioning applications and are classified as propeller, tubeaxial, vaneaxial, and centrifugal. The propeller and tubeaxial fans consist of a propeller or disk-type wheel mounted inside a ring or plate and driven by a belt or direct drive motor.

A vaneaxial fan consists of a disk-type wheel mounted within a cylinder. A set of air-guide vanes is located either before or after the wheel and is belt driven or direct drive. The centrifugal fan is

VENTILATION REQUIREMENTS

a fan rotor or wheel within a scroll-type housing. This type of blower is better known as a squirrel-cage unit. Whenever possible, the fan wheel should be directly connected to the motor shaft. Where fan speeds are critical, a belt drive is employed, and various size pulleys are used.

REVIEW QUESTIONS

1. Define what is meant by the term *ventilation* as applied to a building or space.
2. How do natural and mechanical ventilation differ?
3. What precautions are normally taken to prevent excessive air leakage in residential buildings or homes?
4. What is the function of air filters as installed in an air-conditioning system?
5. State the effect of dust and other impurities on the health of humans.
6. What are the advantages in using throwaway or replaceable type filters?
7. Where are air filters usually placed in residential air-conditioning installations?
8. Describe the function of a humidifier in an air-conditioning system.
9. Name four types of humidifiers.
10. What is the principal function of an air washer?
11. Describe the essential difference between an electrically and air-operated humidifier.
12. What kind of controls are normally employed on electric-type dehumidifiers?
13. What are the factors contributing to heat gains in air ducts?
14. What are the methods used to reduce heat gains in air ducts?
15. How may the total air handling capacity for residential occupancy be calculated?
16. What are the various devices used to supply air circulation in an air-conditioning system?

17. Explain the difference in construction between a propeller and a tubeaxial fan.
18. How is the power input of a fan usually measured?
19. What type of fan or blower is usually employed in air-duct systems?
20. Give a typical attic-fan operating routine for best results.

CHAPTER 5

Room Air Conditioners

A room air conditioner is generally considered to he a unit suitable for placement in any particular room. The room in question may be in an office, a home, or a small shop. Room air conditioners are usually classed according to their design and method of installation, such as window units and portable units.

As the name implies, window units are installed on the window sill. Portable units can be moved from one room to another as conditions and occupancy dictate. The advantages of this type of air conditioner are the relatively low-cost summer cooling in the room selected and the portability and ease of installation. A typical window air-conditioning unit is shown in Fig. 5-1. When properly installed, sized, and serviced, the room air-conditioning unit gives a large measure of comfort, free from irritating exhaust that accompanies sweltering hot spells and extremely high humidity.

The window-type air conditioner may be divided into three functional systems: electrical, refrigerant, and air path. The elec-

AIR CONDITIONING

Fig. 5-1. Exterior view of a typical room air conditioner.

trical system consists of the motor-compressor, unit control switch, fan and fan motor, starting and running capacitors, starting relay (when used), thermostat and necessary wiring. The refrigerant system consists of the compressor, condenser, drier-strainer, capillary tube, evaporator, tubing, and accumulator. In some systems the refrigerant flow is controlled by means of an automatic expansion valve instead of the conventional capillary-tube method. The air path consists of an air discharge and intake grille, air filter, vent door, vent controls, etc. The component parts of a window air conditioner are shown in Fig. 5-2.

OPERATION

Although the various types of room air conditioners differ in cabinet design as well as in arrangement of components, they all operate on the same principle. The evaporator fan draws the recirculated air into the unit through the louvers located on the side. The air passes through the air-filter evaporator and is discharged

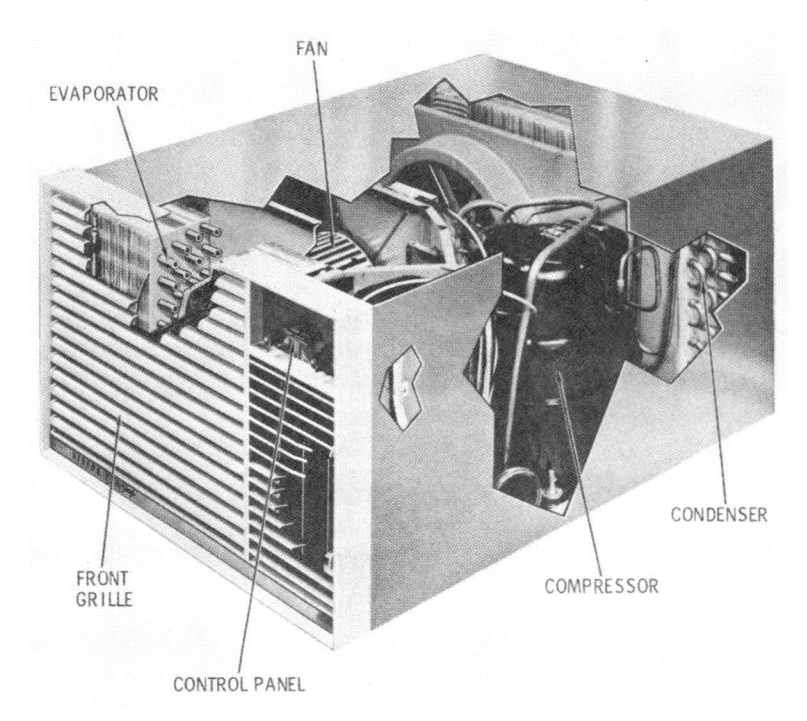

Fig. 5-2. Cutaway view showing component parts of a room air conditioner.

through the grille on the front of the unit into the room. The part of the unit extending into the room is insulated to reduce the transfer of heat and noise.

The condenser and compressor compartment extends outside of the room and is separated from the evaporator compartment by an insulated partition. The condenser air is drawn through the sections of the condenser coil on each side of the condenser fan housing. The air passes through the compressor compartment and is discharged through the center section of the condenser covered by the fan housing. It circulates the air and also disposes of the condensed water from the evaporator, which drops into the base and flows to the condenser end of the unit.

AIR CONDITIONING

COOLING CAPACITY

The cooling capacity of an air-conditioning unit is the ability to remove heat from a room and is usually measured in Btu per hour. The higher the Btu rating, the more heat will be removed. The capacity rating is usually given on the unit's nameplate, together with other necessary information, such as voltage and wattage requirements. It is important that the room air-conditioning units be large enough for the room or rooms to be cooled. When dealing with large units, the term *ton of refrigeration* is most often used. A ton of refrigeration is equivalent to 12,000 Btu/hr. A room air conditioner rated at 10,000 Btu, for example, will supply $^{10}/_{12}$, or 0.83, ton of refrigeration.

CAPACITY REQUIREMENTS

The important variables to keep in mind when estimating the Btu requirements for a room cooling installation are:

1. Room size in square feet of floor area
2. Wall construction, whether light or heavy
3. Heat gain through ceiling
4. The proportion of outside wall area that is glass
5. Exposure to the sun of the walls of the room to be air conditioned.

Additional factors to be taken into account are: room ceiling height, number of persons using the room, and miscellaneous heat loads, such as wattage of lamps, radio and television sets in use in the room, etc.

Table 5-1 should be used to obtain the approximate cooling capacity in Btu. The following steps outline the procedure for using the table, which shows the Btu requirements for various size rooms. In each case, it is assumed that the rooms have an 8-foot ceiling height. The procedure will be as follows:

1. Measure the room to be cooled for square footage (length in feet times width in feet).

ROOM AIR CONDITIONERS

Table 5-1. Room-Cooling Requirements in Btu

Approx. Btu Capacity Required	Space Above Room Being Cooled							
	Occupied Room			Attic			Insulated Flat Roof		
	Area Being Cooled Has Exposed Walls Facing—							
	North or East	South	West	North or East	South	West	North or East	South	West
6,000	400	200	100	200	100	64	250	120	80
7,000	490	250	125	235	130	97	295	155	105
8,000	580	300	150	270	160	130	340	190	130
10,000	750	440	390	340	270	200	470	340	240
12,000	920	580	470	410	320	225	550	375	275
13,000	1000	660	550	450	350	250	600	400	300
16,000	1290	970	790	570	480	390	750	650	540

2. Determine the direction the room faces in order to determine which exposure to use.
3. Determine the condition of the space above the room to be cooled, such as occupied, attic, or insulated flat roof.

Example — To determine the required amount of cooling for a room 23 feet long and 13 feet wide, multiplication of the two figures will give 299 sq ft. It is further assumed that the space above the room to be cooled is occupied and that the room has an exposed wall with windows facing south. A reference to Table 5-1 indicates that the room described will require a unit having a capacity of approximately 8000 Btu. A room of the same size and facing south with an uninsulated attic above it will require a cooling capacity of approximately 10,500 Btu.

Table 5-1 indicates the approximate Btu capacity required to cool different rooms with the approximate area and other conditions as described. The table will also permit checking an installed unit where unsatisfactory performance might be suspected resulting from improper sizing or inadequate unit capacity.

INSTALLATION METHODS

After the unit is removed from the crate, the mounting frame should be located on the sill of the selected window. The window should be on the shady side of the room. If this is not possible and the unit must be exposed to the sun, then some shading of the unit should be used for greater efficiency. Awnings are most effective since they shade both the unit and window at the same time, but the awning must not restrict the free flow of air to and from the unit. The awning top must be held away from the building side so that the hot air can escape. Venetian blinds or shades are a second choice to cut down a great amount of solar heat transmitted to the room through the windows.

Before installing the unit, it should be determined first that it is the correct size for the room and that the electrical power plug is correct and adequate. Whatever type of installation is used, it should be made certain that the location will permit proper distribution of the cooled air throughout the room, as shown in Fig. 5-3, and that there is no obstruction to the outside air flow that could cause restriction or recirculation of outside air back into the unit. The installation instructions provided with the unit should be studied and followed closely. The installation information included here is only general and not intended to replace or substitute the instructions supplied with the unit. The installation

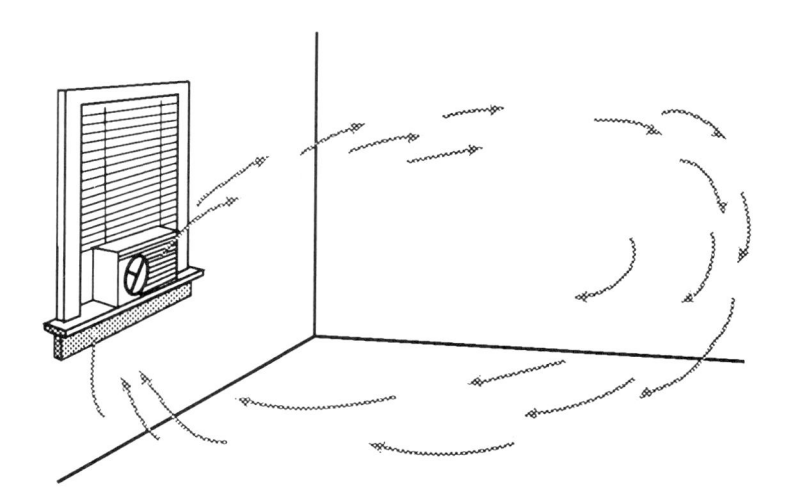

Fig. 5-3. Air circulation with an accompanying cooling effect provided by a typical window-mounted air-conditioning unit.

should be made as neat as possible. Make certain that the unit is properly secured and that the installation is made so there is a good tight seal from the outside. It is very important that no openings are left through which rain or warm air from the outside could enter around the unit.

DOUBLE-HUNG WINDOW INSTALLATION

Most window units are manufactured for installation in sliding windows with free openings from 27 to 40 inches in width. The proper installation of these units is very important to their continued satisfactory performance, and manufacturers' instructions and illustrations should be carefully studied and adhered to. The following is a step-by-step procedure that will assist in reducing installation time:

1. The depth of the unit will determine if it is to be mounted flush to the inside finished wall or at any point between these extremes.
2. Measure and mark the center of the window sill with a line, as shown in Fig. 5-4. Extend the line to the outside sill

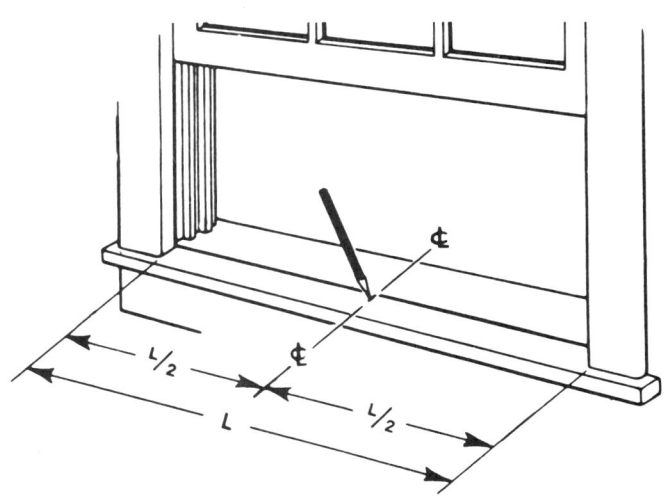

Fig. 5-4. Method of measuring windowsill to establish mounting position of an air-conditioning unit.

AIR CONDITIONING

for use in installing the mounting-frame support. The purpose of the window angles is to hold the side panel in the window channel. Outside of the panel, place the window angle and adjust it for height; outline and drill pilot holes. Screw the angle firmly in place. Repeat the operation with the other angle on the opposite side of the window frame.

3. Place the mounted assembly (Fig. 5-5) in the window,

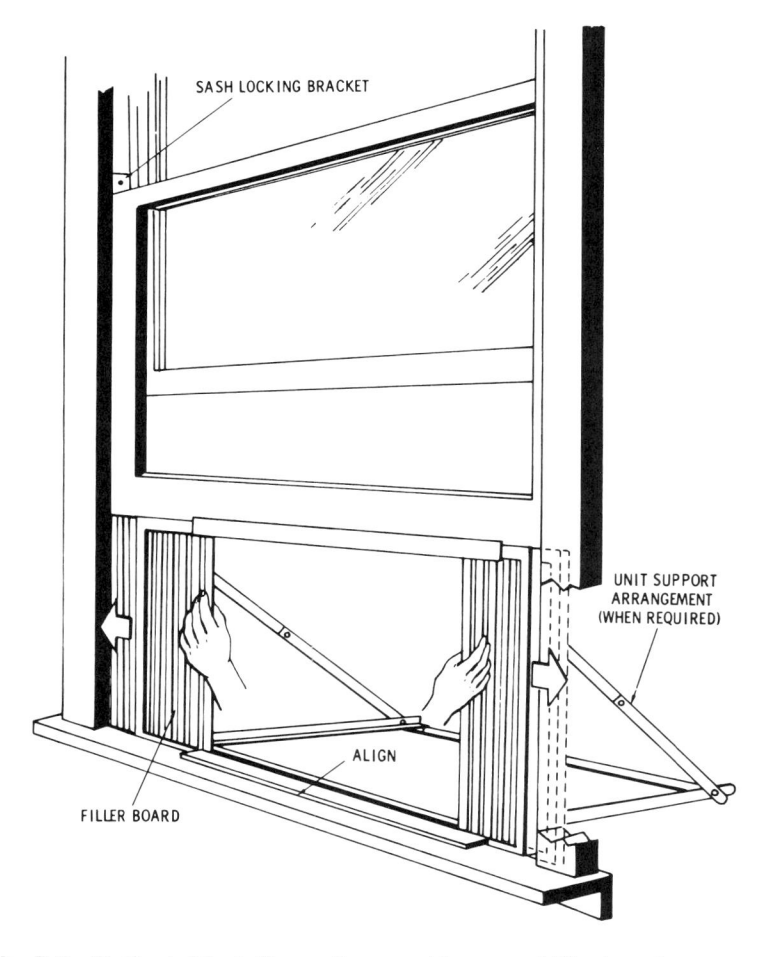

Fig. 5-5. Method of installing unit-support frame and filler boards.

aligning the center hole in the lower frame with the center line mark on the window sill. Lower the window to the top of the frame. Extend sides of the mounted assembly to the full width of the window, as shown by arrows. Install sash locking brackets after seating the window firmly on the assembly frame.
4. Break off excess filler board by pressing outward at the nearest notch beyond the unit opening of the mounting frame, as shown in Fig. 5-6. This operation is easier (especially for wider windows) if a pair of pliers is used.
5. Install the screws as shown in Fig. 5-7 if permanent installation is desired. Use of screws is recommended.
6. Install the bottom unit seal in the groove of the frame, as shown in Fig. 5-8. Stretch the seal material to fill the groove completely.
7. Install the sill clamp in lieu of the sill fastening screws for stone sills or other such materials, as shown in Fig. 5-9.
8. Install the top rail and seal assembly (Fig. 5-10) on top of the unit case in the holes provided for flush mounting only.
9. Remove the unit decorative grille (Fig. 5-11) and install locking screws (two) on flush-mount installations only. Reinstall the grille.

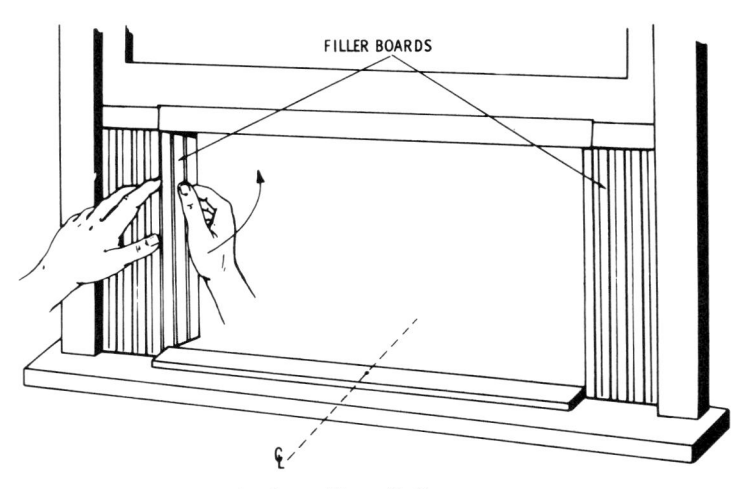

Fig. 5-6. Method of filler-board installation.

AIR CONDITIONING

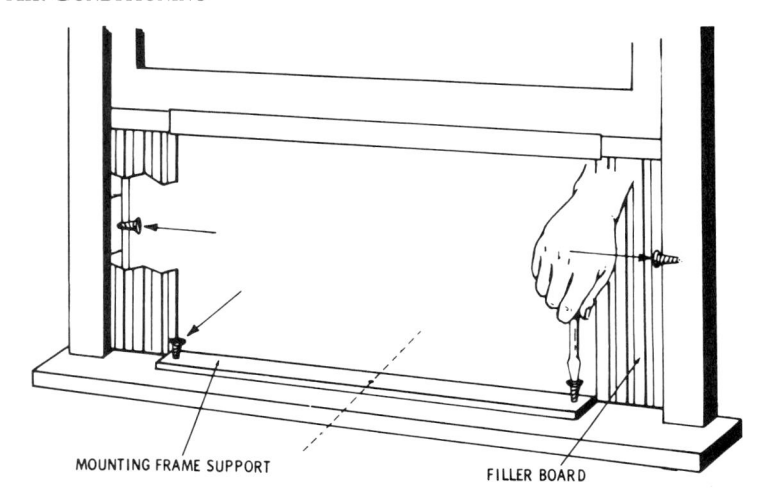

Fig. 5-7. Method of securing sill clamp to wooden windowsill.

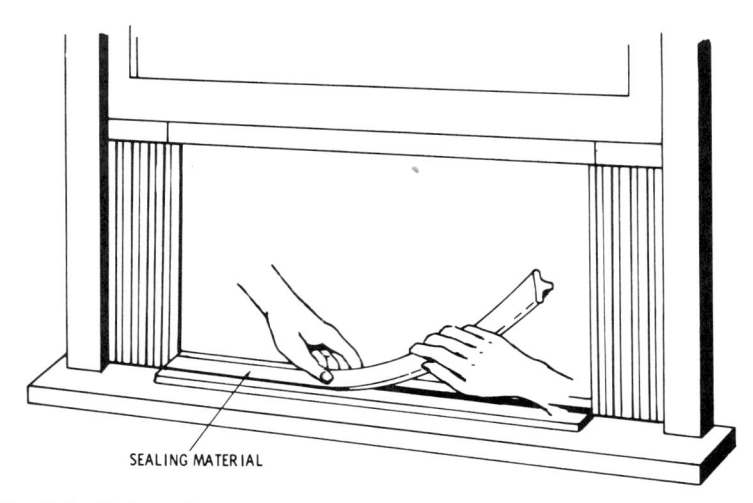

Fig. 5-8. Unit seal arrangement.

ROOM AIR CONDITIONERS

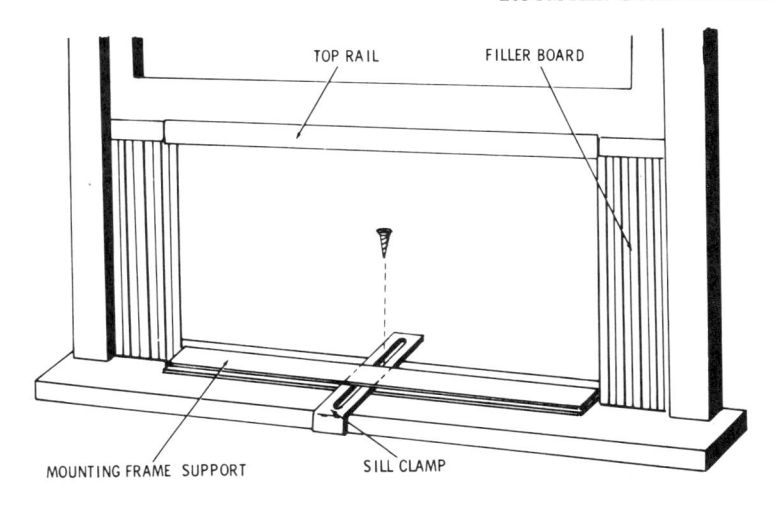

Fig. 5-9. Sill-clamp mounting in stone or masonry sill construction.

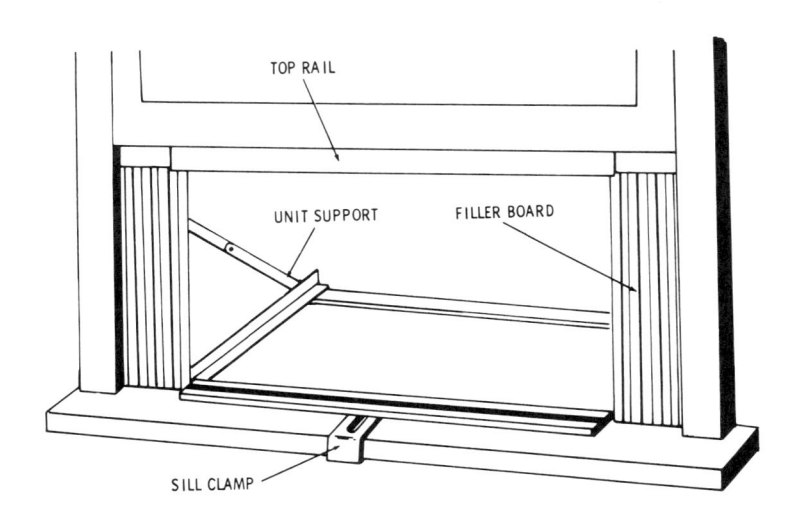

Fig. 5-10. Method of installing top rail and seal assembly.

AIR CONDITIONING

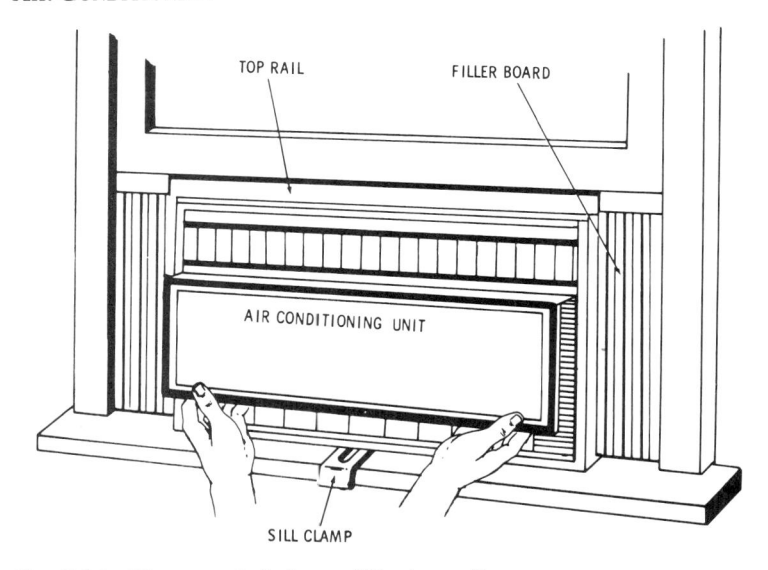

Fig. 5-11. Placement of air-conditioning unit.

10. Install the side seals by forcing them between the unit (filler boards and unit frame).
11. Install the sash-to-glass seal, cutting it to fit the windowglass width, as shown in Fig. 5-12.

CASEMENT WINDOW INSTALLATION

So many different types of casement windows are manufactured that it would be impossible to give a single procedure to cover them all. Each job will require special treatment, thought, and planning. With reference to Fig. 5-13, installation is accomplished by removing sufficient glass and mullions to allow for passage of the unit. It is necessary to build up the windowsill until the top is above the horizontal cross member forming the bottom of the frame on the metal window. If the outer cabinet is allowed to rest on this cross member, any vibration will be transmitted to the window frame and wall and will be amplified. This starting procedure is basic to all casement windows.

Room Air Conditioners

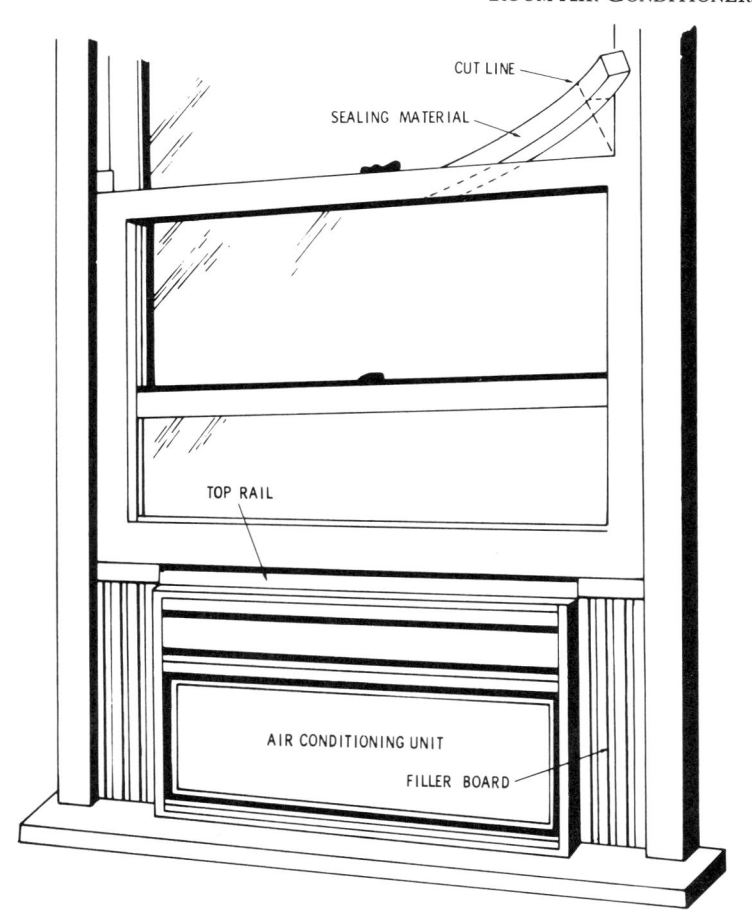

Fig. 5-12. Sealing-installation arrangement.

Measure the height and width of the opening left by the removed glass, and cut a piece of 1/4-inch masonite or equivalent material to fit. This board is referred to as a *filler panel*. Cut out the center of the filler panel to the exact outside dimensions of the outer cabinet. In cutting the height in the board, allow for the height of the bottom cross member on the window frame. Install the outer cabinet, as described previously (under Double-Hung Window Installation). Install the filler panel in the opening, and

129

Air Conditioning

seal the edges to the window frame with putty or caulking compound. The horizontal cross member of the supporting frame may be secured to the temporary sill by using the holes provided for this purpose, or the clamping assembly may be used if it is practical for the installation. Install chassis and inside cabinet.

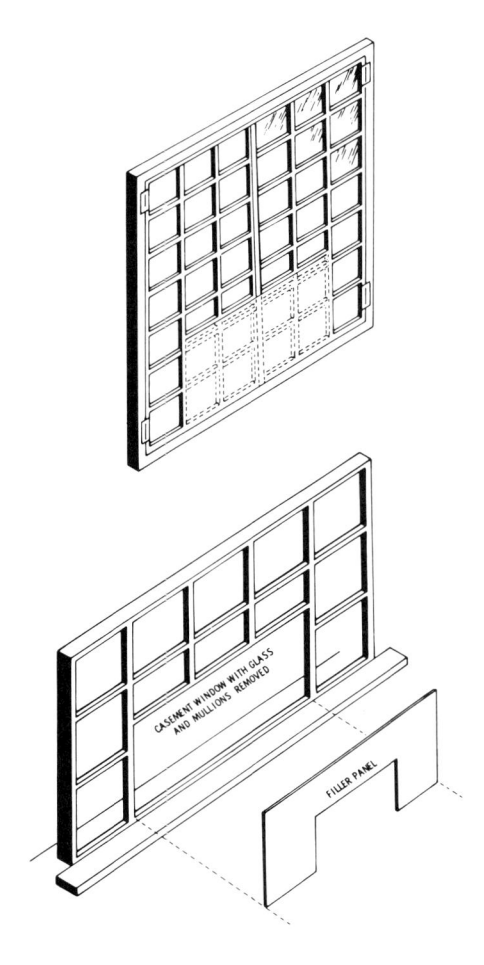

Fig. 5-13. Necessary changes required to permit installation of window-mounted air-conditioning unit in casement-type window.

Alternative Method

Measure height of opening and from this subtract height of the unit. Cut a piece of plywood, Masonite, or Plexiglas equal in height to fit the measurement obtained as a result of the foregoing subtraction and with a width equal to the opening. Install the outer cabinet in the exact center of the opening and secure rigidly. Install the filler board across the top of the unit. Cement these in place at the window frame.

Casement windows with small glass panes will require the cutting out of a horizontal cross member, as well as vertical mullions. In this case it may be advisable to reinstall the cross members at a height equal to the height of the outer cabinet. Cut and reinstall the glass above the cross members. The regular side panels supplied with the unit may be used to fill in the sides, or Plexiglas may be used.

Wall Installations

In certain locations it may be desirable to cut through the existing wall and install the air conditioner in the opening thus provided. When cutting through the existing wall, carefully measure the air conditioner to be installed and cut a hole in the wall large enough to provide for the conditioner as well as the required framing and insulating sleeves.

Frame Construction

Fig 5-14 illustrates a typical framing preparation. After inserting the cabinet into the opening, screw through the cabinet into the framing, as shown. The existing knockouts can be used for either of the two positions; additional holes will have to be drilled along the length of the cabinet for other positions. Also screw the bottom rails to the framing, using #14 screws. After assembling the cabinet securely, the wood-trim strips can be nailed around the cabinet.

Caulking and flashing should be used around the outside to provide a weather-tight seal. Instead of caulking at the bottom, a pan with side flanges and a drip rail may be used to protect the

AIR CONDITIONING

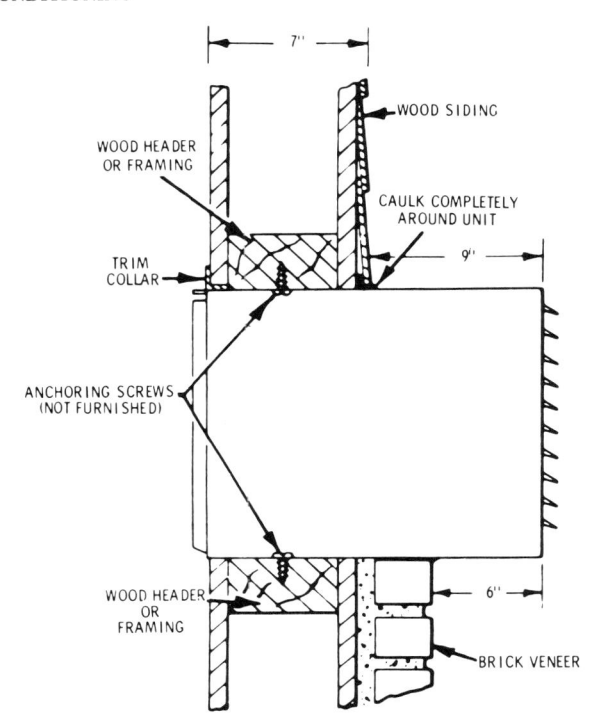

Fig. 5-14. Method of wood-header arrangement to provide support in frame-wall installation.

wall opening. The pan should be sloped to the rear and must also provide a weather-tight seal.

Concrete Block or Brick Construction

To prepare the wall, remove all plaster, furring strips, etc., from the installation area. Be careful not to break the plaster or exterior wall surface beyond the hole dimensions. After obtaining the proper opening, insert the cabinet and secure it with masonry nails. The existing knockouts can be used for either of the two positions; additional holes will have to be drilled along the length of the cabinet for other positions.

After installation, cement the cabinet to the existing masonry

ROOM AIR CONDITIONERS

on the outside. Use caulking and flashing, if necessary, to provide a weather-tight seal. Instead of caulking at the bottom, a pan with side flanges and a drip rail may be used to protect the wall opening. The pan should be sloped to the rear and must also provide a weather-tight seal. Wood-trim strips may be applied around the inside of the cabinet to complete the installation.

OUTSIDE SUPPORT BRACKET

If the air conditioner extends more than 14 inches beyond the outside of the wall opening, the cabinet will require additional support from the outside. This support may be obtained by using the mounting legs and brackets, and corresponding nuts, bolts, and washers as supplied with the mounting kit for the standard window installation. Make sure that the mounting brackets rest firmly against the wall at the wall opening, as shown in Fig. 5-15.

Fig. 5-15. Support-bracket arrangement in typical window installation.

AIR CONDITIONING

CONSOLE-TYPE INSTALLATION

Console-type Room Air Conditioning Systems

The console air conditioner is a self-contained unit, which can come in 2 to 10 hp sizes. These units are used in small commercial buildings, restaurants, stores and banks. They may be water-cooled or air-cooled.

Figure 5-16 shows an air-cooled console conditioner. In installations of this type, the unit should be located so that it can be vented to the outside to get rid of the hot air produced by the compressor and the condenser.

There are also water-cooled console air conditioners. They require connections to the local water supply as well as a water drain and condensate drain (Fig. 5-17). Note the location of the parts in Fig. 5-16. Water is used to cool the compressor. In both models, the evaporator coil is mounted in the top of the unit (Fig. 5-18). Air blown through the evaporator is cooled and directed to the space to be conditioned. In some areas, a water-cooled model is not feasible.

Since the evaporator coil also traps moisture from the air, this condensate must be drained. This dehumidifying action accounts for large amounts of water on humid days. If outside air is brought in, the condensate will be more visible than if inside air is recirculated.

Installation and Service

The console air conditioner is produced by the factory ready for installation. It must be moved to a suitable location and hooked to electrical and plumbing supplies. Once located and connected, it must be checked for level. Electrical and plumbing work must conform to local codes.

Servicing the unit is simple since all the parts are located in one cabinet. Remove the panels to gain access to the compressor, valves, blowers, filter, evaporator and motors. A maintenance schedule should be set up and followed. Most maintenance consists of changing filters and checking pressures. Cleaning the filters, cleaning the inside of the cabinet with a vacuum, and

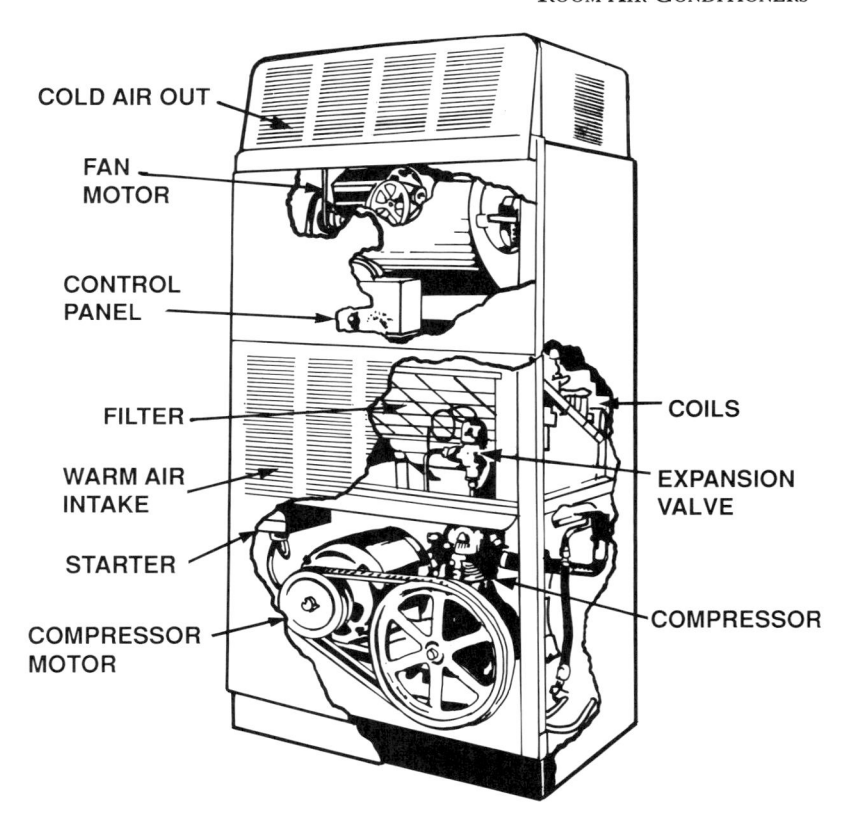

Fig. 5-16. Cut-away view of a self-contained console air conditioner.

cleaning the evaporator fins are the normal service procedures. Water connections and electrical control devices should be checked for integrity. Clean the fan motor. Oil the bearings on the blowers and motors whenever specified by the manufacturers.

Window Mounted Console-type Air Conditioners

It is extremely desirable that a survey of the room is made prior to the actual installation of a console unit. Determine the most favorable location, taking into account the desirable location in the

AIR CONDITIONING

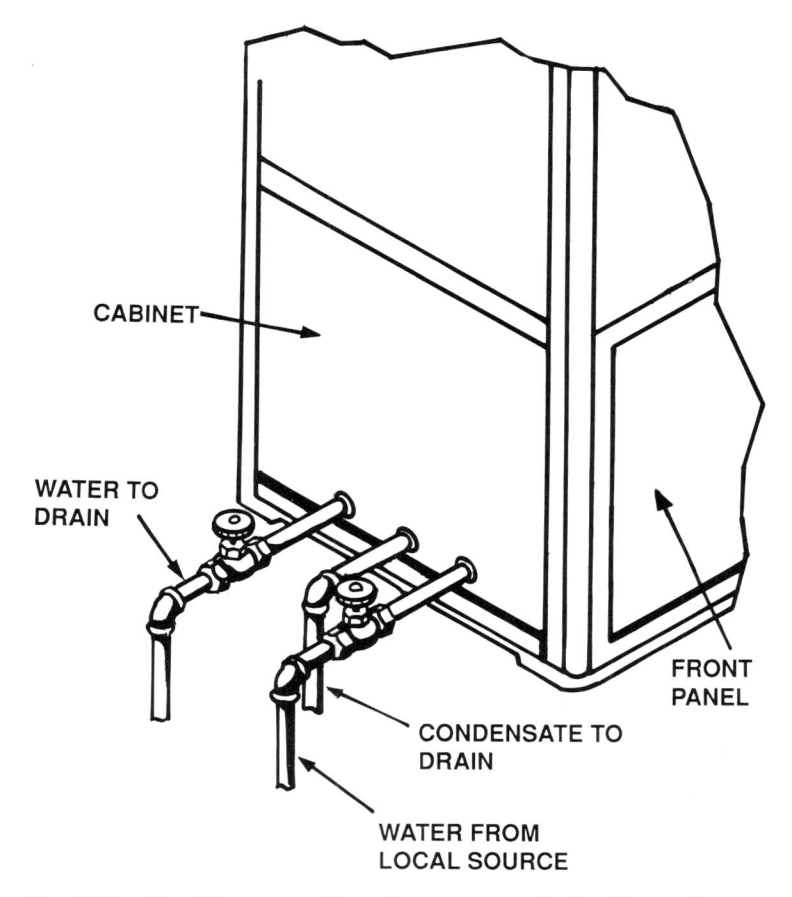

Fig. 5-17. Plumbing connections for a water-cooled, self-contained console air conditioner.

room, exposure of the window, width of the window, its height from the floor, and location of the electrical supply outlet. In this connection, it should be realized that since these types of units usually have a considerably larger cooling capacity than windowsill units, it will be necessary to install a special electrical connection from the meter or distribution panel directly to the location of the unit.

The installation of a typical console room air conditioner in a

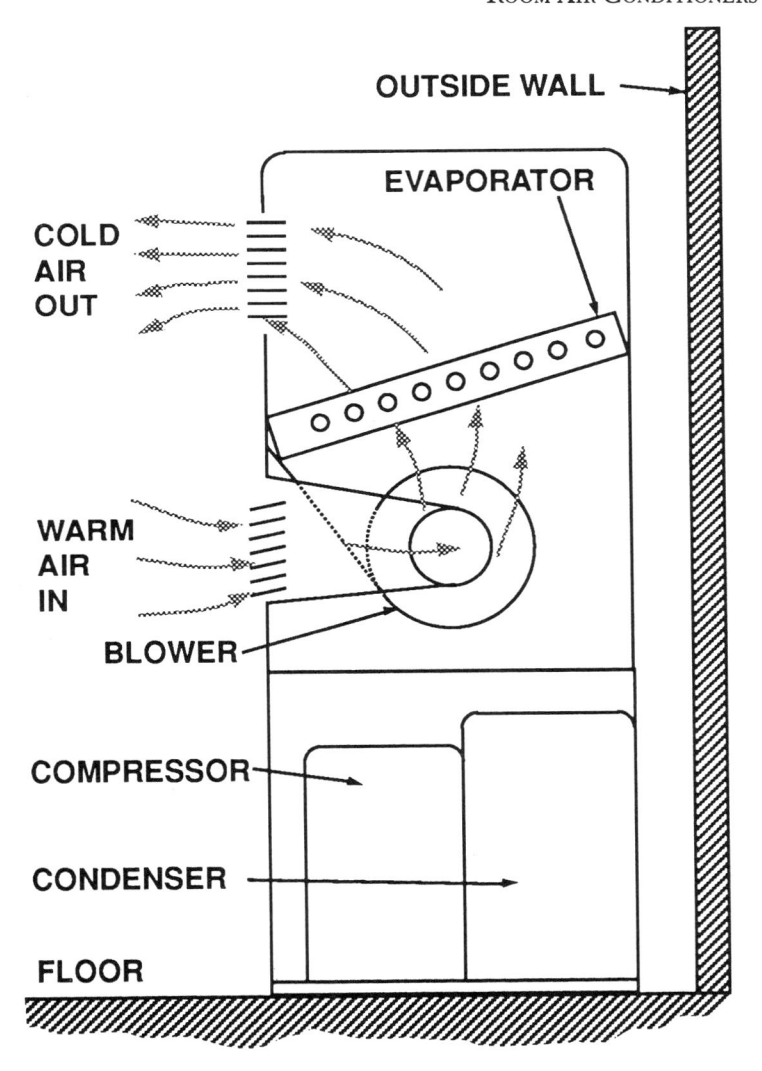

Fig. 5-18. A self-contained console air conditioner, showing air flow over the evaporator.

AIR CONDITIONING

double-hung window is shown in Fig. 5-19. A normal installation of this type will allow the window to be opened or closed without interference from the duct or window filler panels. To completely close the window, the rain hood must be retracted. In order to adjust the height of the unit to obtain the necessary height for the duct outlet (windowsill height may differ by several inches), special wooden bases made up of several sections are usually employed (Fig. 5-20).

Installing Ducts

The standard duct (Fig. 5-21) usually furnished with the console-type unit is approximately 8 inches deep. The unit end has

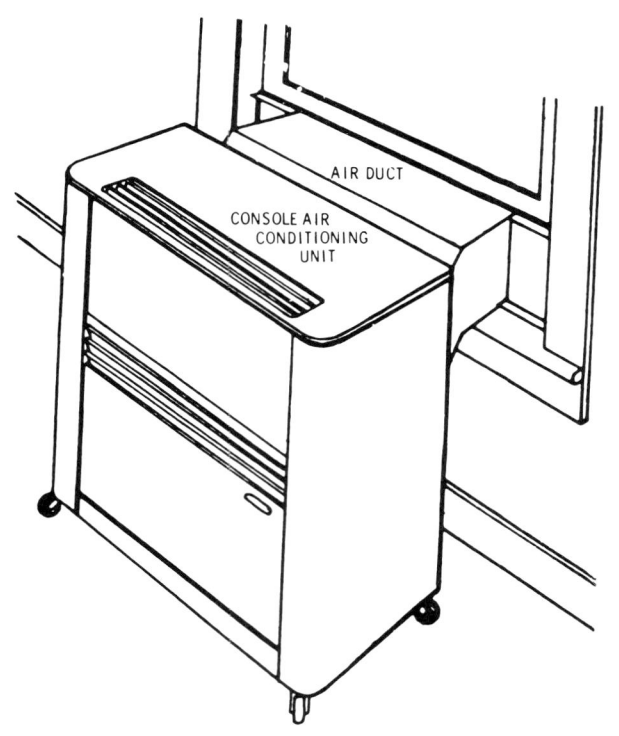

Fig. 5-19. Console-type air-conditioning unit installation in standard double-hung window.

ROOM AIR CONDITIONERS

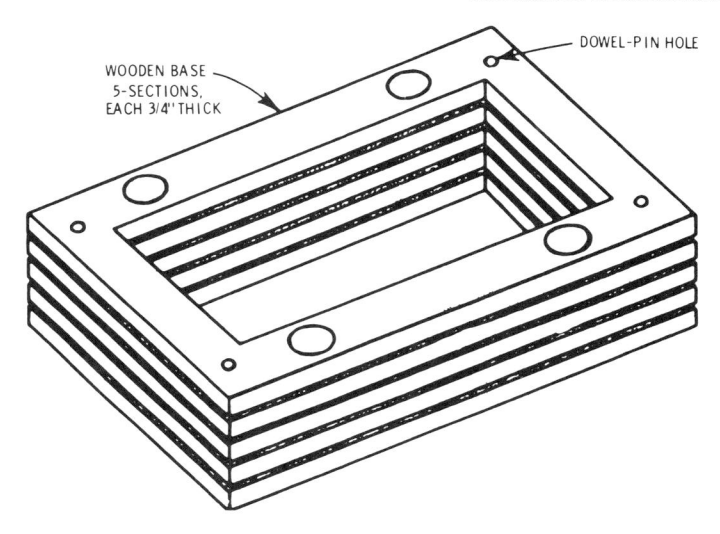

Fig. 5-20. Typical wooden base assembly. Wooden base may be used to increase the height of console units to permit the standard duct to rest on windowsill.

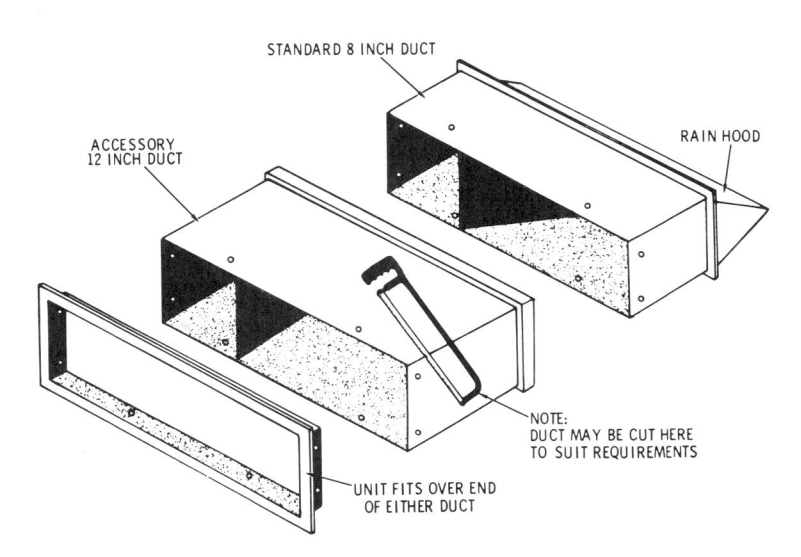

Fig. 5-21. Typical air-duct units as used with console-type air conditioners.

139

AIR CONDITIONING

a removable flange on it which slides on vertical tracks attached to the back of the unit. The window end of the duct has a rain hood fitted in it. When the window is up, the rain hood is pushed out manually and secured in the open position by inserting a screw in each side of the duct after the holes in the duct and rain hood are lined up.

The distance between the window and the nearest permissible location of the unit is measured direct and laid out on the duct, measuring from the window end. Remove the screws holding the removable fitting to the duct; scribe and cut. File off all burrs and break the sharp edges. If the dimension is 8 inches or less, the standard duct is used. If the dimension is greater than 8 inches, the standard and accessory ducts can be connected together to give a total distance of 20 inches. When cutting the accessory duct, be sure to take the measurement from the flared end to allow the extension piece to fit over the standard section. Use the unit fitting to locate new holes in the cut end. Drill with a $1/8$-inch drill and reassemble. Installation of a console air conditioner in casement windows does not differ in any appreciable degree from installation of the windowsill-mounted type.

ELECTRICAL SYSTEM

In order that the room air-conditioning unit may operate properly, it is very important that a check of the available electrical power supply is made before the unit is installed. If an existing branch circuit is to be used, it must be established that it is the proper rating and that the voltage is correct for the unit to be installed (120 or 240 volts, whichever appears on the unit nameplate). It should also be made certain that the electric supply at the outlet is adequate to maintain proper unit operation (Fig. 5-22.)

The following specifications for the operation of room air conditioning units from electric branch circuits are accepted by the National Electric Code and most local codes and ordinances:

1. The unit nameplate amperage should not exceed 80 percent of the branch circuit rating. For example, a No. 14 wire cir-

ROOM AIR CONDITIONERS

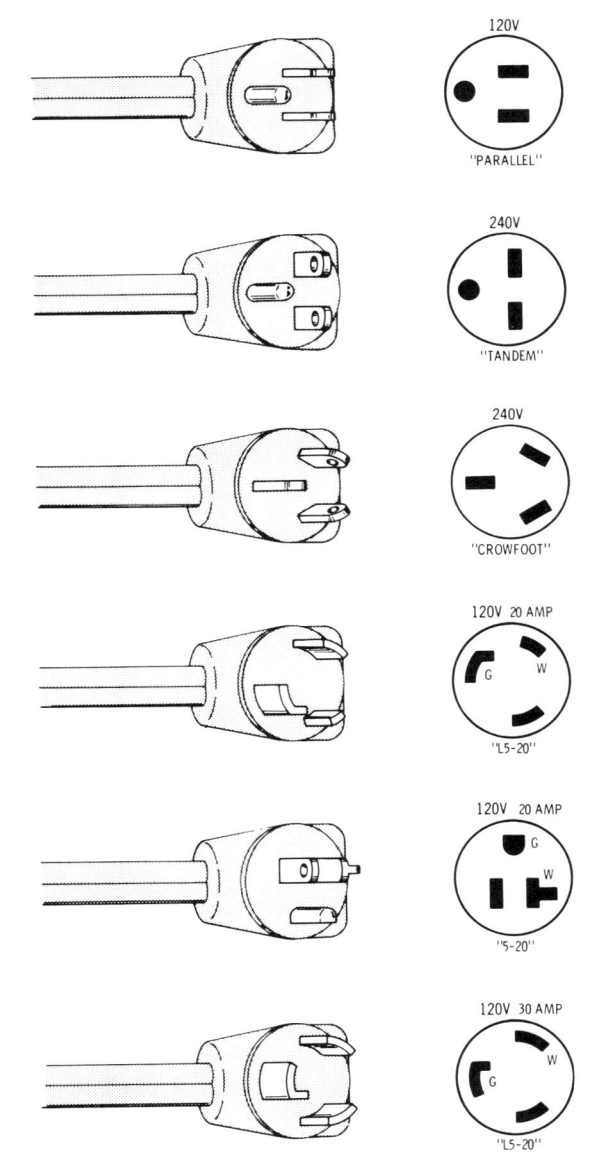

Fig. 5-22. Line cords and plugs for various voltages as used on air conditioning units.

141

cuit is rated at 15 amperes; 80 percent of 15 = 12 amperes, which is maximum loading.
2. If a unit nameplate amperage is greater than 50 percent of the circuit rating, then there should be no other outlet in that circuit available for additional loading by other electrical appliances, lights, etc.
3. If the unit amperage does not exceed 50 percent of the circuit rating, then a multiple-outlet circuit can be used, provided that the additional loading of the circuit combined with the unit amperage does not exceed 80 percent of the circuit rating.

It should be noted also that the determination of a circuit's adequate rating does not necessarily establish that there is adequate voltage available for satisfactory operation of the unit. It is suggested (particularly on 120-volt circuits) that a voltage check be made with the circuit loaded by a current draw comparable to that of the unit nameplate amperage. The voltage under this loading should not drop lower than 10 percent below the unit nameplate voltage. If the voltage drops too low at the outlet but remains satisfactory at the meter, a new branch circuit is necessary.

Electrical load testers are available with selectable wattage that permits simulating a unit current draw. These instruments have a power cord with a wall plug that can be installed directly in the electrical outlet used for the unit. If the voltage does not drop below 90 percent of the unit nameplate voltage with this loading, the circuit should be satisfactory. It may be necessary, however, to install a new circuit for the unit. In some instances, it may be found that there is not enough power supplied to the building, which may require changes in the fuse box and/or the power-supply wiring to the building. It is recommended that the local power company be contacted regarding any questions or problems encountered with either the electric power supply or the electric wiring.

ELECTRICAL COMPONENTS

Figure 5-23 shows the electrical system of a typical room air conditioner, consisting of a motor-compressor with thermal over-

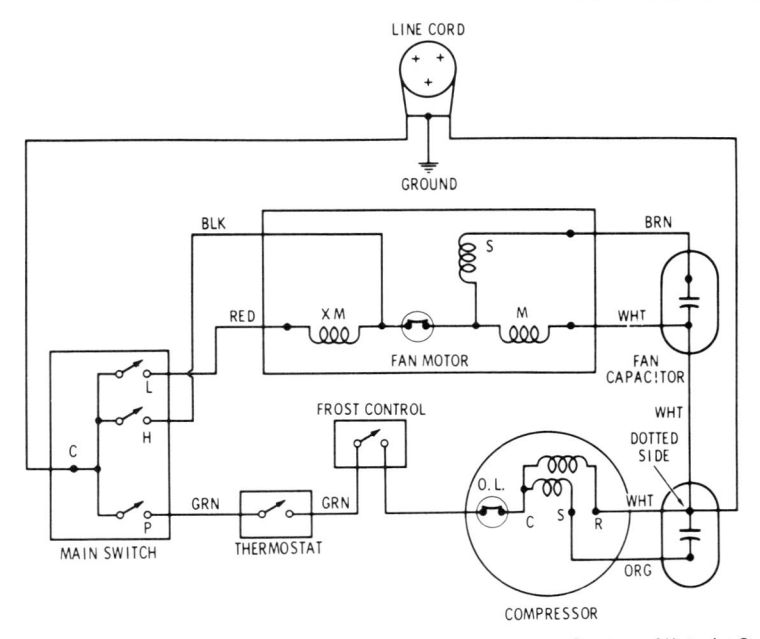

ROOM AIR CONDITIONERS

Fig. 5-23. Schematic diagram showing electrical connection for typical room air-conditioning unit.

load protector, starting relay (when used), starting capacitors, running capacitor, fan motor, fan capacitor (on some models), main thermostat, unit control switch (on-off switch and fanspeed switch combined), reactor (choke coil on some models), and connecting wires.

Compressors

Motor-compressors used on room air conditioners are similar to those used in household refrigeration. In operation, the compressor action necessary for the circulation of the refrigerant is obtained by a piston reciprocating inside a cylinder similar to that of an automobile engine. Half of the shaft rotation is used to draw gas into the cylinder, and the other half is used for compression and discharge. Reciprocating compressors are usually furnished in two styles: a single cylinder (pancake type), with a speed of 3500 rpm, and a twin-cylinder type with a speed of 1750 rpm. Interior and exterior views of a pancake-type compressor are shown in Fig. 5-24.

Starting Relay

The starting-relay coil is in series with the motor run winding. The high current draw on start causes the relay contacts to close, connecting the starting capacitor to the compressor-start winding circuit. As the motor speed increases, current through the run winding drops off and the relay contacts open. This cuts out the starting capacitor, and the compressor continues to run with a running capacitor remaining in the start winding circuit.

The most common faults of a starting relay are:

1. Relay contacts fail to open when the compressor has started.
2. Relay contacts do not close while the compressor is starting.

If the relay contacts fail to open when the compressor has started, starting capacitor failure would likely result and the compressor would draw high amperage. If the relay contacts were not closed when the compressor was starting, difficulty could be encountered.

Overload Protector

The overload protector protects the motor against excessive overload and is usually located inside the compressor terminal cover. It opens the circuit to the compressor motor on abnormally high amperage or objectionally high motor temperature. The overload protector on most models consists of a snap-action, bimetal disk with contacts. The motor current passes through this disk. In the case of overload (failure of the compressor to start or unusual voltage conditions), the current through the disk will be high, thereby increasing its temperature and causing it to automatically open and stop the motor. Due to its direct contact with the com-

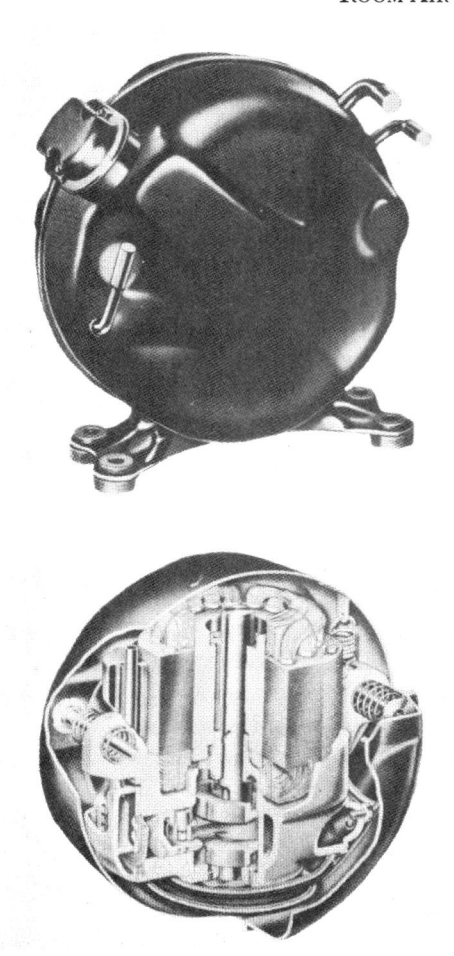

Fig. 5-24. View of single cylinder pancake type compressor. (A) Exterior view; (B) cutaway view.

pressor body, as the motor temperature rises, the heat of the compressor body will increase the overload disk temperature; this lessens the required current to open the circuit.

On some models a heater is combined with the bimetal disk. Excessive current through this heater causes the temperature to rise and heat the disk. When the overload protector trips for any reason, the motor circuit remains open until the compressor cools sufficiently to cause the disk to snap closed, closing the circuit.

Starting Capacitors

All capacitors have two ratings: a microfarad (μF) rating and a voltage rating. The microfarad rating identifies the capacitor's electrical capacitance, while the voltage rating identifies the maximum voltage. A 240-volt capacitor may be used on a 120-volt circuit, but a 120-volt capacitor cannot be directly connected across 240 volts.

Starting capacitors are the electrolytic type and are used in the motor-start winding circuit to affect an increase in starting torque. All starting capacitors are intended for short and infrequent compressor starts. Any operating fault that would cause a starting capacitor to remain in the circuit for more than several seconds could be the cause of its failure. Such causes could be low voltage, faulty relay, some fault of the compressor short-cycling, or a compressor motor failure. The cause of repeated capacitor failure should be investigated as being one the those mentioned above.

If a starting capacitor becomes shorted internally, starting trouble and possible blown fuses would result. If a capacitor should have a broken internal connection to either terminal or any internal open circuit to its terminals, the compressor may not start. If a replacement is available, the simplest way to check a capacitor is to install a new one. If the trouble is corrected, discard the old capacitor.

Running Capacitors

Running capacitors differ from starting capacitors in that they are heavy-duty, oil-filled capacitors that can remain in the circuit continuously. Running capacitors have a much lower microfarad (μF) rating than starting capacitors of comparable size. The running capacitor remains in the motor-start winding circuit at all times during compressor operation. This capacitor also increases starting torque and improves the motor's running abilities. The running capacitor also reduces the running amperage by increasing the power factor.

FAN MOTORS

There are usually two types of fan motors. Some air conditioners use a permanent split-phase capacitor motor, and others use a shaded-pole type. The capacitor type is more efficient electrically, and it is used where there are current limitations. It can be readily identified as it requires a fan capacitor. The number of fan-motor leads will vary from two to four, depending on the type of motor used and its application. Shaded-pole motors will have either two or three leads, depending on whether the speed winding is incorporated in the motor or a reactor is used to reduce the fan speed. Similarly, capacitor-type motors will either have three or four leads for the same reason.

It is very important that the fan-motor leads be connected correctly to prevent damage to the motor. Reversal of the run and phase leads will result in motor rotation reversal on capacitor-type motors. All fan motors have a thermal overload protector embedded in the windings to protect the motor from damage in case of overheating. Certain types of air conditioners employ a *reactor*, sometimes called a *choke coil*, which is used to reduce the speed of the fan motor externally. When connected in the fan-motor circuit, the reactor adds reactance, lowering the voltage to the fan motor and thereby reducing the motor speed.

THERMOSTATS

The thermostat, or temperature control, stops and starts the compressor in response to room temperature requirements. Each thermostat has a charged power element containing either a volatile liquid or an active vapor charge. The temperature-sensitive part of this element (thermostat feeler bulb) is located in the return airstream. As the return air temperature rises, the pressure of the liquid or vapor inside the bulb increases, which closes the electrical contacts and starts the compressor. As the return air temper-

AIR CONDITIONING

ature drops, the reduced temperature of the feeler bulb causes the contacts to open and stops the compressor.

UNIT CONTROL SWITCH

The unit control switch may be located on top of the cabinet or on one of the sides, depending upon the particular design. The control switch is usually of the knob- or rotary-control dial type and normally has four positions, marked OFF, FAN, COOL, and EXHAUST. The damper controls are usually marked SHUT, VENT, and OPEN. To provide cooling, the switch dial is turned to the COOL position, and the damper dial to SHUT or VENT depending on whether or not outside air is desired.

To operate the unit as a ventilator, the switch dial is turned to FAN and the damper dial is turned as far open as desired. In the OPEN position, the unit passes in 100 percent of outside air. To exhaust room air, the switch is turned to the EXHAUST position, and the damper dial is turned to the VENT position. When a thermostat is installed for automatic cooling, the compressor and fans will cycle according to dial requirements.

SERVICE OPERATIONS

Room-air-conditioner servicing is similar to refrigerator servicing. Basically, an air conditioner is a refrigerant system that removes heat from a room. (In a refrigerator, the heat is removed from the food.) An evaporator is employed to remove the heat, and a condenser is used to liquefy the refrigerant. Air movement over the evaporator and condenser surface is accomplished by a fan. Hot air usually contains a greater percentage of moisture than colder air. When the evaporator fan moves hot humid air over the cold surface of the evaporator, a quantity of the moisture in the air will condense (form water droplets) in addition to the air itself being cooled. In this manner, the relative humidity of the recirculated air is also reduced. The condensate water is drained into the condenser section where it is dissipated.

Servicemen who have a good knowledge of refrigeration

combined with the understanding that an air conditioner removes heat and humidity from a room by the process of refrigeration will be able to competently service air conditioners. Since most portable air-conditioning units of present design contain compressors of the hermetic or sealed type, the only parts that can be serviced in the field are the relay, control switch, fan, fan motor, starting and running capacitors, air filters, and cabinet parts. The refrigerating system, consisting of the cooling unit, condensers, compressors, and connecting lines, generally cannot be serviced in the field.

DISMANTLING WINDOW AIR CONDITIONERS

The following discussion lists the procedure and steps necessary to dismantle window air conditioners. The procedure can be used in its entirety or in part, according to the service required. The procedure covers only those parts of the unit that can be serviced in the field and is not a complete disassembly. By presenting a dismantling procedure, it is possible to eliminate repetition of certain steps common to many service operations. Keep in mind that individual manufacturers may recommend a slightly different procedure. In all instances, follow the instructions of the manufacturer when dismantling an air conditioner. When reassembling the unit, the steps should be reversed. To dismantle the air conditioner, proceed as follows:

1. Disconnect the air conditioner from the source of electric supply, and remove the cabinet.
2. Remove the electric control boxes. Remove the screws that secure each box to the partition. Remove the control-box covers, and disconnect the motor leads; remove the controlbox assemblies from the unit.
3. Remove the fan motor and fans. Loosen both fans, and remove them from the shaft. Remove the access panel that fits down over the cooling-unit fan shaft. Loosen and remove the motor cradle supports at each end of the motor, and lift the motor up and out.

ELECTRICAL TESTING

In the event of operating trouble, a thorough check of the electrical system is sometimes necessary. By checking the electrical system a great deal of time can be saved since such a check usually reveals the more obvious troubles. To make a complete electrical test of the unit and its controls, a volt-ohmmeter is the safest and easiest instrument to use. It may be used to check the line voltage to see that it is high enough, and, by using the ohm resistance scales, continuity checks may be made also.

When checking the electrical system, refer to the wiring diagrams (Figs. 5-25 and 5-26). It should be noted, however, that the diagrams shown are only typical; the arrangement and wiring of components may vary, depending on the manufacturer of the particular unit.

Testing for Current Supply

With the volt-ohmmeter set on the proper ac voltage scale, first check the convenience outlet that the unit is plugged into to be certain that current is available. Next, with the unit plugged into the outlet and the volt-ohmmeter on the same setting, place one of the test leads on terminal N and the other on terminal L. If the proper voltage is recorded, current is being supplied to the relay.

Testing Fan Motors

Depending on the size of the unit, it may be equipped with one or two fan motors. In some smaller units, only one fan motor is used, in which case the shaft of the motor extends to operate two fans — one for the evaporator and one for the condenser.

To check for continuity, always pull the plug on the supply cord from the source of supply so as not to damage the ohmmeter portion of the instrument. To check the evaporator fan motor, turn the switch to the running position. Remove the connector from the wires tied together in the switch box. Place the ohmmeter test leads between terminal 1 of the switch and the wire ends from which the connector was removed, which is the lead that goes to

ROOM AIR CONDITIONERS

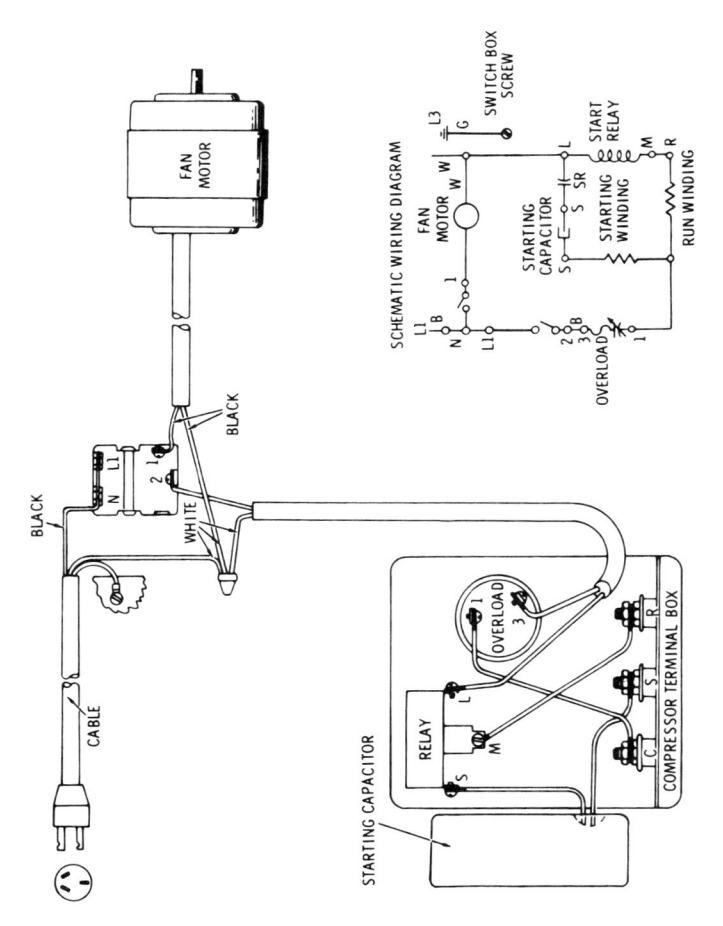

Fig. 5-25. Wiring diagram of typical air-conditioning unit having one fan motor and a sealed-type compressor unit.

the fan motor. If the ohmmeter shows continuity, the circuit to and through the motor is okay.

Most motors are of the high-impedance type, which can stall and not burn out, but some have internal overload protection. If the motor has overload protection and is hot, allow it to cool a few minutes. If the ohmmeter has given an infinity reading, test

AIR CONDITIONING

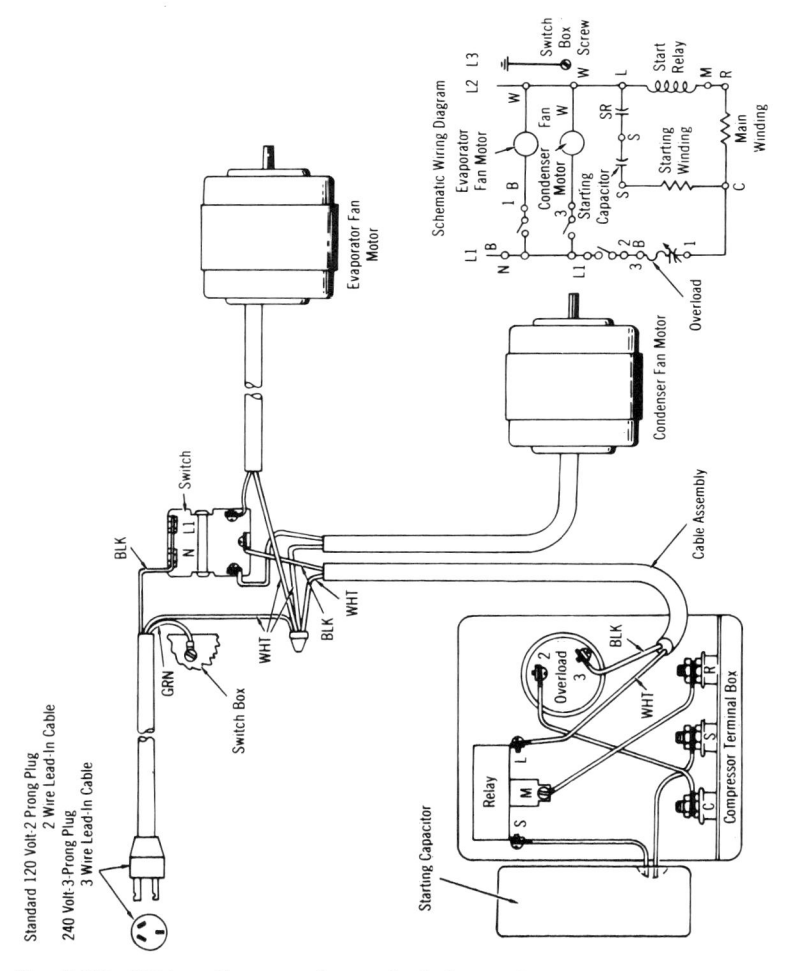

Fig. 5-26. Wiring diagram of a typical air-conditioning unit having two fan motors and a sealed-type compressor unit.

again; if the meter still reads infinity, replace the fan motor. If the fan motor checks out, check the switch by putting one lead of the ohmmeter on N or L_1 (they are tied together), and touch the other lead to terminal 1 and then to terminal 2. If the switch has been turned on, you should get a 0-ohm reading each time. If the

152

switch is defective, replace it with a switch having the same part number.

To check the condenser fan motor, turn the switch to the running position. If the condenser fan motor does not operate, remove the junction box cover and switch, first disconnecting the whole unit from the power source. Remove the connector from the wires tied together. Place the ohmmeter leads on the wire that goes to the condenser motor from the connection where you removed the connector, and connect the other ohmmeter lead to terminal 3 of the switch. If the ohmmeter reads relatively low, the winding is probably okay. If it reads 0 ohm, the winding is probably open in the motor. If the motor is hot, wait a few minutes and read again if the first reading was 0 ohm.

To check the evaporator and condenser fan motor when only one fan motor is used, disconnect the whole unit from the source of supply. Turn the switch to the ON position, and remove the junction-box cover and switch. Remove the connector from the wires tied together in the switch box. Pick out the wire that goes to the motor, and place one ohmmeter test lead on this wire end and the other lead on terminal 1. If the motor is okay, you should get a comparatively low resistance reading. If the meter reads 0 ohm and the motor is hot, let the motor cool a few minutes and recheck. If the meter still reads 0 ohm, the winding is probably open, so replace the motor. Again you should check switch-contact continuity, as explained previously.

Capacitor (Starting)

On most units an electrolytic starting capacitor is used in the starting winding circuit to increase the starting torque. If this capacitor develops a short internally, starting trouble and possible blown fuses will result. If the capacitor develops an open circuit, the compressor will not start. If a replacement is available, the simplest way to check the capacitor is to install a new one. If a replacement is not readily available, a check can be performed in the following manner.

To check a capacitor, disconnect the capacitor leads. Make certain that the capacitor has not retained a charge by placing the blade of an insulated screwdriver across the terminals. Touch the capacitor leads or terminals momentarily with the test probes of an ohmmeter, and observe the meter deflection. A satisfactory capacitor will show at least 100,000-ohm (100-K) resistance. It may take several seconds to arrive at this reading. A shorted capacitor causes the ohmmeter pointer to indicate a continuous low resistance. The pointer moves to the zero end of the scale and remains there as long as the probes are in contact with the terminals. An open capacitor causes no deflection of the pointer, meaning that there is not path or continuity through the capacitor, which should be replaced. It is essential that the capacitor be replaced with an exact duplicate.

Compressor Motor Relay

The relay opens the circuit to the compressor motor starting winding after the compressor has started. The duration of this start or of the time that the starting winding is energized is, under normal conditions, very short — usually 2 or 3 seconds. A defective relay may fail to close, which results in starting trouble; it may fail to open, which results in overload tripouts, capacitor failure, or blown fuses.

Relay Check

Check the capacitor before making any relay-operation checks since a shorted or open capacitor will not allow the motor to operate. Disconnect the entire unit from the supply source. Disconnect lead S from the relay, and check the continuity from S to L. The reading should be 0 ohm. Replace lead S to the relay, and disconnect lead M at the relay, testing the resistance between M and L. The reading should be 0 ohm. These tests indicate that the starting and running contacts of the relay are closed. To doublecheck, in most cases relays may be opened and visual inspection of the contacts made. If the contacts show much burning, even though they are making contact now, replace the relay as a preventive measure.

Compressor Motor Overload

The overload is a protective device used in conjunction with the relay to open the compressor motor circuit at abnormally high currents or dangerously high motor temperatures. The overload consists of a heater and a snap-action bimetal disk on which contacts are mounted. A heater is connected in series with the common terminal of the motor windings. In the case of overloads, the current through the heater is high, thereby increasing its temperature. Thin heats the bimetal disk, thus causing it to automatically snap open and open the motor-winding circuit. Due to the direct contact of the overload to the compressor body, as the motor temperature rises, the heat of the compressor body increases the overload-disk temperature, thereby reducing the current required to open the circuit.

If the overload trips for any reason, the circuit will remain open until the compressor body (or shell) cools sufficiently to cause the disk to snap closed. It will then close the circuit to the motor. This protector is nonadjustable and must be replaced if it fails to function properly. To check the overload, disconnect the power cord from its source, remove leads 2 and 3 from the overload and check the continuity between terminals 2 and 3. The meter should read 0 ohm. If the test shows a high resistance, the points are open and the relay should be replaced.

Often, you will not find an overload protector as shown in Fig. 5-27 because it is a part of the starting relay. Low voltage, shorted capacitor, failure of the compressor to start, and excessive operating pressures or temperatures are some of the causes of overload trip-outs. It is well to have a clamp-on-type ac ammeter to check running currents.

COMPRESSOR AND MOTOR

The compressor motor is a capacitor-type split-phase motor that has two windings. The starting winding has the capacitor in series with it to give great starting torque, with the winding composed of a smaller-size wire than the running winding. Through the use of a capillary-tube method of refrigeration, the starting load is normally rather low. At the start, the running winding is energized, drawing a larger starting current, which causes the relay to close the starting winding contacts, energizing the starting winding almost immediately.

AIR CONDITIONING

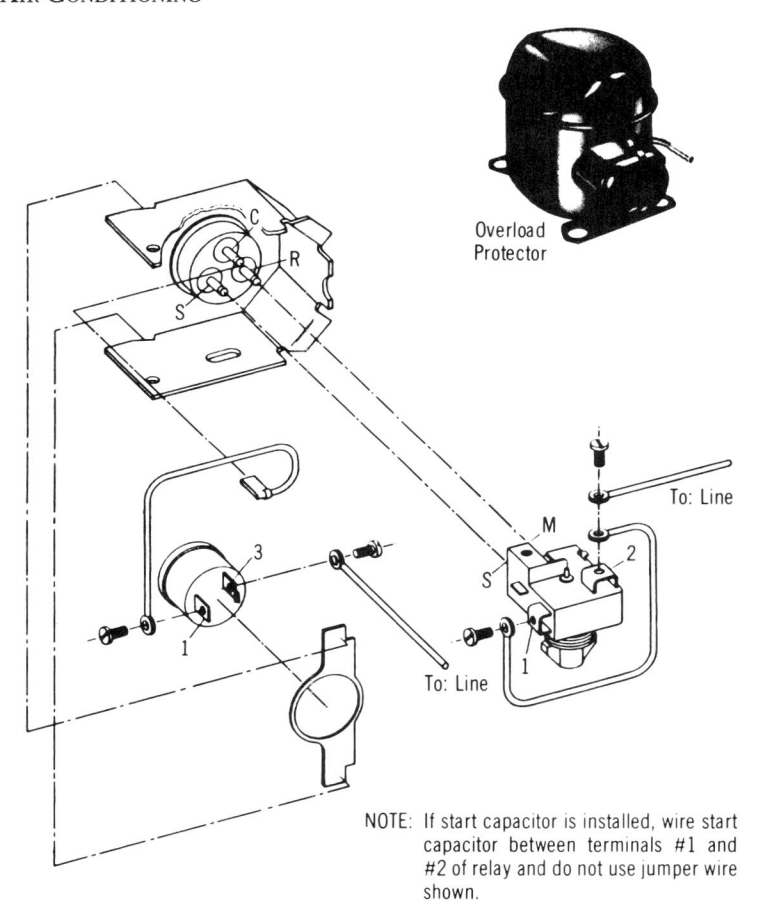

Fig. 5-27. Note location of the overload protector in line that leads to SCR connections on the compressor.

NOTE: If start capacitor is installed, wire start capacitor between terminals #1 and #2 of relay and do not use jumper wire shown.

Courtesy of Tecumseh

As the motor comes up to speed, the current value drops. At about 65 percent of full speed, the current drops low enough for the starting relay to open the circuit to the starting winding and the full load is then carried by the running winding. If the load is too heavy, the current draw will be large and the overload protect-

ROOM AIR CONDITIONERS

ive device will open and stop the motor until the overload protection cools, when it will again attempt to start the motor.

To check the continuity of the compressor motor winding, the three motor leads C, S, and R (Fig. 5-27) should be disconnected, being sure the entire unit is unplugged from the source of electrical supply. The ohmmeter should show continuity between any two of the three terminals, with a larger resistance between terminals C and S than between C and R. If the ohmmeter shows no resistance on any of the tests, an open winding is indicated.

While making the preceding test, you should test the compressor motor for grounding. This is done by putting the ohmmeter on its highest scale and touching one lead to the compressor housing and the other to each of the motor terminals, one at a time. If the winding is not grounded, the ohmmeter should read full scale, or the maximum resistance, in ohms.

It must be remembered that faults other than motor trouble may be the cause of compressor failure or of a motor drawing high current. A stuck compressor, high head pressure, low voltage, or a plugged capillary tube are some of the causes of compressor failure.

OILING OF MOTORS

The fan motors should be oiled at the start of each cooling season or every six months if the unit is operated all year. Use a good grade of electric motor oil or SAE No. 20 automobile oil. A few drops in each oil hole is sufficient.

LEAKS IN SYSTEM

In the event that any part of the condensing unit develops a leak, it will be necessary to repair the leak before the unit can be put into operation. A leak is usually discovered by the presence of oil around the point at which the leak developed. It must not be assumed, however, that the presence of oil on any part of the unit is a positive indication of a leak. Always check the suspected area with an approved leak detector.

AIR CONDITIONING

FILTERS

The filters should be inspected at regular intervals. Refer to individual manufacturer's instructions for filter removal and cleaning. Periodic cleaning of filters ensures maximum air delivery by the air conditioner at all times. Best results can be obtained if filters are replaced every year and cleaned often between replacements.

INTERIOR CLEANING

The interior of the unit should be cleaned periodically of all dust, grease, and foreign matter. Special attention should be given to the condenser and evaporator coils. Regular cleaning will ensure continuous good service of the unit.

WINTER CARE

In many parts of the United States, service of the cooling unit will not be required during the winter months. Such a unit can be easily removed from the window and stored in a convenient place. This prevents condensation of moisture in the unit and also makes available a greater window area. Before the unit is stored, however, it should be checked for dirt in the evaporator and condenser and cleaned when necessary. Cleaning the evaporator once a year is recommended to avoid the development of objectionable odors and possible accumulation of dirt.

The coils can be cleaned by the use of a stiff bristle brush and a strong solution of soap and water. All parts should be rinsed off after cleaning. It is also advisable to wash out the drain pan and retouch with an asphalt paint. When storing the unit, it should be put on blocks to take the weight off the sponge rubber mounting. At the beginning of the cooling season when the unit is reinstalled, the fan motors should be oiled. After the unit has been reinstalled and started, it should be checked after a few minutes of operation to ascertain that the temperature drop across the evaporator is 10°F or more.

REFRIGERANT-SYSTEM SERVICE

When a unit operates but does not have its normal cooling capacity, the refrigerant system should be checked. Check for some obvious symptom, such as partial icing of the evaporator, an abnormally warm suction line, or other unusual condition, which could indicate a loss of refrigerant or a restricted capillary tube.

The refrigerant system of most room air-conditioning units is a sealed-type circuit. Should trouble develop that requires opening the refrigerant circuit, the usual caution and requirements for repair of a capillary-tube-type system are necessary. This requires equipment for evacuating and leak testing, as well as a means of accurate refrigerant charging. Observe all appropriate safety precautions specific to the refrigerant.

Evaporator Temperature

Under average operating conditions of outdoor and indoor temperatures, all evaporator return bends should be cold and at the same approximate temperature. On most models, this check can be easily made by removing the front grille. If the first evaporator passes are iced over or if there is a noticeable temperature increase of the return bends of the last several passes, it could be an indication of either restriction or loss of refrigerant. It must be remembered, however, that with higher outdoor temperatures, the very last evaporator pass will not have as much refrigerant to effect cooling as it would have at lower temperatures. Therefore it will be at a slightly higher temperature if the outside temperature is high.

Frequently, servicemen check the dry-bulb-temperature drop across the evaporator to determine if the unit is cooling satisfactorily. The temperature drop across the evaporator is not the same on all units. Also, higher outdoor dry-bulb temperatures and high inside relative humidity tend to reduce the dry-bulb-temperature drop across the evaporator. On most units with an outdoor temperature between 80 and 90°F and with an indoor relative humidity between 40 and 50%, the temperature drop across the evaporator should be between 17 and 20°F. Where proper evaporator air flow is maintained, a 20° dry-bulb-temperature drop

AIR CONDITIONING

would normally indicate satisfactory cooling. When the room air relative humidity is considerably higher than 50%, a smaller dry-bulb temperature drop across the evaporator will be obtained. When the relative humidity is considerably below 40%, a higher temperature will result. It is therefore important that room air relative humidity be checked and considered when checking air temperature drop across the evaporator as a check for proper cooling.

Compressor Amperage

With normal evaporator loading and average outdoor temperatures, if proper cooling is being accomplished, the compressor motor will draw the approximate amperage shown on the unit nameplate. If a unit is drawing considerably below compressor nameplate amperage, the refrigerant system may not be producing proper cooling, assuming that there is a normal flow of air over the evaporator and that the evaporator loading is not abnormally low.

Refrigerant Leaks

Whenever a shortage of refrigerant is suspected, this condition can be confirmed by identifying the source of leakage. A leaktest operation can be performed with a halide leak detector. The end of the exploring hose is passed around areas where a refrigerant leak is suspected. The air drawn into the hose passes through a copper reactor plate and through a small flame. If any refrigerant is contained in this air, the color of the flame turns slightly green. If a considerable quantity of refrigerant is leaking, the flame turns purple. When the location of leak has been determined, it is sometimes difficult to pinpoint the exact leak location; soapsuds can be used to pinpoint the exact source of leak.

Automatic Expansion Valves

Automatic expansion valves are used on some room air conditioners instead of the capillary tubes to control the flow of refrigerant into the evaporator. These valves are set to maintain a normal suction pressure. They are factory adjusted and sealed, and no field adjustment is required. A small bleed port in the valve permits pressure equalization from the high to low side on shut down. A small liquid-line strainer is located in the liquid line ahead of the expansion valve.

TROUBLESHOOTING GUIDE
(Window mounted units)

The troubleshooting guide shown on the following pages lists the most common operating faults and suggests possible causes and remedies. This guide provides a quick reference to the cause and correction of a specific fault.

ROOM AIR CONDITIONER TROUBLESHOOTING GUIDE

Symptom and Possible Cause *Possible Remedy*

Unit Will Not Run

(a) Blown fuse
(b) Broken or loose wiring
(c) Low voltage

(d) Defective unit starting switch

(a) Check for an electrical short at wall receptacle.
(b) Replace fuse.
(c) Check voltage. Voltage should be within 10 percent of that shown on the unit nameplate.
(d) Check voltage at switch. If there is proper voltage at switch but no continuity through it, replace switch.

Fan Runs, but Compressor Will Not Operate

(a) Inoperative thermostat

(a) If turning thermostat to its

AIR CONDITIONING

Symptom and Possible Cause   *Possible Remedy*

    coldest setting does not start compressor (and room temperature is above 75°F) but shorting across its terminals does, change thermostat.

(b) Loose or broken wiring
(b) Check unit wiring and wiring connections at the unit starting switch and at the compressor.

(c) Starting capacitor faulty (if used)
(c) Check starting capacitor (if used).

(d) Running capacitor faulty
(d) Check running capacitor.

(e) Relay faulty (if used)
(e) Check relay (if used).

(f) Off on overload or overload fault
(f) Check for overheated compressor or for defective overload.

(g) Low voltage
(g) Check voltage.

Unit Blows Fuses

(a) Shorted or incorrect unit wiring
(a) Check unit wiring.

(b) Shorted starting or running capacitor
(b) Check capacitors.

(c) Shorted or stuck compressor
(c) Check compressor.

(d) Compressor starting difficulty
(d) Check for low voltage; check starting capacitor and relay.

(e) Incorrect fuse
(e) Check fuse size.

Compressor Cycles Off and On

(a) Low voltage
(a) Check voltage and amperage with unit operating. If low, check other appliance loads.

Symptom and Possible Cause	Possible Remedy
(b) Restricted air across the condenser	(b) Check condenser fan and check for any restriction of condenser air.
(c) Recycling of condenser discharge air back into unit	(c) Check for inadequate clearance for proper discharging of condenser air; check for any cause of hot discharge air reentering the unit.
(d) Dirty or plugged condenser	(d) Check and clean all dirt or lint from condenser coil fins. Clean air filter.
(e) Thermostat feeler bulb out of position	(e) Check bulb for proper location in return air. Bulb should not touch the evaporator coil.
(f) Faulty or incorrect overload	(f) Check if overload has tripped. Refer to Capacitor, Overload and Relay Chart for correct overload. Check compressor temperatures. If compressor is not overheated, and if amperage is not high and overload trips out, change overload.

Unit Vibrates or Rattles

(a) Compressor shipping block not removed or shipping bolts not removed or properly loosened	(a) Remove shipping bolt and shipping block on externally spring-mounted compressors only.
(b) Discharge or suction tube striking metal surface	(b) Bend tube slightly for clearance where striking.
(c) Loose compressor junction-box cover or capacitor	(c) Tighten.
(d) Loose or bent fan blades	(d) Replace fan.

AIR CONDITIONING

Symptom and Possible Cause *Possible Remedy*

(e) Fan motor out of alignment or loose on mounting
(e) Check alignment and tighten mounting.

Water Drips from Unit

(a) Unit not properly leveled
(a) Level unit. Most models can be tipped slightly toward outside.
(b) Condensate drain to condenser pan plugged
(b) Clean condensate drain.
(c) Condenser-fan slinger ring out of position
(c) Check motor mounting and alignment for proper positioning of condenser-fan slinger ring in water sump.
(d) Evaporator drip pan, shroud, or other source of water leaking and requires sealing
(d) Use Permagum or other sealing compound to seal leak.

Noisy Fan Operation

(a) Fan striking
(a) Check for proper positioning and clearance of fans.
(b) Loose fan or fan motor
(b) Check for fan tightness on shaft. Check for loose fan blade, fan loose at rubber hub, and loose fan-motor mounting.
(c) Bent fan blades
(c) Replace fan blades.

Unit Not Cooling Properly

(a) Clogged air filter
(a) Check air filter for dirt. Clean or replace.
(b) Outside air entering the unit
(b) Check for proper closing of outside air doors.

Symptom and Possible Cause   *Possible Remedy*

(c) Air across condenser restricted
(d) Dirty or blocked condenser

(e) Hot condenser discharge air reentering unit

(f) Compressor or refrigerant system faulty

(c) Check condenser air.

(d) Check condenser for blocked dirt or lint. Clean or replace.

(e) Check for restriction or obstruction to condenser discharge air that could cause it to reenter unit.

(f) If compressor runs but the evaporator does not get cool, this would indicate either an inefficient compressor, restricted capillary tube, or lost refrigerant charge.

Evaporator Ices Over

(a) Clogged or dirty air filter
(b) Evaporator fan motor tripping on overload

(c) Unit operating at too low room temperature

(d) Outside temperature too low

(a) Clean or replace air filter.
(b) Check for overheated evaporator fan motor and tripping off on its internal overload.

(c) If room temperature drops below 70°F, the evaporator may ice over.

(d) Check for unit being operated when it is unusually cool outside.

SUMMARY

A room air conditioner is usually considered to be a unit suitable for placement in any particular room. The room may be an office, a home or a small shop. Room air conditioners are usually classed according to their design and method of installation, such as window units and portable units.

Although the various types of room air conditioners differ in cabinet design as well as in arrangement of components, they all operate on the same principle. The evaporator fan draws the recirculated air into the unit through the louvers located on the side. The air passes through the air-filter evaporator and is discharged through the grill on the front of the unit into the room.

A ton of refrigeration is 12,000 Btu/hr. Room air conditioners are rated in Btu/hr. The amount of cooling capacity can be determined by checking a table of values for various Btu ratings.

A number of installation methods may be employed in mounting air conditioners in the window of a desired room. Some units come with filler boards to make the unit fit any size window. A console type air conditioning unit can be made to fits a standard double-hung window.

Line cords and plugs vary according to the voltage and current the unit will draw. At least six different standard type plugs are available for window or room air conditioners.

The electrical components of a room air conditioner consist of a motor-compressor (with thermal overload protector), starting relay (when used), starting capacitor, running capacitor, fan motor, fan capacitor on some models, main thermostat, unit control switch (made up of the on-off and speed selection switches).

Compressors may be of single or double cylinder type. Rotary compressors are used in some units to reduce the noise inherent with the reciprocating types.

Room air conditioner servicing is similar to refrigerator servicing. Basically an air conditioner is a refrigerant system that removes heat from a room whereas a refrigerator removes the heat from the food compartment and from the food inside it.

When checking the electrical system, refer to the wiring diagram. Keep in mind that the compressor motor is a capacitor-type split-phase motor that has two windings. Fan motors should be oiled at the start of each cooling season or every six months if the unit is operated all year. Use a good grade of electrical oil or SAE No. 20 automobile oil. A few drops in each oil hole is sufficient.

Filters should be inspected at regular intervals. Periodic cleaning of filters ensures maximum air delivery by the air conditioner at all times.

Before the unit is stored for the winter it should be checked for dirt in the evaporator and condenser and cleaned when necessary. Cleaning the evaporator once a year is recommended to avoid the development of objectionable odors and possible accumulation of dirt. The coils should be cleaned by the use of a stiff bristle brush and a strong solution of soap and water. All parts should be rinsed off after cleaning.

A troubleshooting guide should be consulted to check the symptoms and possible causes of trouble along with the possible remedy.

REVIEW QUESTIONS

1. Where can a room air conditioner be most useful?
2. List at least 6 parts of a room air conditioner.
3. List 5 important variables to keep in mind when estimating the Btu requirements for a room cooling installation.
4. What affect does a west facing wall have on a room when figuring the room cooling requirements in Btus?
5. Why are filler boards sometimes needed in an air conditioner installation?
6. How does the console type air conditioner differ from that of the window type?
7. What are the three specifications that the National Electrical Code states for operation of room air conditioning units?
8. Draw the plug arrangement for a "crowfoot" 240-volt plug.
9. What is the symbol used in the schematic diagram of a typical air conditioning unit for representing the compressor motor?
10. Where is the start-relay located in the electrical circuit of a room air conditioner?
11. In what circuit is the overload protector located?
12. What is the difference between a run-capacitor and a start-capacitor?
13. Name two types of fan motors.

AIR CONDITIONING

14. What is another name for a choke coil?
15. What is the difference between an air conditioner and a refrigerator?
16. How do you test fan motors for proper operation?
17. What does CSR stand for on a compressor motor housing?
18. How often should fan motors be oiled?
19. What is one indication of a refrigerant leak in an air conditioner system?
20. What happens to the color of a flame when this method is used to detect a refrigerant leak?

CHAPTER 6

Refrigerants

Refrigeration is a process of removing heat from a substance or a space. A desirable refrigerant should possess chemical, physical, and thermodynamic properties that permit its efficient application in refrigerating systems. In addition, there should be no danger to health or property in case of its escape due to leaks or other causes in a refrigerating system.

DESIRABLE PROPERTIES

The following are requirements of a good refrigerant for commercial use:

1. Low boiling point
2. Safe and nontoxic
3. Easy to liquefy at moderate pressure and temperature
4. High latent heat value

AIR CONDITIONING

5. Operation on a positive pressure
6. Not affected by moisture
7. Mixes well with oil
8. Noncorrosive to metal

In the past, refrigerants were selected for use principally for their boiling points, pressures, and stability within the system or unit, regardless of other important necessary properties, such as nonflammability and nontoxicity. There are, of course, many factors that must be taken into account when selecting a chemical compound for use as a refrigerant other than the boiling point, pressure, stability, toxicity, and flammability. These factors must include molecular weight, density, compression ratio, heat value, temperature of compression, compressor displacement, and design or type of compressor, to mention only a few of the major considerations.

CLASSIFICATIONS

The refrigerants listed in Table 6-1, for all practical purposes, are no longer used, but it is possible that you may run across them in old installations. The refrigerants used most at this time are: Freon-12, Freon-22, Freon-502, and ammonia (NH_3). (Some information will be given on the others [for purposes of information only].) Table 6-2 gives refrigerant pressure and temperature for the commonly used refrigerants.

Refrigerants may be divided into three classes according to their manner of absorption or extraction of heat from the substances to be refrigerated.

Class 1 — This class includes refrigerants that cool by the absorption or extraction of latent heat from the substances to be refrigerated. These refrigerants and their characteristics are listed in Table 6-1.

Class 2 — These refrigerants cool substances by absorbing their sensible heats. They are air, calcium chloride brine, sodium chloride (salt) brine, alcohol, and similar nonfreezing solutions. The purpose of Class 2 refrigerants is to receive a reduction of temper-

Table 6-1. Characteristics of Typical Refrigerants

Name*	Boiling Point, °F	Heat of Vaporization at Boiling Point, Btu/lb (at 1 atm)
Sulfur dioxide (SO_2)	14.0	172.3
Methyl chloride (CH_3Cl)*	-10.6	177.80
Ethyl chloride (C_2H_5Cl)	55.6	177.00
Ammonia (NH_3)	-28.0	554.70
Carbon dioxide (CO_2)	-110.5	116.00
Freezol (isobutane) [$(CH_3)3CH$]	10.0	173.50
Freon-11 (CCl_3F)*	74.8	78.31
Freon-12 (CCl_2F)*	-21.7	71.04
Freon-13 ($CClF_3$)*	-114.6	63.85
Freon-21 ($CHCl_2F$)*	48.0	104.15
Freon-22 ($CHClF_2$)*	-41.4	100.45
Freon-113 (CCl_2F—$CClF_2$)*	117.6	63.12
Freon-114 ($CClF_2$)*	38.4	58.53
Freon-115 ($CClF_2CF_3$)*	-37.7	54.20
Freon-502	-50.1	76.46

*Note: The Freon family of refrigerants, originally designated as Freon-12, Freon-13, etc., are presently listed under various tradenames, such as Ucon-12, Ucon-22, or simply Refrigerant-12, Refrigerant-22, sometimes abbreviated R-12, R-22, R-113, etc., depending on the particular refrigerant characteristics desired.

ature from Class 1 refrigerants and convey this lower temperature to the area to be air conditioned. By doing this, a lesser amount of Class 1 refrigerant is required (which is more expensive than Class 2 refrigerant); the final results are the same, and less of the system is subject to the problems that accompany Class 1 refrigerant.

Class 3 — This group consists of solutions that contain absorbed vapors of liquefiable agents or refrigerating media. These solutions function by nature of their ability to carry liquefiable vapors, which produce a cooling effect by the absorption of their latent heat. Examples are ammonia and lithium bromide, with water as the absorbing agent. See Chapter 13 for a discussion of absorption-type air conditioning.

Class 1 refrigerants are employed in the standard compressor type of refrigeration systems; Class 2 refrigerants are used in conjunction with Class 1 refrigeration, in some cases; and Class 3 refrigerants are used with Class 2 systems in the same manner as Class 1 systems.

Table 6-2. Refrigerant Pressure Versus Temperature

Temperature, °F	Pressure at Sea Level psi Gage			
	R-12	R-22	R502	R717 (NH$_3$)
−40	11.0*	0.5	4.1	8.7*
−35	8.4*	2.6	6.5	5.4*
−30	5.5*	4.9	9.2	1.6*
−25	2.3*	7.4	12.1	1.3
−20	0.6*	10.1	15.3	3.6
−15	2.4	13.2	18.8	6.2
−10	4.5	16.5	22.6	9.0
−5	6.7	20.1	26.7	12.2
0	9.2	24.0	31.1	15.7
5	11.8	28.2	35.9	18.6
10	14.6	32.8	41.0	23.8
15	17.7	37.7	46.5	28.4
20	21.0	43.0	52.5	33.5
25	24.6	48.8	58.8	39.0
30	28.5	54.9	65.6	45.0
35	32.6	61.5	72.8	51.6
40	37.0	68.5	80.5	58.6
45	41.7	76.0	88.7	66.3
50	46.7	84.0	97.4	74.5
55	52.0	92.6	106.6	83.4
60	57.7	101.6	116.4	92.9
65	63.8	111.2	126.7	103.1
70	70.2	121.4	137.6	114.1
75	77.0	132.2	149.1	125.8
80	84.2	143.6	161.2	138.3
85	91.8	155.7	174.0	151.7
90	99.8	168.4	187.4	165.9
95	108.2	181.8	201.4	181.1
100	117.2	195.9	216.2	197.2
105	126.6	210.8	231.7	214.2
110	136.4	226.4	247.9	232.3
115	146.8	242.7	264.9	251.5
120	157.6	259.9	282.7	271.7
125	169.1	277.9	301.4	293.1
130	181.0	296.8	320.8	
135	193.5	316.6	341.3	
140	206.6	337.2	362.6	

*In.Hg (mercury) standard atmosphere.

Properties of Some Refrigerants

Sulfur Dioxide — Sulfur dioxide (SO_2) is a colorless gas or liquid that is toxic and has a very pungent odor. It is obtained by burning sulfur in air. It is not considered a safe refrigerant, especially in quantities. It combines with water and forms sulfurous and sulfuric acid, which is corrosive to metal and has an adverse effect on almost anything that it comes in contact with. It boils at about 14°F (standard conditions) and has a latent heat value of 166 Btu/lb.

Sulfur dioxide has the disadvantage that it operates on a vacuum to give temperatures required in most refrigeration work. Should a leak occur, moisture-laden air will be drawn into the system that will eventually corrode the metal parts and stick up the compressor. Also in relation to Freon or methyl chloride, approximately one-third more vapor must be pumped in order to get the same amount of refrigeration. This means that either the condensing unit will have to be speeded up to give the desired capacity or the size of the cylinders increased proportionately. Sulfur dioxide does not mix well with oil. The suction line must be on a steady slant to the machine; otherwise the oil will trap out, making a constriction in the suction line. On many installations it is not possible to avoid traps, and on these jobs sulfur dioxide is not satisfactory.

Due to its characteristic pungent odor, comparatively small leaks are readily detected. Even the smallest leaks are readily located by means of an ammonia swab. A small piece of cloth or sponge may be secured to a wire and dipped into strong aqua ammonia or household ammonia and then passed over points where leaks are suspected. A dense white smoke forms where the sulfur dioxide and ammonia fumes come in contact. Where no ammonia is available, leaks may be located by the usual soap-bubble or oil test.

Methyl Chloride — Methyl chloride is a good refrigerant, but due to the fact that it will burn under some conditions and is slightly toxic, it does not conform to some city codes. Roughly speaking, the average relative concentration by weight of refrigerant vapor in a room of a given size that produces an undesirable effect on a

AIR CONDITIONING

person breathing the air thus contaminated may be specified approximately as follows:

Carbon dioxide	100
Methyl chloride	70
Ammonia	2
Sulfur dioxide	1

In other words, methyl chloride is 35 times safer than ammonia and 70 times safer than sulfur dioxide. To produce any serious effects from breathing methyl chloride, a considerable quantity is required. For instance, in a room 20 × 20 × 10 feet, it is necessary to liberate about 60 lb of methyl chloride to produce a dangerous condition.

Methyl chloride has a low boiling point. Under standard atmospheric pressure, it boils at $-10.6°F$. It is easy to liquefy and has a comparative high latent-heat value of approximately 178 Btu/lb under standard conditions. It will operate on a positive pressure as low as $-10°F$ and mixes well with oil. In its dry state, it has no corrosive effect on metal; but in the presence of moisture, copper plating of the compressor parts results and, in severe cases of moisture, a sticky black sludge is formed, which is detrimental to the working parts of the system.

Leaks are readily detected by means of a special leak-detecting halide torch. Some torches use alcohol as a fuel, normally producing a colorless flame that turns green when the leak-detector tube picks up very minute concentrations of methyl chloride, and brilliant blue with stronger concentrations. The space where the torch is being used should be well ventilated to prevent possible harmful effects of combustion products. Since the pressure in all parts of the system is above atmospheric (except in low-temperature applications), a soap-bubble or oil test also effectively locates leaks. Methyl chloride is not irritating, and, consequently, does not serve as its own warning agent in case of leaks to anywhere near the degree that sulfur dioxide does. In some cases, a warning agent is added, such as a small percentage (1 percent) of acrolein.

Ethyl Chloride—Due to the low pressure at which it evaporates, ethyl chloride is not commonly used in domestic units. In many respects, ethyl chloride is quite similar to methyl chloride. It has a boiling point of 55.6°F at atmospheric pressure, with a critical temperature of 360.5°F at a pressure of 784 lb absolute. Ethyl chloride is a colorless liquid or gas with a pungent ether-like odor and sweetish taste. It is neutral toward all metals, which allows the use of iron, copper, and even tin and lead in the construction of the unit. Since ethyl chloride softens all rubber compounds or gasket material, it is best to employ only lead as gasketing.

For the lower temperatures required for the preservation of foodstuffs, ethyl chloride must be evaporated at pressures below atmospheric. This is somewhat of a disadvantage because it is quite difficult to detect leaks in the low or evaporating side and large leaks will allow air or brine to be filtered or sucked into the system. On the other hand, the low pressure similarly encountered on the high or condensing side (ranging from 6 to 20 psig) has resulted in the use of rotary compressors. Since the displacement required per unit of refrigeration in the compressor cylinder is excessively large, the rotary method of compression serves to keep down the size of the unit itself, whereas a reciprocating compressor usually presents a rather bulky appearance. Rotary compressors operate with less noise and do not have the pounding found in the reciprocating type, especially in starting up where liquid refrigerant may enter the suction line.

Like methyl chloride, ethyl chloride presents the same difficulties of lubrication and leak detection, with the added disadvantages of operating under low pressures. In order to detect leaks, a small quantity of some liquid having a powerful or penetrating odor is sometimes added to the system. The most common method of leak detection, however, is to put pressure on the system and apply a soapy solution on the suspected points. In locations where there is no danger of explosion, a halide torch may be used. The flame will turn green when held under the leak if ethyl chloride is present.

Ammonia—Ammonia is used in absorption types of refrigeration equipment, which are operated by heat instead of by electric motors. It is a colorless gas with a pungent characteristic odor. Its boiling temperature at normal atmospheric pressure is −28°F, and its freezing point is −107.86°F. It is very soluble in water, 1 volume of water absorbing 1.148 volumes of ammonia at 32°F. It is combustible or explosive when mixed with air in certain proportions (about 1 volume of ammonia to 2 volumes of air) and much more so when mixed with oxygen. Due to its high latent-heat value (555 Btu at 18°F), large refrigeration effects are possible with relatively small-sized machinery. It is very toxic and requires heavy steel fittings. Pressures of 125 to 200 psi are not uncommon, and water-cooled units are essential.

There are two common methods for the detection of small ammonia leaks. One is the employment of a sulfur candle, which will give off a very thick, white smoke if it comes in contact with escaping ammonia. The other method is to test with phenolphthalein paper. The faintest trace of ammonia vapor will cause the moistened paper strip to turn pink, while a large amount of continued exposure will turn it a vivid scarlet color. Phenolphthalein paper usually may be secured from ammonia producers. Both of the preceding tests are convenient and accurate. Ammonia is also labeled R-717, while water is labeled R- 718.

Carbon Dioxide — Carbon dioxide is a colorless gas at ordinary temperatures, with a slightly pungent odor and acid taste. It is harmless to breathe except in extremely large concentrations when the lack of oxygen will cause suffocation. It is nonexplosive, nonflammable, and does not support combustion. The boiling point of carbon dioxide is so extremely low that at 5°F, a pressure of well over 300 psi is required to prevent its evaporation. At a condenser temperature of 80°F, a pressure of approximately 1000 psi is required to solidify the gas. Its critical temperature is 87.8°F, and its triple point is − 69.9°F.

Due to its high operating pressure, the compressor of a carbon dioxide refrigeration unit is very small even for a comparatively large refrigerating capacity. However, because of its low efficiency, as compared with other common refrigerants, it is seldom used in household units. However, it is used in some industrial applications and aboard ships.

Leakage of carbon dioxide gas can be detected by making sure that there is pressure on the part to be tested and then use a soap solution at the suspected points. Leakage into the condenser water can be tested with the use of bromothymol blue. Both the water entering and leaving the condenser should be tested at the same time because of the sensitivity of the test. When carbon dioxide is present, the normal blue color of the water containing the test chemicals will change to yellow.

Freezol (Isobutane) — Freezol is a colorless gas with a characteristic natural-gas odor. It is stable and does not react with water. It does not decompose to form foreign gases in refrigerating systems and has no corrosive action on metals. Freezol has a boiling point of 10°F, which permits the production of the desired temperature at atmospheric pressure with a positive back pressure. Its condensing pressure is approximately 49 lb gage at 90°F. When used as a refrigerant, any good grade of refrigerating oil may be employed as a lubricant in the system. The gas has a specific volume of 6.3 cu ft/lb at 60°F and at normal atmospheric pressure. For the detection of leaks, put pressure on the system and apply a soapy solution to the suspected points.

Calcium Chloride ($CaCl_2$) — Calcium chloride brine is used extensively in commercial plants as a simple carrying medium for refrigeration. In the indirect system of refrigeration, some refrigeration medium, such as brine, is cooled down by the direct expansion of the refrigerant and is then pumped through the material or space to be cooled, where it absorbs its sensible heat. Brine systems are used to advantage in large installations where there is danger due to leakage of the large amount of refrigerant present and where the temperature fluctuates in the space to be refrigerated.

For small refrigerated rooms or spaces where it is desired to operate the refrigerating machine only part of each day, the brine coils are supplemented by holdover or congealing tanks. A holdover tank is a steel tank containing strong brine in which direct expansion coils are immersed. During the period of operation of the refrigeration machine, the brine is cooled down. During the shut-down period, the refrigerant is capable of absorbing heat, the amount depending on the quantity of brine, its specific heat, and the temperature head.

Congealing tanks serve the same general purpose but operate on a different principle. Instead of storing brine, they contain a comparatively weak brine solution that freezes or congeals to a slushy mass of crystals during the period of operation. This mass of congealed brine, in addition to its sensible heat, is capable of absorbing heat equivalent to its latent heat of fusion. For the same refrigerating effect, congealing tanks can be made much smaller than holdover tanks because advantage is taken of the latent heat of fusion of the brine. This results in a considerable saving in construction and maintenance costs.

In modern refrigeration plants, however, the tendency is to operate with brine at low temperature, which permits the use of less brine, less piping or smaller-diameter pipe, smaller pumps, and lower pumping cost. In other words, instead of cooling a large volume of brine to a given temperature, the same number of refrigeration units is utilized in cooling a smaller volume of brine to a lower temperature. This practice results in greater economy of installation and operation. The use of an extremely low freezing brine, such as can be obtained with calcium chloride, is particularly desirable in the case of shell-type coolers. Unexpected low temperatures sometimes may occur in the tubes of such a cooler, either through excess vacuum on the cold side of the refrigerating unit or from shutting down a brine agitator, or both. Salt brine with a minimum possible freezing point of $-6°F$ may solidify under such conditions and cause considerable trouble and loss of operating time. In some cases where this has occurred, the cooler has been ruined completely.

The refrigerants we have just discussed, with the exception of ammonia and brine, will seldom be found in the field today. They are the early refrigerants used before the Freons.

FREON REFRIGERANTS

The Freon family of refrigerants has been one of the major factors responsible for the impressive growth of not only the home refrigeration and air-conditioning industry, but also the commercial refrigeration industry. The safe properties of these products have permitted their use under conditions where flammable or more toxic refrigerants would be hazardous.

Refrigerants

Freon-11 — Freon-11 has a boiling point of 74.8°F and has wide usage as a refrigerant in indirect industrial and commercial air-conditioning systems employing single or multistage centrifugal compressors with capacities of 100 tons and above. Freon-11 is also employed as a brine for low-temperature applications and provides relatively lob operating pressures with moderate displacement requirements.

Freon-12 — The boiling point of Freon-12 is −21.7°F. It is the most widely known and used of the Freon refrigerants. It is used principally in household and commercial refrigeration and air-conditioning units, for refrigerators, frozen-food cabinets, ice cream cabinets, food-locker plants, water coolers, room and window air-conditioning units, and similar equipment. It is generally used in reciprocating compressors, ranging in size from fractional to 800 hp, and in rotary-type compressors in the smaller sizes. The use of centrifugal compressors with Freon-12 for large air-conditioning and process-cooling applications is increasing.

Freon-13 — The boiling point of Freon-13 is −144.6°F. It is used in low-temperature specialty applications employing reciprocating compressors and generally in cascade with Freon-12 or Freon-22.

Freon-21 — Freon-21 has a boiling point of 48°F. It is used in fractional-horsepower household refrigerating systems and drinking-water coolers employing rotary vane-type compressors. Freon-21 is also used in comfort-cooling air-conditioning systems of the absorption type where dimethyl ether or tetraethylene glycol is used as the absorbent.

Freon-22 — The boiling point of Freon-22 is −41.4°F. It is used in all types of household and commercial refrigeration and air conditioning applications with reciprocating compressors. The outstanding thermodynamic properties of Freon-22 permit the use of smaller equipment than is possible with similar refrigerants, making it especially suitable where size is a problem.

Freon-113 — The boiling point of Freon-113 is 117.6°F. It is used

Air Conditioning

in commercial and industrial air-conditioning and process water and brine cooling with centrifugal compression. It is especially useful in small-tonnage applications.

Freon-114 — The boiling point of Freon-114 is 38.4°F. It is used as a refrigerant in fractional-horsepower household refrigerating systems and drinking-water coolers employing rotary vane-type compressors. It is also used in indirect industrial and commercial air-conditioning systems and in industrial process water and brine cooling to $-70°F$ employing multistage centrifugal-type compressors in cascade of 100 tons refrigerating capacity and larger.

Freon-115 — The boiling point of Freon-115 is $-37.7°F$. It is especially stable, offering a particularly low discharge temperature in reciprocating compressors. Its capacity exceeds that of Freon-12 by as much as 50 percent in low-temperature systems. Its potential applications include household refrigerators and automobile air conditioning.

Freon-502 — Freon-502 is an azeotropic mixture composed of 48.8% Freon-22 and 51.2% Freon-115 by weight. It boils at $-50.1°F$. Because it permits achieving the capacity of Freon-22 with discharge temperatures comparable to Freon-12, it is finding new reciprocating compressor applications in low-temperature display cabinets and in storing and freezing of food.

PROPERTIES OF FREONS

The Freon refrigerants are colorless and almost odorless, and their boiling points vary over a wide range of temperatures. Those Freon refrigerants that are produced are nontoxic, noncorrosive, nonirritating, and nonflammable under all conditions of usage. They are generally prepared by replacing chlorine or hydrogen with fluorine. Chemically, Freon refrigerants are inert and thermally stable up to temperatures far beyond conditions found in actual operation. However, Freon is harmful when allowed to escape into the atmosphere. It can deplete the ozone layer around earth and cause more harmful ultraviolet rays to reach the surface of the earth.

Physical Properties

The pressures required to liquefy the refrigerant vapor affect the design of the system. The refrigerating effect and specific volume of the refrigerant vapor determine the compressor displacement — the heat of vaporization and specific volume of the liquid refrigerant affect the quantity of refrigerant to be circulated through the pressure-regulating valve or other system device.

Flammability

Freon is nonflammable and noncombustible under conditions where appreciable quantities come in contact with flame or hot metal surfaces. It requires an open flame at 1382°F to decompose the vapor. Even at this temperature, only the vapor decomposes to form hydrogen chloride and hydrogen fluoride, which are irritating but are readily dissolved in water. Air mixtures are not capable of burning and contain no elements that will support combustion. For this reason, Freon is considered nonflammable.

Amount of Liquid Refrigerant Circulated

It should be noted that the Freon refrigerants have relatively low heat values, but this must not be considered a disadvantage. It simply means that a greater volume of liquid must be circulated per unit of time to produce the desired amount of refrigeration. It does not concern the amount of refrigerant in the system. Actually, it is a decided advantage (especially in the smaller- or low-tonnage systems) to have a refrigerant with low heat values for the reason that the larger quantity of liquid refrigerant to be metered through the liquid-regulating device will permit the use of more accurate and more positive operating and regulating mechanisms of less sensitive and less critical adjustments. The quantities of liquid refrigerant metered or circulated per minute under standard ton conditions are listed in Table 6-3.

Table 6-3. Quantities of Refrigerant Circulated Per Minute Under Standard Ton Conditions

Refrigerant	Pounds Expanded per Minute	Cu Ft/lb Liquid 86°F	Cu In. Liquid Expanded per Minute	Specific Gravity Liquid 86°F (Water-1)
Carbon dioxide	3.528	0.0267	162.8	0.602
Freon-22	2.887	0.01367	67.97	1.177
Ammonia	0.4215	0.02691	19.6	0.598
Freon-12	3.916	0.0124	83.9	1.297
Methyl chloride	1.331	0.01778	40.89	0.898
Sulfur dioxide	1.414	0.01184	28.9	1.358
Freon-114	4.64	0.01112	89.16	1.443
Freon-21	2.237	0.01183	45.73	1.360
Freon-11	2.961	0.01094	55.976	1.468
Methylene chloride	1.492	0.01198	30.88	1.340
Freon-113	3.726	0.01031	66.48	1.555

$$\text{Cu in. liquid refrig./min} = \frac{200 \text{ Btu/min}}{\text{Btu refrig. effect/lb}} \times \text{vol. liquid/lb } 86°F \times 1728$$

Volume (Piston) Displacement

For reason of compactness, cost of equipment, reduction of friction and compressor speed, the volume of gas that must be compressed per unit of time for a given refrigerating effect, in general, should be as low as possible. Freon-12 has a relatively low volume displacement, which makes it suitable for use in reciprocating compressors, ranging from the smallest size to those of up to 800-tons capacity, including compressors for household and commercial refrigeration. Freon-12 also permits the construction of compact rotary compressors in the commercial sizes. Generally, low-volume displacement (high-pressure) refrigerants are used in reciprocating compressors; high-volume displacement (low-pressure) refrigerants are used in large- tonnage centrifugal compressors; intermediate-volume (intermediate-pressure) refrigerants are used in rotary compressors. There is no standard rule governing this usage.

OPERATING PRESSURES

Condensing Pressure

Condensing (high-side) pressure should be low to allow construction of lightweight equipment, which affects power consumption, compactness, and installation. High pressure increases the tendency toward leakage on the low side as well as the high side when pressure is built up during idle periods. In addition, pressure is very important from the standpoint of toxicity and fire hazard.

In general, a low volume displacement accompanies a high condensing pressure, and a compromise must usually be drawn between the two in selecting a refrigerant. Freon-12 presents a balance between volume displacement and condensing pressure. Extra-heavy construction is not required for this type of refrigerant, and so there is little or nothing to be gained from the standpoint of weight of equipment in using a lower-pressure refrigerant.

Evaporating Pressure

Evaporating (low-side) pressures above atmospheric are desirable to avoid leakage of moisture-laden air into the refrigerating systems and permit easier detection of leaks. This is especially important with open-type units. Air in the system will increase the head pressures, resulting in inefficient operations, and may adversely affect the lubricant. Moisture in the system will cause corrosion and, in addition, may freeze out and stop operation of the equipment.

In general, the higher the evaporating pressure, the higher the condensing pressure under a given set of temperatures. Therefore, in order to keep head pressures at a minimum and still have positive low-side pressures, the refrigerant selected should have a boiling point at atmospheric pressure as close as possible to the lowest temperature to be produced under ordinary operating conditions. Freon-12, with a boiling point of $-21.7°F$, is close to ideal in this respect for most refrigeration applications. A still lower boiling point is of some advantage only when lower operating temperatures are required.

PRESSURE-TEMPERATURE CHART

A logarithmic-scaled chart giving the relation between pressure and corresponding temperatures in degrees Fahrenheit for common refrigerants is shown in Fig. 6-1. The axis of the abscissa shows the temperature, and the axis of the ordinate shows the pressure in pounds per square inch gage and absolute, respectively.

Thus, to ascertain the pressure of a refrigerant at any particular temperature, follow the desired temperature until the curve of that particular refrigerant is reached. The corresponding pressure is found on the pressure axis. For example, the temperature corresponding to a pressure of 25 psi for sulfur dioxide is approximately 60°F, and the corresponding temperature at the same pressure for dichlorodifluoromethane (Freon-12) is approximately 27.2°F, etc.

REFRIGERANT CHARACTERISTICS

The freezing point of a refrigerant should be below any temperature that might be encountered in the system. The freezing point of all refrigerants, except water (32°F) and carbon dioxide (−69.9°F, triple point), are far below the temperatures that might be encountered in their use. Freon-12 has a freezing point of −247°F.

Critical Temperature

The critical temperature of a refrigerant is the highest temperature at which it can be condensed to a liquid, regardless of a higher pressure. It should be above the highest condensing temperature that might be encountered. With air-cooled condensers, in general, this would be above 130°F. Loss of efficiency caused by superheating of the refrigerant vapor on compression and by throttling expansion of the liquid is greater when the critical temperature is low.

All common refrigerants have satisfactorily high critical temperatures, except carbon dioxide (87.8°F) and ethane (89.8°F).

REFRIGERANTS

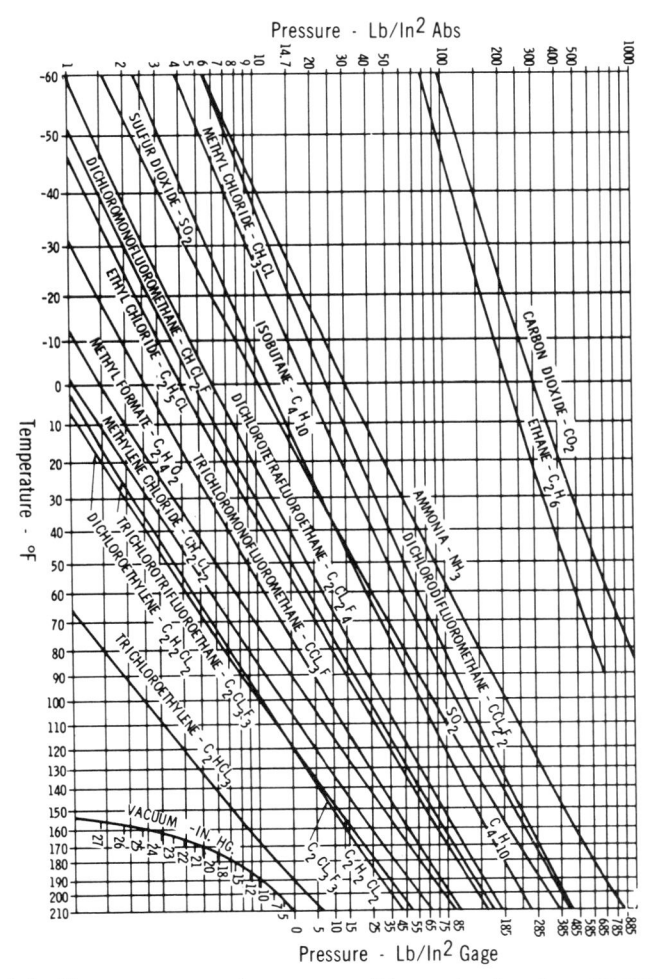

Fig. 6-1. Vapor pressure of common refrigerants at various temperature ranges.

These two refrigerants require condensers cooled to temperatures below their respective critical temperatures, thus generally requiring water.

Latent Heat of Evaporation

A refrigerant should have a high latent heat of evaporation per unit of weight so that the amount of refrigerant circulated to produce a given refrigeration effect may be small. Latent heat is important when considering its relation to the volume of liquid required to be circulated and the net refrigerating effect. Since other factors enter into the determination of these, they are discussed separately.

The refrigerant effect per pound of refrigerant under standard ton conditions determines the amount of refrigerant to be evaporated per minute. The refrigerating effect per pound is the difference in Btu content of the saturated vapor leaving the evaporator (5°F) and the liquid refrigerant just before passing through the regulating valve (86°F). While the Btu refrigerating effect per pound directly determines the number of pounds of refrigerant to be evaporated in a given length of time to produce the required results, it is much more important to consider the volume of the refrigerant vapor required rather than the weight of the liquid refrigerant. By considering the volume of refrigerant necessary to produce standard ton conditions, it is possible to make a comparison between Freon-12 and other refrigerants so as to provide for the reproportioning of the liquid orifice sizes in the regulating valves, sizes of liquid refrigerant lines, etc.

A refrigerant must not be judged only by its refrigerating effect per pound, but the volume per pound of the liquid refrigerant must also be taken into account to arrive at the volume of refrigerant to be vaporized. Although Freon-12 has relatively low refrigerating effect, this is not a disadvantage because it merely indicates that more liquid refrigerant must be circulated to produce the desired amount of refrigeration. Actually, it is a decided advantage to circulate large quantities of liquid refrigerant because the greater volumes required will permit the use of less sensitive operating and regulating mechanisms with less critical adjustment.

Refrigerants with high Btu refrigerating effects are not always desirable, especially for household and small commercial installations, because of the small amount of liquid refrigerant in the system and the difficulty encountered in accurately controlling its

REFRIGERANTS

flow through the regulating valve. For household and small commercial systems, the adjustment of the regulating-valve orifice is most critical for refrigerants with high Btu values.

Specific Heat

A low specific heat of the liquid is desirable in a refrigerant. If the ratio of the latent heat to the specific heat of a liquid is low, a relatively high proportion of the latent heat may be used in lowering the temperature of the liquid from the condenser temperature to the evaporator temperature. This results in a small net refrigerating effect per pound of refrigerant circulated and, assuming other factors remain the same, reduces the capacity and lowers the efficiency. When the ratio is low, it is advantageous to precool the liquid before evaporation by heat interchange with the cool gases leaving the evaporator.

In the common type of refrigerating systems, expansion of the high-pressure liquid to a lower-pressure, lower-temperature vapor and liquid takes place through a throttling device such as an expansion valve. In this process, energy available from the expansion is not recovered as useful work and, since it performs no external work, reduces the net refrigerating effect.

Power Consumption

In a perfect system operating between 5 and $-86°F$ conditions, 5.74 Btu is the maximum refrigeration obtainable per Btu of energy used to operate the refrigerating system. This is the theoretical maximum *coefficient of performance* on cycles of maximum efficiency (e.g., the Carnot cycle). The minimum horsepower would be 0.821 hp/ton of refrigeration. The theoretical coefficient of performance would be the same for all refrigerants if they could be used on cycles of maximum efficiency.

However, because of engineering limitations, refrigerants are used on cycles with a theoretical maximum coefficient of performance of less than 5.74. The cycle most commonly used differs in its basic form from (1) the Carnot cycle, as already explained in employing expansion without loss or gain of heat from an outside source, and (2) in compressing adiabatically (compression without

187

gaining or losing heat to an outside source) until the gas is superheated above the condensing medium temperature. These two factors, both of which increase the power requirement, vary in importance with different refrigerants; but it so happens that when expansion loss is high, compression loss is generally low, and vice versa. All common refrigerants (except carbon dioxide and water) show about the same overall theoretical power requirement on a 5 to $-86°F$ cycle. At least the theoretical differences are so small that other factors are more important in determining the actual differences in efficiency.

The amount of work required to produce a given refrigerating effect increases as the temperature level to which the heat is pumped from the cold body is increased. Therefore, on a 5 to $-86°F$ cycle, when gas is superheated above $86°F$ temperature on compression, efficiency is decreased and the power requirement increased unless the refrigerating effect due to superheating is salvaged through the proper use of a heat interchanger.

Volume of Liquid Circulated

The volume of liquid required to be circulated for a given refrigerant effect should be low to avoid fluid-flow (pressure-drop) problems and to keep down the size of the required refrigerant change. In small capacity machines, the volume of liquid circulated should not be so low as to present difficult problems in accurately controlling its flow through expansion valves or other types of liquid metering devices.

With a given net refrigerating effect per pound, a high density of liquid is preferable to a low volume. However, a high density tends to increase the volume circulated by lowering the net refrigerating effect. The volumes of liquid required to be circulated for the various refrigerants are listed in Table 6-1.

SIMPLE REFRIGERATION SYSTEM

The principle of using the latent heat of vaporization of a liquid, such as sulfur dioxide, for producing refrigeration can be illustrated very easily by thinking of a refrigerator of very simple

REFRIGERANTS

design, similar to the one shown in Fig. 6-2. The refrigerator is made up of a box that is completely insulated on all six sides to prevent as much head transfer by conduction, convection, and radiation as possible. A series of finned coils with one end connected to the cylinder charged with sulfur dioxide is located in the top of the cabinet. Two pounds of sulfur dioxide are charged into the coil, after which the compressed cylinder is sealed and disconnected from the line, with the charging end of the pipe open to the atmosphere.

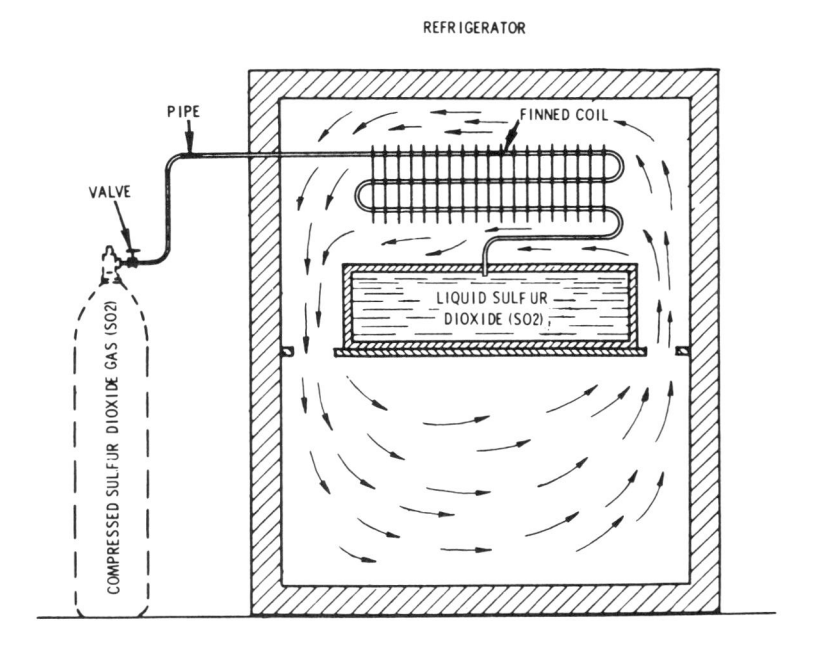

Fig. 6-2. Typical refrigeration cycle.

Since the liquid sulfur dioxide is exposed to the air, the only pressure to which the liquid is subjected is atmospheric pressure, which is approximately 14.7 psi absolute pressure, or 0 psig. At this pressure, the liquid sulfur dioxide will boil or vaporize at any temperature higher than 14°F. For example, if the temperature of the room in which the refrigerator is located is 70°F, the temperature of the cabinet at the time the sulfur dioxide liquid is added will also be 70°F. The liquid sulfur dioxide in the coils therefore will start boiling and vaporizing immediately because the surrounding temperature is above the boiling point (14°F) of the liquid. As the liquid boils away, it will absorb heat from the cabinet because for every pound of sulfur dioxide liquid vaporized, 168 Btu of heat will be extracted from the cabinet. As soon as the temperature of the cooling coil is reduced to a point lower than the cabinet temperature, the air in the cabinet will start circulating in the direction shown by the arrows in Fig. 6-2 because heat will always flow from the warmer to the colder object.

With this method, however, the 2 lb of sulfur dioxide liquid will soon vaporize and the colorless gas will be given off to the air outside the cabinet. It has a sharp odor. Refrigeration will then stop until a new charge is placed in the cooling coil. Sulfur dioxide is expensive and difficult to handle, and so some means must be used to reclaim the vapor in order to use the original charge continuously and indefinitely. The inconvenience of recharging the coil must also be prevented, and the refrigerator must be built so that it will automatically maintain the proper food preservation temperatures at all times, with absolutely no inconvenience to the customer. This is accomplished by the compressor pulling the warm sulfur dioxide gas from the cooling unit and pumping it into the condenser where it is changed to a liquid, ready to return to the cooling unit.

HANDLING REFRIGERANTS

It has been pointed out previously that one of the requirements of an ideal refrigerant is that it must be nontoxic. In reality, however, all gases (with the exception of pure air) are more or less toxic or asphyxiating. It is therefore important that wherever gases or highly volatile liquids are used, adequate ventilation be provided because even nontoxic gases in air produce a suffocating effect.

Vaporized refrigerants, especially ammonia and sulfur dioxide, bring about irritation and congestion of the lungs and bronchial organs, accompanied by violent coughing, vomiting, and, when breathed in sufficient quantity, suffocation. It is therefore of the

REFRIGERANTS

utmost importance that the serviceman subjected to a refrigerant gas find access to fresh air at frequent intervals to clear his lungs. When engaged in the repair of ammonia and sulfur dioxide machines, approved gas masks and goggles should be used. Carrene, Freon (F-12), and carbon dioxide fumes are not irritating and can be inhaled in considerable concentrations for short periods without serious consequences.

It should be remembered that liquid refrigerant will refrigerate or remove heat from anything it comes in contact with when released from a container. In the case of contact with refrigerant, the affected or injured area should be treated as if it has been frozen or frostbitten.

STORING AND HANDLING REFRIGERANT CYLINDERS

Refrigerant cylinders should be stored in a dry, sheltered, and well-ventilated area. The cylinders should be placed in a horizontal position, if possible, and held by blocks or saddles to prevent rolling. It is of the utmost importance to handle refrigerant cylinders with care and to observe the following precautions:

1. Never drop the cylinders nor permit them to strike each other violently.
2. Never use a lifting magnet nor a sling (rope or chain) when handling cylinders. A crane may be used when a safe cradle or platform is provided to hold the cylinders.
3. Caps provided for valve protection should be kept on the cylinders at all times except when the cylinders are actually in use.
4. Never overfill the cylinders. Whenever refrigerant is discharged from or into a cylinder, weigh the cylinder and record the weight of the refrigerant remaining in it.
5. Never attempt to mix gases in a cylinder.
6. Never use cylinders for rollers, supports, or for any purpose other than to carry gas.
7. Never tamper with the safety devices in valves or on the cylinders.

8. Open the cylinder valves slowly. Never use wrenches or tools except those provided or approved by the gas manufacturer.
9. Make sure that the threads on regulators or other unions are the same as those on the cylinder valve outlets. Never force a connection that does not fit.
10. Regulators and gages provided for use with one gas must not be used on cylinders containing a different gas.
11. Never attempt to repair or alter the cylinders or valves.
12. Never store the cylinders near highly flammable substances, such as oil, gasoline, or waste.
13. Cylinders should not be exposed to continuous dampness, salt water, or salt spray.
14. Store full and empty cylinders apart to avoid confusion.
15. Protect the cylinders from any object that will produce a cut or other abrasion on the surface of the metal.

CYLINDER CAPACITY

To determine the allowable capacity of any cylinder or other authorized container, first find the weight of water it will hold by weighing it full of water and subtracting the container weight. This water capacity in pounds multiplied by 1.25 will give its allowable shipping capacity for sulfur dioxide. The water capacity multiplied by 0.75 will give its allowable shipping capacity for methyl chloride. Thus, for example, a cylinder weighs 120 lb empty and 243 lb when completely filled with water. Its water capacity is therefore 243 − 120, or 123 lb. It will hold 1.25 × 123, or 154 lb of sulfur dioxide, and 0.75 × 123, or 92 lb of methyl chloride.

FIRST AID

The following procedures are recommended for persons suffering from exposure to refrigerants.

REFRIGERANTS

Ammonia Fumes

Take the patient into fresh air and call a physician promptly. If the patient is unconscious, apply artificial respiration. Ammonia gas is lighter than air. Keep the head as low as possible in the presence of ammonia. Keep near the floor in going to the rescue of anyone overcome by ammonia, and keep a wet sponge or cloth over the mouth and nostrils.

Sulfur Dioxide

Liquid sulfur dioxide on any part of the body produces freezing which should be treated as any ordinary frostbite. Freezing any part of the body constricts the blood vessels in that area. The natural restoration of circulation will hold in check unnecessary swelling and destruction of the tissues.

Every eye freezing should immediately have the service of an eye specialist to determine the presence of ulcers and to prevent later infection. *Caution: Do not rub the affected eye.*

Methyl Chloride

Excessive exposure to methyl chloride develops the characteristic symptoms of drowsiness, mental confusion, nausea, and possible vomiting. In any case of excess exposure, the patient should be removed immediately to fresh air, and a physician should be called. Like other refrigerant gases, methyl chloride vaporizes rapidly when it escapes, and a person coming in contact with liquid methyl chloride will be subject to a freezing action. The treatment is the same as for any freeze and is of relatively minor importance except when the eyes are the affected part.

In case of an accident involving the eyes, the person should be taken to an eye specialist as soon as possible.

Freons

Should the liquid gas of Freons come in contact with the skin, the injury should be treated in the same manner as though the skin had been frostbitten or frozen. An eye injury should be treated the same as that described for methyl chloride. It is best to wear gloves when handling Freon cylinders. The eyes should be protected by goggles.

SUMMARY

Refrigeration, regardless of the means by which it is obtained, may be defined as a process of removing heat from a substance or from space. Refrigerants, being the prime source without which mechanical refrigeration systems would not be possible, have been fully described in this chapter.

Refrigerants are generally divided into three classes according to their manner of absorption or extraction of heat from the substance to be refrigerated. Refrigerants are selected principally for their boiling points, pressures, and stability within the system. This includes molecular weight, density, compression ratio, heat value, temperature of compression, and compressor displacement.

The Freon family of refrigerants has been one of the major factors responsible for the large growth in the refrigeration and air-conditioning field. This is largely because the properties of Freon permit their use under conditions where flammable or more toxic refrigerants would be hazardous to use. Freon has been determined to harm the environment.

REVIEW QUESTIONS

1. Name the desirable properties of a good refrigerant for commercial use.
2. How are commercial refrigerants classified?
3. What is the purpose of calcium chloride brine as used in commercial refrigeration plants?
4. Why is ammonia employed in preference to other refrigerants in commercial plants?
5. State the boiling point and heat of vaporization at the boiling point for Freon-12.
6. In what respect does the Freon family of refrigerants differ from such refrigerants as methyl or ethyl chloride?

7. Name the chemical and physical properties of Freons.
8. Why is Freon-12 preferred in both commercial and household refrigeration units?
9. What is meant by the critical temperature of a refrigerant?
10. Why does a high latent heat of evaporation per unit of weight affect the desirability of a refrigerant?
11. Describe the methods of leak detection in a refrigeration system.
12. What precautions should be observed in the storage and handling of refrigerant cylinders?
13. Enumerate the procedures recommended when suffering from exposure to various refrigerants.

CHAPTER 7

Compressors for Air Conditioners

The components necessary for operating a gas-compression refrigeration system may be divided into two parts: mechanical and electrical. The mechanical components are the compressor, condenser, receiver, evaporator, and various types of refrigerant controls. In addition, suitable pipelines with the necessary shutoff valves, strainers, dryers, heat interchangers, fans, pumps, etc., are required, depending on the size of the system and other factors. The electrical components consist of such items as the compressor motor, compressor contactor or relay, compressor starter, overload protector, capacitor, potential relay, pressure switch, transformers (when used), evaporator fan, condenser fan, evaporator-fan relay, thermostat, etc., as shown in Fig. 7-1.

Compressors

Compressors are manufactured in various sizes, depending on requirements, and are classified according to their method of op-

AIR CONDITIONING

eration as *reciprocating*, *rotary*, and *centrifugal*. The function of any compressor is to provide the necessary pumping action to remove the refrigerant gas from the evaporator, compress the gas, and finally discharge the gas into the condenser where it is liquefied. The pumping action of the compressor provides the means of lifting the gas from the condition of *low* to that of *high* pressure, and the volume of gas to be compressed is dependent on the compressor piston displacement. The compressor and component parts are shown in Fig. 7-2.

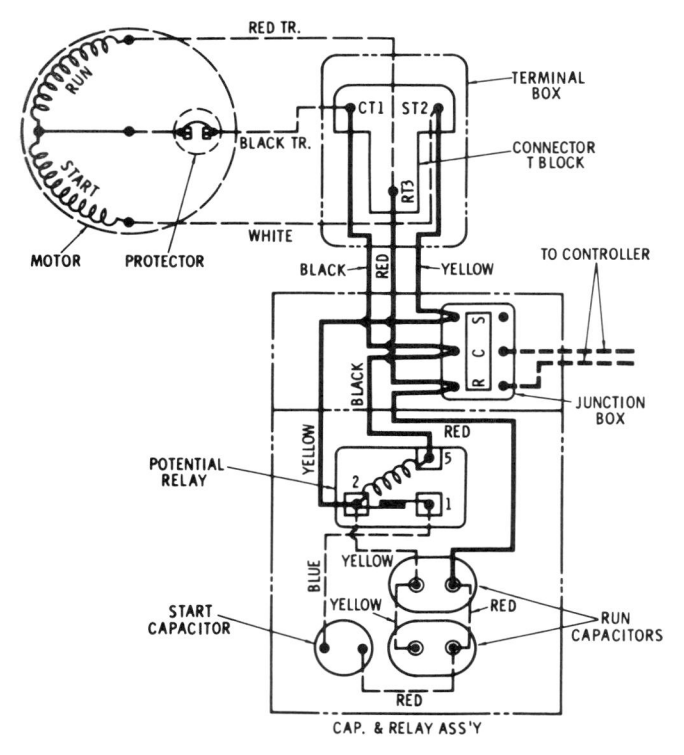

Courtesy Copeland Refrigeration Corporation

Fig. 7-1. Diagram for a compressor with a single-phase capacitor-start motor using a potential relay and motor-overload protector.

COMPRESSORS FOR AIR CONDITIONERS

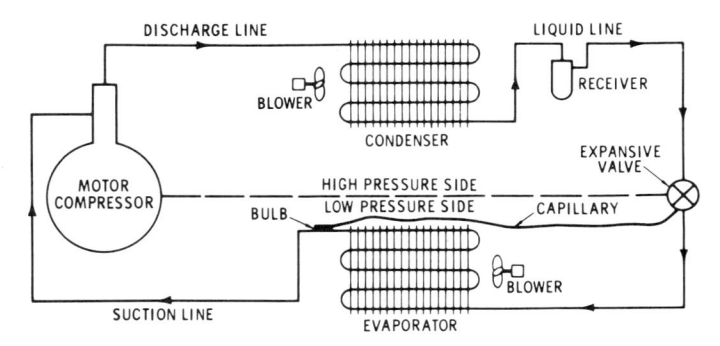

Fig. 7-2. Typical compression-refrigeration system.

Condensers

When the refrigerant gas leaves the compressor and enters the condenser, it contains the full load of heat, which consists of the heat picked up in the evaporator, the heat of cooling the liquid from the condenser to the evaporator temperature, the heat picked up in the suction line and cylinder chamber, and the heat of compression. The function of the condenser is to remove this heat and return the refrigerant to the refrigerant control, thus enabling the system to repeat its cycle.

Several types of condensers are used, the most common being the water-cooled types used on commercial and larger systems: the double-pipe, flooded atmospheric; the bleeder shell-and-tube; and the shell-and-coil. Air-cooled condensers are most universally used on household and smaller air-conditioning units. They are usually of the natural draft or fan-cooled types, furnished with plain or finned tubes or plates, and either series pass or parallel pass.

Receivers

The receiver is the reservoir for excess liquefied refrigerant that is not being used in the system. Unless other conditions prevent it, the receiver should have sufficient capacity to hold the total amount of refrigerant in the system. Liquid receivers are most commonly made of drawn steel shells welded together to provide maximum strength.

Evaporators

The evaporator is that part of the system where the useful cooling is accomplished by removing the heat from the area to be cooled. This is done by bringing the temperature of the liquid refrigerant below the temperature of the surrounding medium. The heat passes into the liquid refrigerant and is absorbed as latent heat, changing the state of the refrigerant from a liquid into a vapor. It is then withdrawn by action of the compressor.

There are two general types of evaporators: *dry* and *flooded*. In the dry type, the refrigerant enters in the liquid state, and the design provides for complete evaporation with the vapors slightly superheated. In the flooded type, not all of the refrigerant is evaporated, and the liquid/vapor mixture leaving the evaporator flows into a surge drain from which the vapors are drawn into the compressor suction line; the liquid is recirculated through the evaporator.

Refrigerant Controls

In addition to the compressor, condenser, receiver, and evaporator, several auxiliaries are required for proper operation of an air-conditioning system. Certain devices must be furnished for the controlled expansion of the refrigerant from the high condenser pressure to the low evaporator pressure. Controls are also necessary for the on-off operation of the compressor, flow of the condensing medium, and safety purposes. Proper piping is required for interconnection of the various parts of the system.

Expansion Valves

To control the rate of flow of the liquefied refrigerant between the high- and low-side pressure of the system, some form of expansion device must be provided. This device is usually an expansion valve. Automatic expansion valves are pressure-controlled devices that operate to maintain a constant pressure in the evaporator. Such an expansion valve is usually applied to evaporators of the direct-expansion type but is not satisfactory for fluctuating loads such as those encountered in air-conditioning installations.

COMPRESSORS FOR AIR CONDITIONERS

Thermostatic expansion valves are similar in construction to automatic expansion valves but, in addition, incorporate a power element responsive to changes in the degree of superheat of the refrigerant gas leaving the evaporator coil. This power element consists of a bellows connected by means of a capillary tube to a feeler bulb fastened to the suction line from the evaporator. (For further details of air-conditioning components, see Chapters 5, 8 to 11, and 13.)

REFRIGERATION CYCLE

In the great majority of air-conditioning applications and almost exclusively in the smaller-horsepower range, the vapor-compression system is used. In larger applications, however, absorption systems and steam-jet vacuum systems are frequently used. The centrifugal system, which is largely an adaption of the vapor-compression cycle, is used in large-capacity industrial applications.

The vapor-compression refrigeration cycle consists of a series of changes of state needed to restore the refrigerant to a condition in which it will possess the ability to extract heat from the space to be cooled. For all vapor-compression-type systems, the cycle consists of the following four processes:

1. Heat gain in the evaporator (the vaporizing process)
2. Pressure rise in the compressor (the compressing process)
3. Heat loss in the condenser (the condensing process)
4. Pressure loss in the expansion valve (the pressure-reducing process)

Briefly, and with reference to Fig. 7-2, the cycle of operation that is the cooling process consists of four separate steps as follows:

1. *Vaporizing process* — As the low-temperature, low-pressure refrigerant in the evaporator absorbs the heat from the material and space that is being refrigerated, it is transformed from a liquid to a vapor in the process.
2. *Compression process* — The compressor draws the heat-laden

vapor from the evaporator through the return piping; it then compresses this vapor until the temperature is above that of the condensing medium.

3. *Condensing process* — When the compressor has raised the temperature of the vapor to a temperature above that of the condensing medium (usually 15 to 25°F), the heat of vaporization will flow from the vapor to the condensing medium and so condense the refrigerant to a high-pressure liquid. This high-pressure liquid then flows to the receiver where it is stored until it is supplied to the cooling unit through the expansion valve.

4. *Pressure-reducing process* — As the compressor withdraws the refrigerant vapor from the evaporator, the cooling unit must be supplied with more low temperature, low-pressure refrigerant capable of absorbing heat. This is accomplished by a liquid control valve known as an *expansion valve*. This valve has the following functions: It reduces the pressure of the high-pressure vapor from the receiver to a low-pressure liquid capable of absorbing heat; it maintains a constant supply of liquid in the evaporator; and it acts as a dividing point between the high- and low-pressure sides of the system.

REFRIGERANTS

By definition, a refrigerant is a substance that produces a refrigeration effect by its absorption of heat while expanding or vaporizing. A desirable refrigerant should possess chemical, physical, and thermodynamic properties that permit its efficient application in refrigerating systems. In addition, there should be little or no danger to health or property in case of leaks in the refrigeration system.

The selection of a practical refrigerant is determined by the type of condensing unit and system in which it is to be used. Differences in units and systems are to allow for differences in characteristics of the available refrigerants — they also occur because of the type of application, initial cost, compactness, etc. Reciprocating, rotary, and centrifugal condensing units are used extensively.

Dry expansion and flooded systems with high-side floats, low-side floats, expansion valves, and capillaries as refrigerant controls employed in conjunction with condensing-unit systems have proved ideal for all applications, and each varies from the other in refrigerant requirements. (For further refrigerant information, see Chapter 6.)

COMPRESSORS

The function of any compressor in a mechanical refrigeration system is to establish a pressure difference in the system to create a flow of the refrigerant from one part of the system to the other. It is this difference in pressure between the high side and the low side that forces liquid refrigerant through the control valve and into the cooling unit. The vapor pressure in the cooling unit must be above the pressure at the suction side of the condensing unit to force the low-side vapor to leave the cooling unit and flow to the condensing unit.

Compressors employed in air conditioning may be divided into several classes with respect to their construction and operation as

1. Single cylinder
2. Multicylinder

With respect to the method of compression, they are classified as

1. Reciprocating
2. Rotary
3. Centrifugal

With respect to the drive employed, they are classified as

1. Direct drive
2. Belt drive

With respect to location of the prime mover, they may be classified as:

1. Independent (belt-drive)
2. Semihermetic (direct drive; motor and compressor in separate housings)
3. Hermetic (direct drive motor and compressor in same housing)

Reciprocating Compressors

Cylinder design in reciprocating compressors may vary as to the number, arrangement, and action (such as single- and double-acting). See Figs. 7-3 to 7-6. Reciprocating units are manufactured with from 1 to 16 cylinders, with the arrangement being V-, W-, or radial-type, depending on requirements.

Drive Methods

Reciprocating compressors are usually motor driven and may be subdivided as to whether they are of open or hermetic design. The division of compressors into the open or closed design is dependent on whether the compressor is driven by an externally mounted motor or the motor is sealed within the compressor housing. If the motor is enclosed within the compressor housing, the compressor is classified as *closed*, or *hermetic*. The hermetic type of compressor eliminates the necessity of shaft seals, thus preventing refrigerant leakage and, in addition, reducing operating noise.

Compressor Valves

The function of compressor valves is to direct the flow of refrigerant through the compressor. These are termed *suction* or *discharge* valves, according to the function they perform. The suction and discharge valves (Fig. 7-7) may be arranged with both located in the compressor head or with the suction valve on top of the piston and the discharge valve in the compressor head. The valves are usually classified according to their construction as *poppet*, *ring-plate*, or *reed*.

Pistons and Piston Rings

The piston is an important part of the compressor mechanism. It consists essentially of a long cylindrical casting closed at the top

COMPRESSORS FOR AIR CONDITIONERS

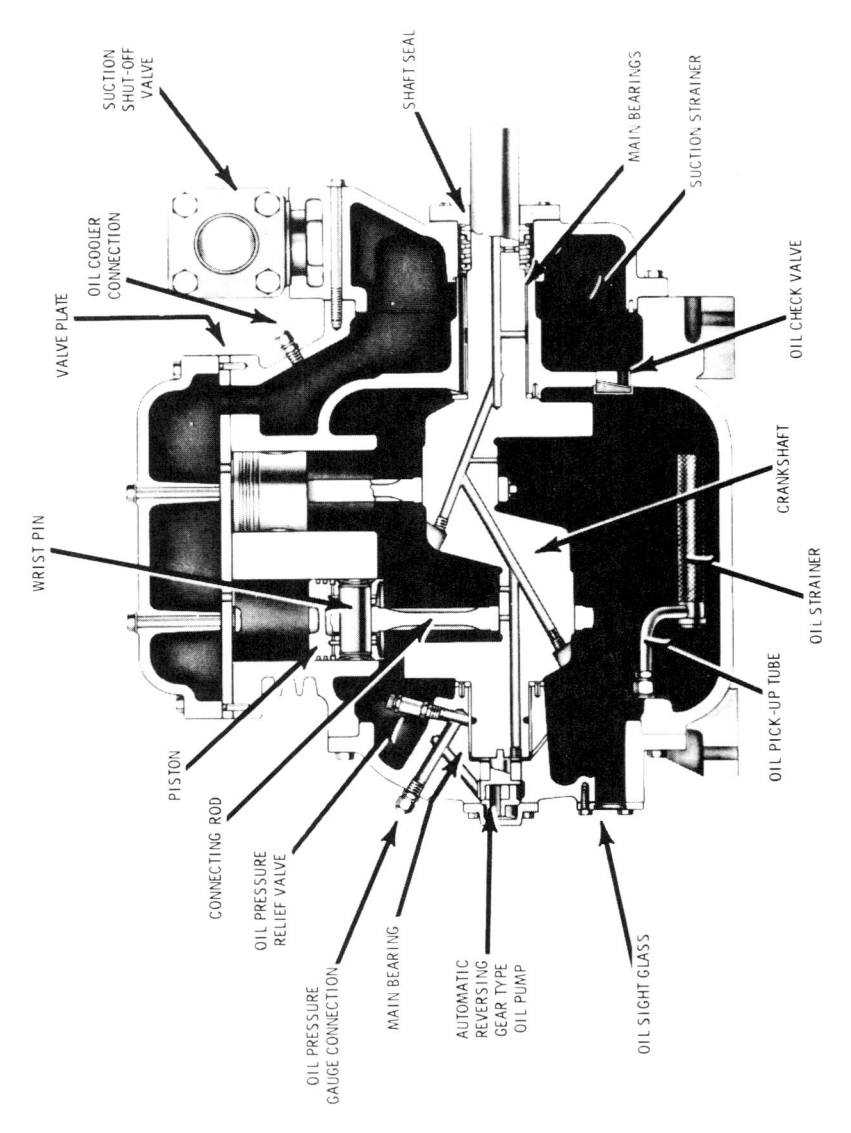

Courtesy Dunham-Bush, Inc.

Fig. 7-3. Direct-drive reciprocating compressor with arrangement of component parts.

205

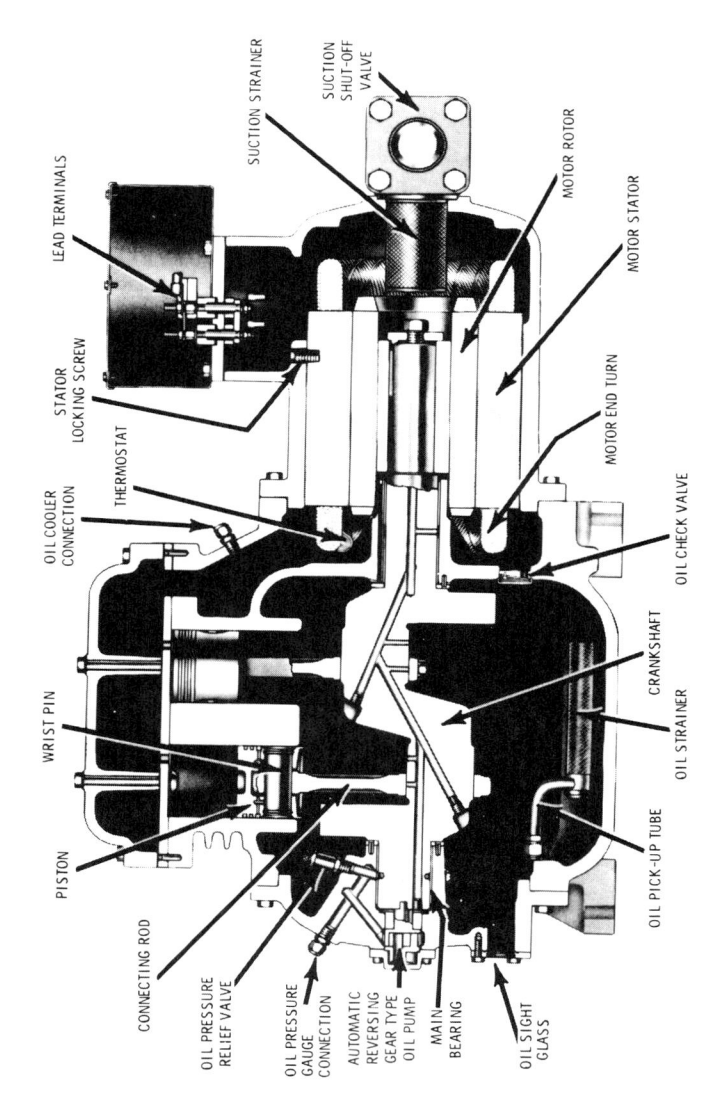

Fig. 7-4. Typical reciprocating hermetic compressor showing arrangement of parts with a motor terminal for across-the-line or part-winding starting and built-in thermal protection.

COMPRESSORS FOR AIR CONDITIONERS

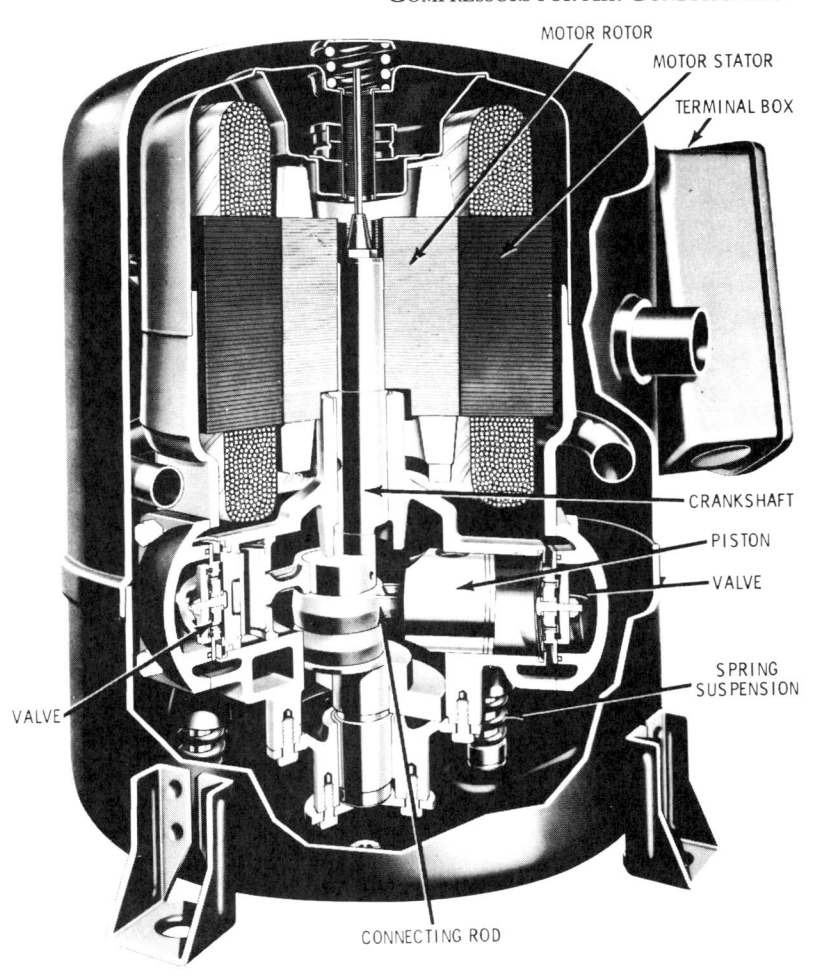

Fig. 7-5. Sectional view of vertical hermetic compressor.

and open at the bottom end, having attached at an intermediate point a wrist pin that transmits the pressure to the refrigerant gas by means of the connecting rod attached to the crank-shaft. In this manner the pumping action of the compressor provides the necessary circulating feature of the refrigerant in a mechanical refrigeration system.

Air Conditioning

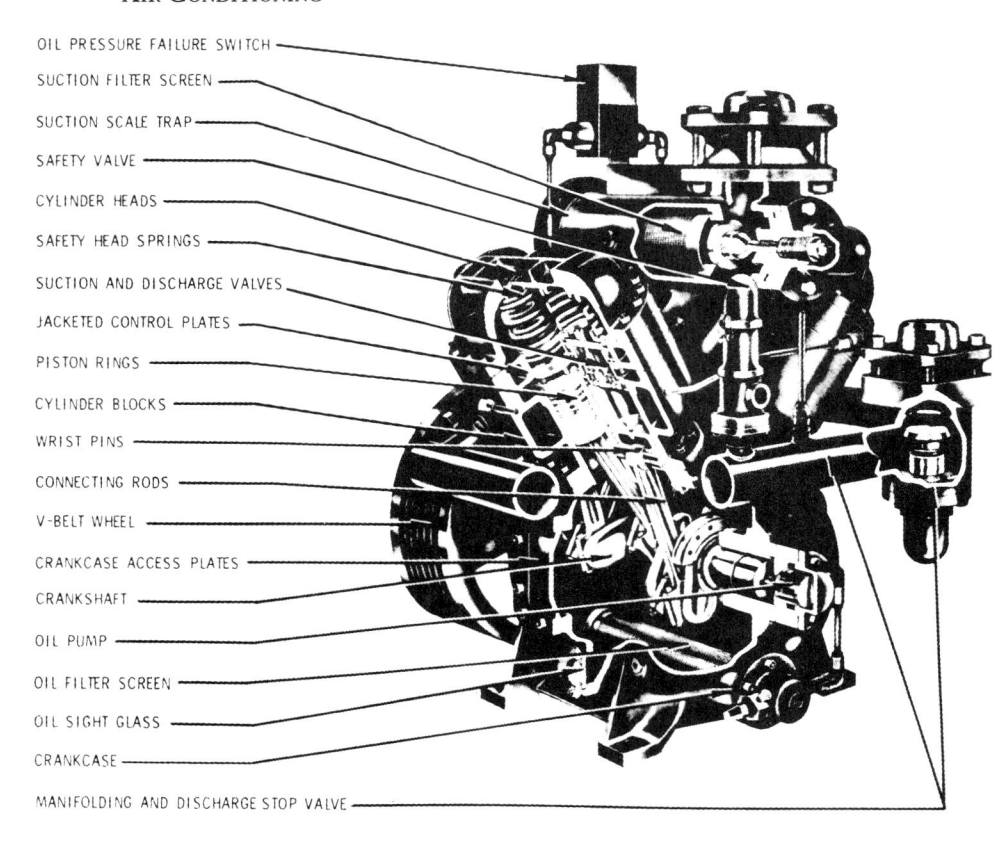

OIL PRESSURE FAILURE SWITCH
SUCTION FILTER SCREEN
SUCTION SCALE TRAP
SAFETY VALVE
CYLINDER HEADS
SAFETY HEAD SPRINGS
SUCTION AND DISCHARGE VALVES
JACKETED CONTROL PLATES
PISTON RINGS
CYLINDER BLOCKS
WRIST PINS
CONNECTING RODS
V-BELT WHEEL
CRANKCASE ACCESS PLATES
CRANKSHAFT
OIL PUMP
OIL FILTER SCREEN
OIL SIGHT GLASS
CRANKCASE
MANIFOLDING AND DISCHARGE STOP VALVE

Fig. 7-6. Cutaway view of a Frick ellipse reciprocating compressor equipped for automatic operation.

Pistons employed in the majority of compressors are made from the best grade of cast iron. They are carefully machined, polished, and fitted to the cylinders at close tolerances. There are several different designs of pistons. Some early models were not equipped with piston rings but were lapped to the cylinder in a very close fit. At present, however, two or more piston rings are employed, their function being to ensure proper lubrication and sealing of the cylinder walls.

Fig. 7-7. Sectional view through a Garland valve showing arrangement of parts.

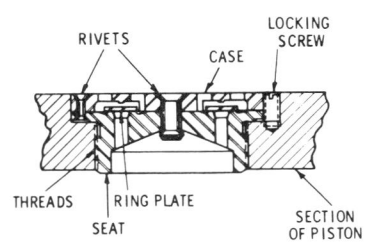

Courtesy the Frick Company

When assembled, pistons should fit the cylinder walls so that when inserted they do not drop of their own weight but must be urged through with the fingers. This fit is usually obtained by allowing 0.0003 inch (three ten-thousandths of an inch) of clearance per inch of cylinder diameter. Thus, a $3^{1}/_{2}$-inch cylinder would have a 0.00105-inch clearance between piston and cylinder wall (Fig. 7-8).

If cylinders are scored, they must be honed after reboring to ensure a high degree of wall finish that maintains cylinder proportions over long periods. After rings have been fitted, the compressor should be operated overnight to permit the rings to wear in. The compressor should then be drained and flushed with petroleum spirits or dry-cleaning fluid and dehydrated before being returned to service.

Connecting Rods

The connecting rod, as the name implies, serves as the connecting link between the piston and the crankshaft. It transforms the rotary motion of the crankshaft into the reciprocating motion of the piston. One end of the connecting rod is connected to the piston by means of a hardened, ground, and highly polished steel wrist pin, and the other end is connected to the crankshaft (Fig. 7-9). Several different methods are used to secure the piston wrist pin and rod. In some designs, the wrist pin is tightly clamped to the connecting rod with the moving bearing surface in the piston, while others have a bushing in the connecting rod.

AIR CONDITIONING

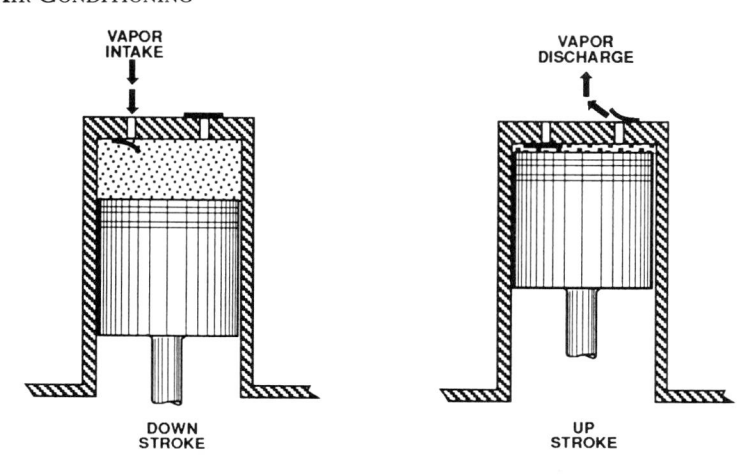

Fig. 7-8. Operation of the valves in a reciprocating compressor.

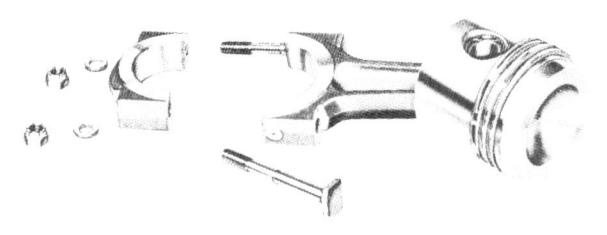

Courtesy the Trane Company
Fig. 7-9. Compressor piston and connecting-rod arrangement.

Crankshafts

The duty performed by this part of the compressor is to convert the rotating motion of the crankshaft into the reciprocating motion of the piston. In its simplest form it consists essentially of a shaft having a lever or arm, known as a *crank integral*, attached at right angles to the shaft. The outer end of the lever has a pin that engages with the large bearing on the connecting rod, as shown in Fig. 7-10. Crankshafts are commonly made of steel forgings,

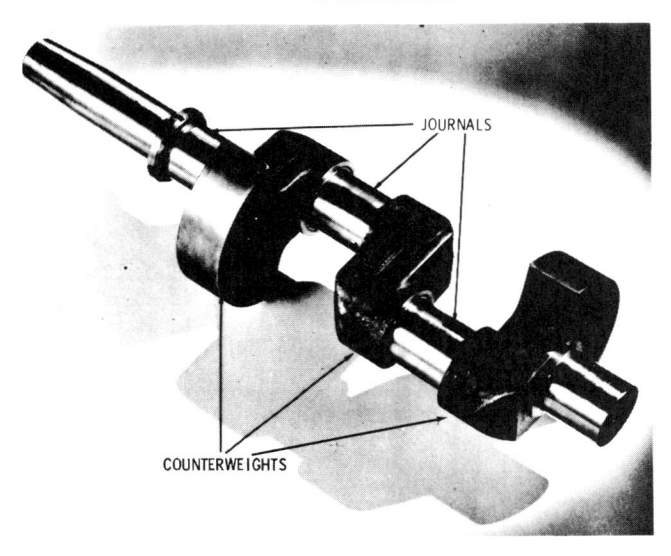

Fig. 7-10. Typical one-piece crankshaft.

which are machined to the proper tolerances for the main and connecting rod bearings. The double crank is employed in all two-cylinder compressors, while the single crank is used on the single-cylinder type.

Crankshafts are equipped with counterweights and are carefully balanced to ensure smooth and vibrationless compressor operation. Some types of compressors are equipped with what is known as the *eccentric shaft*. This is a different application of the crankshaft principle. It employs an assembly consisting of the main shaft (over which an eccentric is fitted) and an outer eccentric strap.

Shaft Seals

The function of the shaft seal in a refrigerating compressor is to prevent the gas from escaping from the compressor at the point where the shaft leaves the compressor housing. This problem was difficult to solve in the early stages of the refrigeration industry.

AIR CONDITIONING

The shaft, of necessity, must revolve, and yet the refrigerant must not be allowed to escape (when there is a pressure within the crankcase), nor must air be allowed to enter when the pressure of the crankcase is below 0 lb pressure. A typical compressor shaft seal is shown in Fig. 7-11.

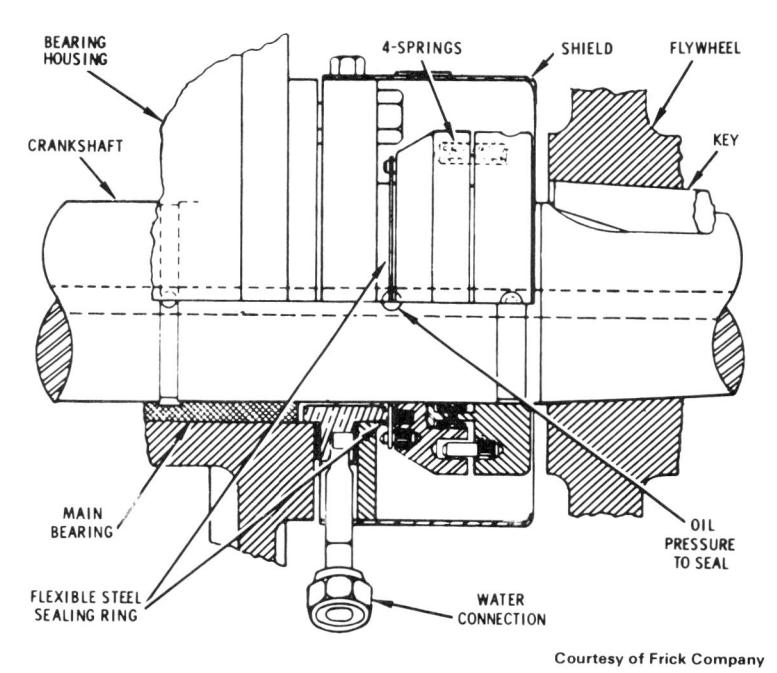

Courtesy of Frick Company

Fig. 7-11. Compressor shaft seal showing cooling-water connection.

The solution to the problem of crankshaft leaks was found in the development of a bellows-type seal, which is now a part of most compressors. The seal is an assembly of parts consisting of a special bronze seal ring soldered to a bellows, which, in turn, is soldered to a brass seal flange. A spring surrounds the bellows, with one end resting against the seal ring and the other against the seal flange. The face of the seal ring is lapped flat and smooth. When assembled in the compressor, the seal ring presses against a shoulder on the shaft (which has also been lapped), forming a gas-

and oil-tight joint at this point. The bellows provides the degree of flexibility necessary to keep the seal ring in perfect contact with the shaft, and the seal flange is clamped against a lead gasket around an opening in the side of the compressor.

Lubrication and Cooling

The method of lubrication depends largely on the size of the compressor; forced lubrication is universally used on large compressors, the splash method is used on smaller units. Large compressors are usually water-cooled, with the water jacket cooling either only the cylinder walls or both the cylinder walls and compressor head. Small compressors are usually air-cooled with an extended finned surface cast on the cylinder exterior. In this connection, it should be noted that, although water cooling is more effective than air cooling, cylinder cooling removes only a portion of the superheat in the refrigerant gas. The removal of heat from the cylinder will also result in a decrease in the work of compression, with an accompanied reduction in the condenser load.

ROTARY COMPRESSORS

Since they are hermetically sealed, resulting in an improved drive method (the motor and compressor being located in a common enclosure), rotary compressors are presently manufactured in large quantities, particularly in fractional-tonnage applications. By definition, a hermetic compressor is one in which the electric motor and compressor are contained within the same pressure vessel, with the motor shaft integral with the compressor crankshaft; the motor is in constant contact with the refrigerant.

In operation, the rotary-type compressor performs the same function as the reciprocating compressor; the compression of the gas establishes a pressure difference in the system to create a flow of refrigerant from one part of the system to the other. The rotary-type compressor, however, employs a slightly different method of accomplishing the compression of the gas. The rotary compressor

AIR CONDITIONING

compresses gas due to the movement of the rotor in relation to the pump chamber.

There are two types or configurations of rotary compressors. The *stationary blade* rotary compressor and the *rotating blade* rotary compressor. Figure 7-12 is an example of the stationary blade type. Note how the sliding barrier operates inside the steel cylinder.

The only moving parts in a stationary blade rotary compressor

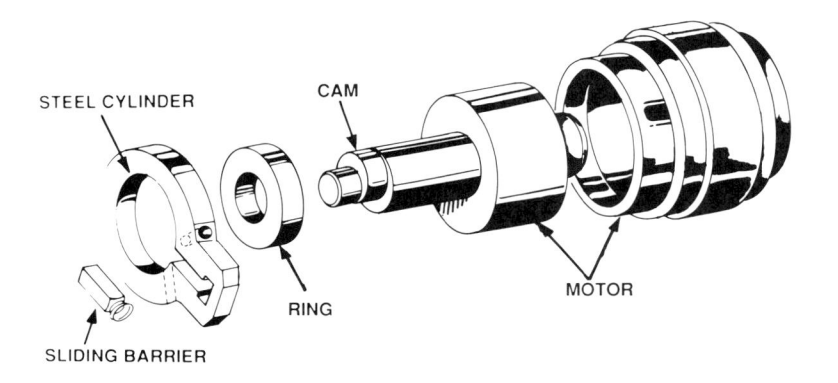

Fig. 7-12. Parts of a stationary blade rotary compressor.

are a steel ring, an eccentric or cam and a sliding barrier. Fig. 7-13 labels the parts of this type of operation while Fig. 7-14 shows how the rotation of the off-center cam compresses the gas refrigerant in the cylinder of the rotary compressor. The cam is rotated by an electric motor. As the cam spins it carries the ring with it. The ring rolls on its outer rim around the wall of the cylinder.

The gas must have a pathway to be brought into the chamber. Note that in Fig. 7-15 the vapor comes in from the evaporator and goes out to the condenser through holes that have been drilled in the compressor frame. Note that the gas is compressed by an offset rotating ring. Fig. 7-16 shows how the refrigerant vapor in the compressor is brought from the evaporator as the exit port is opening. When the compressor starts to draw the vapor from the evap-

Compressors for Air Conditioners

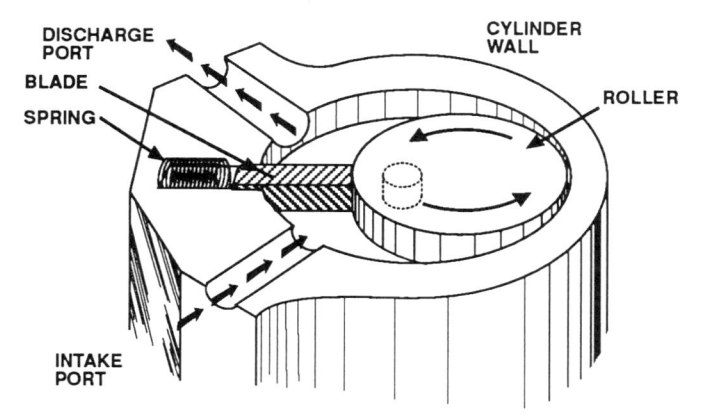

Fig. 7-13. Parts labeled for a stationary blade rotary compressor.

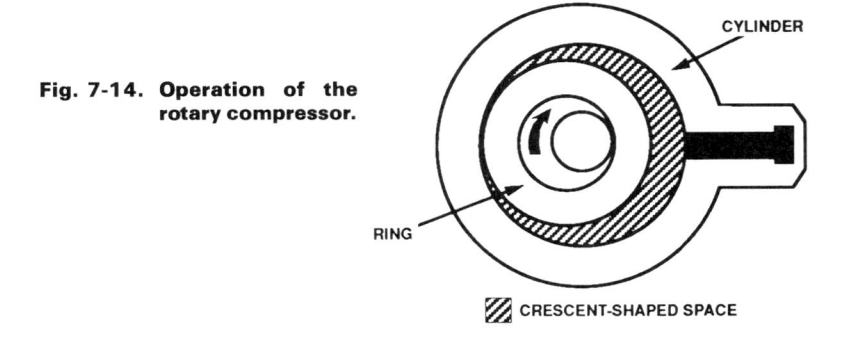

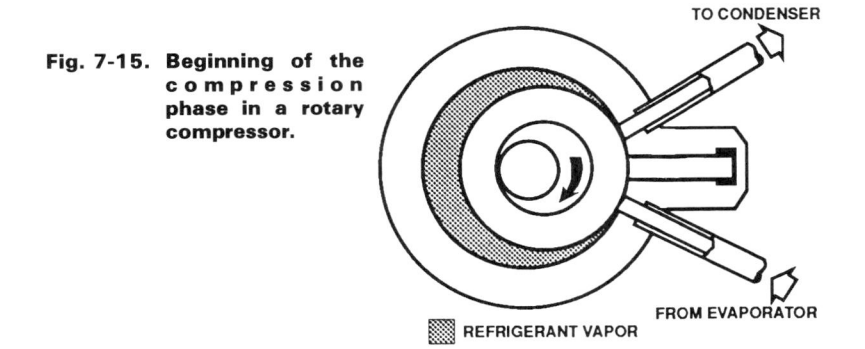

Fig. 7-14. Operation of the rotary compressor.

Fig. 7-15. Beginning of the compression phase in a rotary compressor.

215

AIR CONDITIONING

orator the barrier is held against the ring by a spring. This barrier separates the intake and exhaust ports. As the ring rolls around the cylinder, it compresses the gas and passes it on to the condenser (Fig. 7-17). The finish of the compression portion of the stroke or operation is shown in Fig. 7-18. The ring rotates around the cylinder wall, and is held in place by the spring tension of the barrier's spring and the pressure of the cam being driven by the electric motor. This type of compressor is not used as much as the reciprocating hermetic type of compressor, as there are some problems with lubrication. Better seals are still being developed. This type of compressor is used in some home refrigerators and air conditioners; however, they have been known to become overworked in very hot climates and have been replaced by the reciprocating type.

The other type of rotary compressor is the rotating blade type (Fig. 7-19). This type has a roller centered on a shaft that is eccentric to the center of the cylinder. Usually, there are two spring-loaded roller blades, mounted 180° apart. The blades sweep the sides of the cylinder. The roller is so mounted that it touches the cylinder at a point between the intake and discharge ports. As the roller rotates, vapor moves into the cylinder through the intake port and is trapped in a space between the cylinder wall, the blade, and the point of contact between the roller and the cylinder. Vapor is compressed as the next blade passes the contact point and the vapor space becomes smaller and smaller. When the vapor has been compressed to a certain pressure, it leaves by way of the exhaust port on its way to the condenser. This type of rotary compressor, like the stationary blade model, requires extremely close tolerances.

Some manufacturers make rotary blade compressors for commercial applications. They are used primarily with ammonia. Thus, there is no copper or copper alloy tubing or parts. Most of the ammonia tubing and working metal is stainless steel.

CENTRIFUGAL COMPRESSORS

Centrifugal compressors are essentially high-speed machines and best suited for steam-turbine drives. Because they are de-

Compressors for Air Conditioners

Fig. 7-16. **Beginning of the intake phase in a rotary compressor.**

■ REFRIGERANT VAPOR IN COMPRESSION
▒ REFRIGERANT VAPOR FROM EVAPORATOR

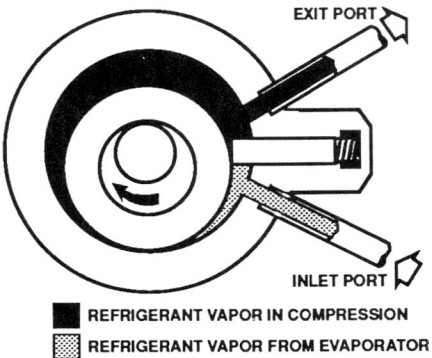

Fig. 7-17. **Compression and intake phases half completed in a rotary compressor.**

■ REFRIGERANT VAPOR IN COMPRESSION
▒ REFRIGERANT VAPOR FROM EVAPORATOR

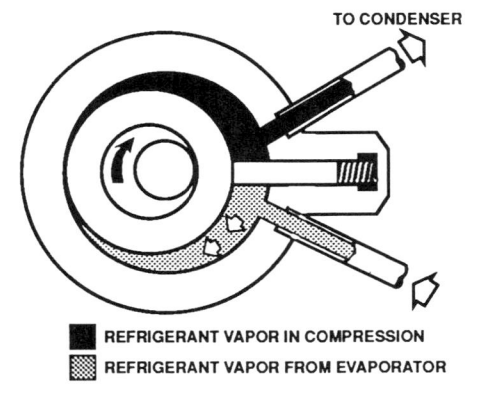

Fig. 7-18. **Finish of the compression phase of the rotary compressor.**

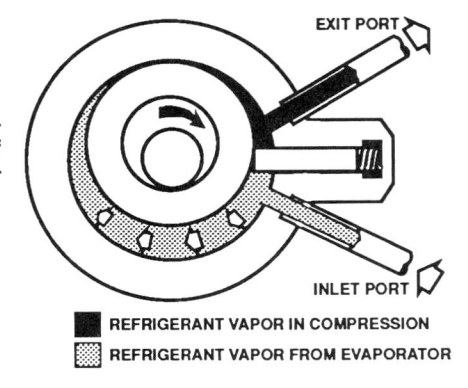

■ REFRIGERANT VAPOR IN COMPRESSION
▒ REFRIGERANT VAPOR FROM EVAPORATOR

AIR CONDITIONING

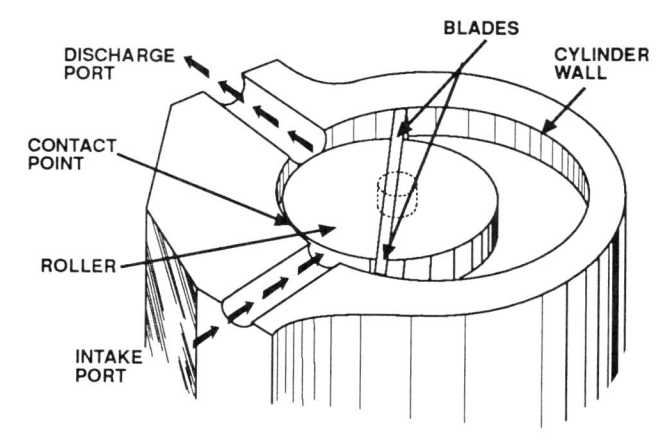

Fig. 7-19. Rotating blade compressor.

signed for the same high speed as the turbine, they may be directly connected. Where high-pressure steam is generated, the turbine can act as a reducing valve, and the low-pressure steam that leaves the turbine can be used for heating and other purposes. In a great many applications, particularly in the smaller sizes, however, they are driven by electric motors and equipped with gear-type speed increasers. Centrifugal compressors (Fig. 7-20) are used with low-pressure refrigerants, and both evaporator and condenser usually work below atmospheric pressure.

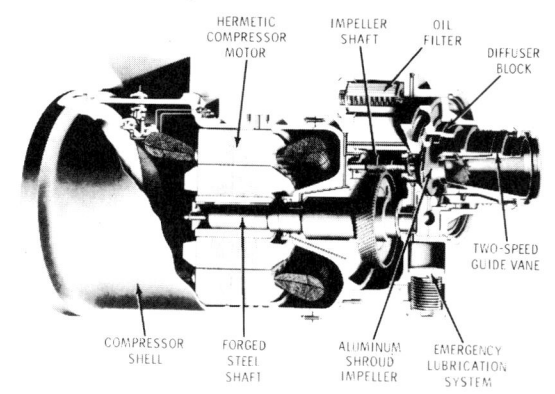

Courtesy Westinghouse Electric Corporation—Air Conditioning Division

Fig. 7-20. Cutaway view of a single-stage centrifugal compressor.

Compressors for Air Conditioners

Compression of the refrigerant is accomplished by means of centrifugal force; and, because of this, centrifugal compressors are best suited for large refrigerant volumes and low-pressure differentials. They are well suited to low-temperature refrigerating cycles, especially those using petroleum or halogenated hydrocarbons as a refrigerant.

Centrifugal-compressor installations are economically advantageous where steam-turbine drive is permitted, since the installation and labor required for such an installation is relatively small compared to that required for a comparable gas-engine-driven compressor plant. This is mainly because of the compactness and lightness of the unit compared to the amount of horsepower involved; in addition, a centrifugal compressor will occupy only a fraction of the space ordinarily required for refrigerating equipment. Refrigerating units of the centrifugal type are available in various sizes varying from 100 to 2500 tons capacity and are manufactured for use with electric motors, steam turbines, or internal-combustion engines.

Construction Features

Because the centrifugal compressor differs radically in construction from the familiar reciprocating and rotary types, we shall give a brief review of the principal parts. With reference to Fig. 7-20, it will be noted that its operating parts consist of the following:

1. The compressor motor, consisting of a dependable double-squirrel-cage type for operation from a standard 50- or 60-Hertz power source. The compressor is effectively cooled by controlled liquid injection and is protected against damage from abnormal operation.
2. The overhung steel motor shaft smoothly transmits power to the impeller shaft through matched, hardened, ground, and crowned helical gearing.
3. The impeller shaft design cushions the impeller and gearing against unusual stresses that could be set up by abnormal system operation.
4. Emergency lubrication is provided by a built-in emergency

system. Should a power failure occur, oil under pressure is available for the coast-down period.
5. Oil is supplied to the bearings and is cleaned by a 10-μ oil filter. A check valve on the discharge side of the filter reservoir allows easy filter replacement without the need for system pumpdown.
6. The movable diffuser block is positioned by the guide-vane piston. As loads vary, the diffuser block is moved to match the diffuser passage to the gas flow and to allow quiet, stable operation from 1 to 100 percent of unit capacity.
7. A two-speed, hydraulically operated guide-vane activating piston positions the vanes at speed one (the normal response to variations in the chiller water temperature leaving the unit). A built-in compensating control allows automatic override of normal operation and quick response at speed two to close the valves for low suction pressure or current-limiting duty.
8. The inlet guide vanes modulate the compressor capacity and direct the refrigerant gas into the impeller in a manner that minimizes refrigerant pressure losses.
9. An aluminum closed-shroud impeller is precision cast. It incorporates an unusual noise-abatement system and is mounted on the impeller shaft by means of a rugged self-centering type of friction mounting.

Lubrication System

A separately driven hermetic oil-pump assembly supplies oil at controlled temperature and pressure to all bearing surfaces and is the source of hydraulic pressure for the capacity-control system. The oil pump is automatically started and must run for a fixed time period before the compressor is started and after it has stopped. In addition, temperature and pressure controls prevent the compressor from operating unless oil pressure and temperature are at satisfactory levels.

Oil is collected from a well located at the bottom of the gear casing and is returned to the oil pump through an external oil line. Any refrigerant returning to the pump with the oil is purged from the pump housing after entrained oil has been separated. Oil is separated as the refrigerant gas passes through an unusual centrifuge type of oil separator that is mounted on the pump motor rotor, as shown in Fig. 7-21.

Oil from the pump is supplied to the compressor and control system after it has been cooled. All bearing surfaces are pressure lubricated. Drive gears are operated in a controlled oil-mist atmosphere, which efficiently cools and lubricates them. Oil is made available under pressure to the unit-control system and is used to position the inlet guide vanes in response to changes in temperature of the chiller water leaving the unit. Should a power failure occur, an emergency oil system guarantees adequate oil flow under pressure and prevents damage that could occur during a spindown period with the oil pump stopped.

INSTALLATION AND MAINTENANCE

The selection of space in which the condensing unit is to be located is very important. If the unit is to be installed within an enclosure or room having only natural ventilation and a volume of less than 2500 cu ft., an exhaust fan or a suitable water coil and fan must be installed to cool the air.

Foundation

Due to the weight of the unit, a concrete foundation that is level and strong must be provided. The foundation must be constructed and the unit so installed that it sets level in all directions (Fig. 7-22). Sufficient space should be provided around the foundation to permit any necessary service operation and especially at the end of the unit in order to provide room for cleaning the condenser tubes.

Motor and Belts

In condensing-unit installations having belted compressors, the procedure is generally as follows:

1. Align the motor pulley with the compressor flywheels by stretching a string from the outer edge of one flywheel to the

Air Conditioning

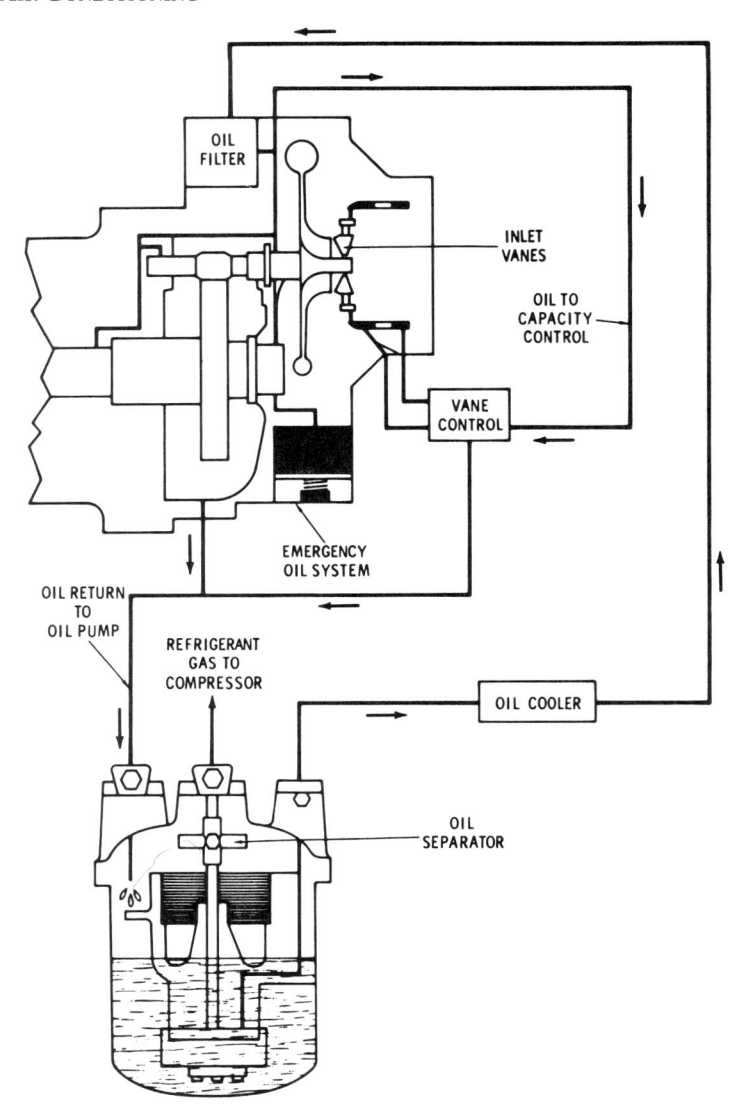

Fig. 7-21. Oil-lubrication system of a single-stage centrifugal refrigerant compressor.

COMPRESSORS FOR AIR CONDITIONERS

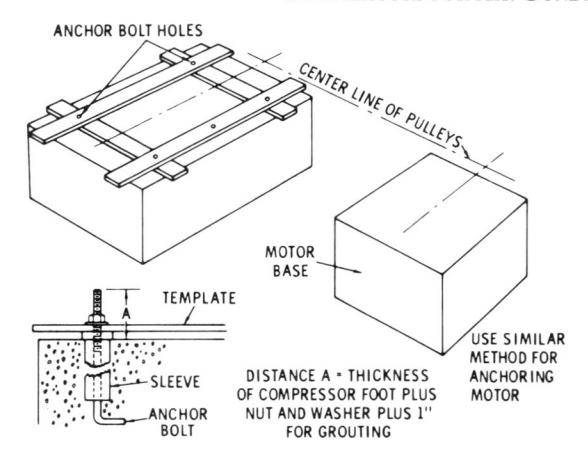

Fig. 7-22. Method of setting anchor bolts and template prior to pouring concrete foundation.

outer edge of the other, or by the use of a long rigid straightedge. It may be necessary to loosen the motor mounting bolts and slide the motor forward or backward to get the pulley properly aligned. Do not slide the pulley further out or closer in on the motor shaft for this initial alignment; a slight adjustment may be made in this manner, if necessary, after the condensing unit is assembled and operating.

2. To install the belts, start by placing a belt in the pulley groove nearest the motor. Then place the belt in the flywheel groove that lines up with this rear groove in the pulley. With the motor mounting base in its lowest position, the belt may be put on using only the hands. Never use a pry bar, screwdriver, or wood strips, because this will result in broken fibers and shorten the belt life.

3. Adjust the belt tension by raising the motor mounting base by means of the nuts on the adjusting screws. Proper tension will allow some give in the belt when depressed with the hand. Be sure that all four corners of the motor base are the same distance above the condensing-unit frame because the motor must be level. It may be necessary to move the motor slightly at this time in order to equalize the tension in the belt to the compressor.

Compressor Flowback Protection

In most air-conditioning and refrigeration systems, it is possible to control the refrigerant feed so that liquid refrigerant does not return to the compressor during operation. Some systems, by reason of poor expansion-valve operation, sudden changes in loading, etc., periodically return liquid refrigerant to the compressor, although not in sufficient quantities to do real damage. However, there are certain systems which, by their very design, will periodically flood the compressor with excessive amounts of liquid refrigerant. Principally, these prove to be systems in which there is a periodic reversal of the refrigerant cycle to do one of the following:

1. Provide hot gas for defrosting an evaporator
2. Furnish heat for release purposes in ice-making apparatus
3. Change from cooling to heating cycle, or vice versa, in heat-pump systems

These systems are almost identical in operation and in the effect on the compressor. Consider an air-source heat pump for a typical analysis of what happens. The flowback occurs when reversing the cycle in either direction. During the cooling cycle, the outdoor coils act as condensers and have warm liquid continually draining from them. The lines to which the thermal expansion-valve bulbs are strapped are now hot-gas lines and are hot. When the cycle reverses to heating, the outdoor coils change from condensers to evaporators. The liquid that has been draining from them during the condensing cycle is now dumped into the suction line. Of even greater significance, the thermal expansion valves also open wide as the result of their bulbs being on warm lines (lines that were previously hot-gas lines). Therefore these valves will flow through until control is reestablished at the bulbs. The total of these two effects constitutes a substantial liquid flowback through the suction line.

When there is no way to control the hazard of periodic flowback of substantial proportions through the suction line, it is necessary to take measures to protect the compressor against it; otherwise compressor life will be materially shortened. The most satisfactory method appears to be a trap arrangement that catches the liquid flowback and may do one of the following: meter it slowly into the suction line when it is cleaned up with a liquid-suction heat interchanger; or evaporate 100 percent of the liquid in the trap itself and automatically return the oil to the suction line.

Valve Maintenance

Sometimes the suction-valve reed is held off its seat by corrosion and dirt lodged under it, thus leaving the cylinder open to the low side. In this case, there will be no refrigeration, the condenser will not get warm, and the power consumption by the drive motor will be low. To determine if the inlet valve is causing these conditions, connect the compound and pressure gage, and start the unit. The pressure gage will not show an increase in pressure, and the compound gage will not show any decrease in pressure and will fluctuate. Flushing the valve may wash the obstruction away and permit it to function normally. If the discharge valve is held open by something lodged under it, such as corrosion or dirt, the cylinder will remain open to the high side and there will be no refrigeration.

To determine if the discharge valve leaks, connect the compound and pressure gage and operate the unit until the low-side pressure is reduced to normal. Shut off the unit. Immediately place your ear close to the compressor. If the valve leaks, a hissing noise produced by escaping gas can be heard. The low-side pressure will rise rapidly, and the high-side pressure will fall rapidly, equalizing the pressures. The rapidity of this action will determine how badly the valve leaks.

Shaft-Seal Leaks

To determine whether a compressor is leaking at the seal, proceed as follows: close the shut-off valves by turning the stem clockwise as far as possible. To ensure adequate refrigerant pressure in the compressor crankcase and on the seal bearing face, attach a refrigerant drum containing the correct refrigerant to the suction shutoff valve outlet port. In making this connection, there should be a gage in the line from the drum to the compressor so as to de-

AIR CONDITIONING

termine accurately the pressure in the compressor.

Test pressures for this purpose should be approximately 70 to 80 lb. Should the compressor be located in a cool location, it may be necessary to raise the pressure in the drum by adding heat. In this process, care must be taken not to exceed 100 psi since greater pressure may damage the bellows assembly of the seal. With this pressure on the crankcase of the compressor, test for leaks with a halide torch, moving the finder tube close around the seal nut, crankshaft, seal-plate gasket, and where the seal comes in contact with the seal plate. If this does not disclose a leak, turn the flywheel over slowly by hand, holding the finder tube close to the aforementioned parts.

After a leak has been detected, locate the exact place where it is leaking, if possible. If the leak is around the seal-plate gasket, replace the gasket; if the leak should be at the seal seat or nut, replace with a new seal plate, gasket, and seal assembly. To replace the seal, observe the following instructions:

1. Remove the compressor from the condensing unit in the usual manner.
2. Remove the flywheel, using a puller. Leave the flywheel nut on the crankshaft so that the wheel puller will not distort the threads.
3. Remove the seal guard, seal nut, and seal assembly.
4. Remove the seal plate and gasket.
5. When assembling the seal, put a small quantity of clean compressor oil on the seal face (both plate and seal).
6. To reassemble, reverse these operations, making sure that the seal plate is bolted in place and that the seal guard is at the top.

Compressor Knocks

A knock in the compressor may be caused by a loose connecting rod, eccentric strip and rod, eccentric disk, piston pin, crankshaft, or too much oil in the system. A compressor knock can be determined by placing the point of a screwdriver against the crankcase and your ear against the handle. A knock can then very easily be heard. It will not be possible to determine what causes the knock before the compressor is disassembled. Sometimes it may be possible to determine a looseness of the aforementioned parts without completely disassembling the compressor.

First, remove the cylinder head and valve plate to expose the head of the piston. Now, start the motor and press down on the top of the piston with the finger. Any looseness can be felt at each stroke of the piston. The loose part should be replaced. It is always well to check the compressor oil level first before analyzing and determining the compressor repairs. Oil knocks are usually caused by adding too much oil in servicing.

It should never be necessary to add oil to a system unless there has been a leakage of oil. A low charge is sometimes diagnosed as lack of oil. Always make sure that a low oil level is actually due to lack of oil rather than to a low charge before adding oil to the system.

Stuck or Tight Compressor

The reason for a compressor being *stuckup* is usually the result of moisture in the system or a lack of lubrication. When this occurs, the compressor should be thoroughly cleaned out. The compressor should be completely disassembled and the parts thoroughly cleaned and refitted. New oil and refrigerant should be put into the cleaned system. A tight compressor will result when a cylinder head, seal cover, or similar part has been removed and not replaced carefully, or when the screws have been tightened unevenly.

Compressor Removal

To remove a compressor from the unit, proceed as follows:

1. Attach a compound gage to the suction-line valve. Close the suction-line shutoff valve and run the compressor until 20 to 25 inches of vacuum is obtained; then close the discharge shutoff valve. Before removing any fittings, crank the suction-line valve to bring the gage reading back to zero.
2. Before removing the service valves, loosen the pressure gage and relieve the pressure in the head of the compressor. Re-

move the belt and cap screws holding the suction, and discharge valves to the compressor. If the compressor is to be taken away from the premises for repairs, place service valves over both the discharge and suction-line openings to prevent air and moisture from entering and oil from leaking out. Loosen the four nuts that hold the compressor to the base, and bend the tubing away from the compressor just enough to permit the assembly to be lifted out. Care should be taken not to loosen the mounting pads and washers on which the compressor rests.

Compressor Replacement

The reinstallation of a compressor is roughly a reversal of the process of removing it. First, place the mounting pads and washers in position. Place the compressor carefully on the base in the same position it occupied before removal. Bolt the compressor in place. Replace the suction and discharge valves, using new gaskets. After the compressor and valves are bolted in place, the compressor must be evacuated to remove the air.

Compressor Startup

When starting a water-cooled compressor after winter shutdown, the following procedure is suggested:

1. Install a pressure gage on the condenser, and check for pressure. If there is no pressure in the condenser, determine how the charge was lost and repair the leak before opening any valves.
2. Open the liquid-line valves, and allow the refrigerant to flow through the system.
3. Check the entire system for refrigerant leaks, using a halide leak detector.
4. If the system is gas tight, connect the water lines, open the hand valve in the water lines, and open the discharge and suction valves. *Caution: Be sure the condenser has not been frozen before turning on the water.*

5. Install the belts and check for correct tension after a short period of operation.
6. Check the motor for oiling; if old oil is heavy or dirty, replace with new oil of the proper viscosity.

Winter Shutdown of Water-Cooled Compressors

Where water-cooled refrigeration and air-conditioning systems are not operated in the winter, it is necessary to service them before discontinuing their operation. They should be pumped down, water drained from the condensers and piping, belts loosened, and motors oiled. The compressor should be thoroughly tested for leaks, and any found should be corrected.

To pump down and discontinue service, attach head and suction gages, and close the liquid shutoff valve at the receiver. Run the compressor until a suction pressure of from 2 to 5 psig is obtained. If, on stopping the compressor, the suction pressure rises, repeat until this pressure is steady. Close the suction and discharge shutoff valves to confine the refrigerant charge in the receiver. Open the main switch and *be sure* to tag it with a warning not to start the compressor without first opening the discharge and suction stop valves, as damage to the compressor may result. Tighten all packing nuts on valve stems. Cap all valve stems tightly.

The condenser is usually of sufficient capacity to hold the complete charge; however, if it is not, the overcharge must be pumped into an empty drum. Never completely fill a receiver with a liquid refrigerant (indicated by a rise in head-gage pressure above the safety cutout pressure) so that no gas space exists. Should the compressor room temperature increase, there is danger of rupturing the receiver. Shut off the water supply to the compressor, and disconnect the outlet water line from the condenser. This will allow the water to drain from the condenser. After the water has all drained out, connect the air line or an automobile tire pump to the blowout plug, and proceed to force the remaining water, if any, out of the condenser. Condensers with removable heads can be effectively drained by shutting off the waste supply and removing the condenser heads, thus allowing all the water to drain out.

On compressors of large capacity, it will be necessary to arrange for a connection to accommodate the air line or pump. After all water has been forced from the condenser, it is recommended that all water lines be disconnected from the condenser and the water valves drained by removing the drain plug from the bottom of the water valves. For frequent and quick draining, it is suggested that a tee and drain valve be installed between the water-regulating valve and condenser inlet. The safest, surest way to prepare condensers for winter shutdown is to flush them out with an antifreeze solution. After the condenser has been treated with the antifreeze solution and drained, the condenser inlet and outlet water connections should be plugged in order to maintain the remaining antifreeze solution in the condenser. Remove the belts from the compressor motor, but not from the flywheel, so that they will not take a permanent set. It is also advisable to oil the motor in order to protect the bearings from corrosion, etc.

Compressor Performance—The performance of a compressor is measured by its ability to provide the maximum refrigeration effect with the least power input. Other desirable factors are maximum trouble-free life expectancy at the lowest possible cost and a wide range of operating conditions. The performance factor of a compressor may be written:

$$\text{Performance factor} = \frac{\text{capacity, Btu}}{\text{power input, watts}}$$

Compressor Capacity—Capacity of a compressor means the useful refrigeration effect that can be accomplished. It is equal to the difference in the total enthalpy between the refrigerant liquid at a temperature corresponding to the pressure leaving the compressor and the refrigerant vapor entering the compressor. It is measured in Btu.

Mechanical Efficiency—Mechanical efficiency of a compressor is defined as the ratio of the work delivered to the gas, as obtained by an indicator diagram, to the work delivered to the compressor shaft.

Volumetric Efficiency—Volumetric efficiency is defined as the ratio of the actual volume of a refrigerant gas pumped by the com-

COMPRESSORS FOR AIR CONDITIONERS

pressor to the volume displaced by the compressor piston. The compression ratio of a compressor is the ratio of the absolute head pressure to the absolute suction pressure. Factors determining compressor efficiency include the clearance volume above the piston, clearance between the piston and the cylinder walls, valve-spring tension, and valve leakage. For a given compressor, the effect of compressor efficiency because of design factors is fairly constant, and volumetric efficiency will vary almost inversely with the compression ratio.

Compressor Efficiency Test—An efficiency test generally means a check on the relative amount of useful work that the compressor will accomplish. Strictly speaking the efficiency of any machine is taken as the ratio of the power output to the power input in the same units. This is usually written:

$$\text{Efficiency} = \frac{\text{output}}{\text{input}}$$

Factors that determine the efficiency of a compressor are:

1. The degree to which the piston valve holds closed on the up stroke.
2. The degree to which the discharge valve holds tight when the piston is on the down stroke.

To test the compressor efficiency, proceed in the following manner:

1. Stop the compressor, and install a compound gage on the suction-line valve.
2. Close the suction valve, and operate the compressor until about a 25-inch vacuum is obtained on the compound gage; then stop the unit, and note the gage readings. If the compressor will not pull a 25-inch vacuum or better, it is an indication that air is leaking back by the discharge valve and also leaking by the piston valve.
3. If the head pressure drops and the vacuum-gage reading remains practically constant, this indicates that there is an ex-

AIR CONDITIONING

ternal leak at one of the points on the head of the compressor or at the gage and shutoff-valve connection.
4. To repair either a leaky suction or discharge valve, remove the head of the compressor and carefully remove the discharge-valve plate. In disassembling the discharge valve, caution should be used not to disturb the actual conditions prevailing during the test. It is possible for some dirt, scale, or other foreign matter to get under the valve disk and on the seat, causing poor performance. If there is no evidence of dirt or foreign matter, check the seat on both the discharge-valve and piston assembly for low spots or scratches. If these are found, replace the disks or the complete valve plate.

Generally the trouble can be found at either the suction valve or the discharge valve. Badly scored seats on either the discharge or suction valve require replacing with a new assembly. The removal of deep scores changes the valve lift, further endangering the efficiency. After repairs are made, thoroughly clean the parts with gasoline or carbon tetrachloride, reassemble them (using new gaskets), and repeat the efficiency test to ensure that the trouble has been eliminated. Caution: carbon tetrachloride (CCl_4) can cause permanent liver damage if the fumes are breathed and can be *fatal* if proper ventilation is not used.

SERVICE GUIDE

The following service guide will assist in troubleshooting air conditioning systems. It should be noted, however, that although the trouble may differ at times from that given in the service guide, experience will usually quickly reveal the exact cause, after which the remedy and its correction may be carried out.

AIR CONDITIONING TROUBLESHOOTING GUIDE

Symptom and Possible Cause   *Possible Remedy*

Compressor Will Not Start

(a) Control circuit not functioning properly.
(a) Locate open control and determine cause.

Compressors for Air Conditioners

(b) Overload tripped or fuse blown.

(b) Reset overload, replace fuses, and examine for cause of condition and correct.

(c) Switch open.
(d) No charge of gas in system operated on low pressure control.

(c) Close switch.
(d) With no gas in system, there is insufficient pressure to throw in low-pressure control. Recharge system; correct leak.

(e) Solenoid valve closed.

(e) Examine holding coil; if burned out or defective, replace.

(f) Dirty contact points on controls.
(g) Out on safety controls, high-pressure cutout, oilpressure safety switch.

(f) Clean contacts.

(g) Discharge service valve may be partially closed. Check oil level. Reset control button. Check for faulty wiring.

(h) Burned out or defective motor.
(i) Compressor stuck.

(h) Repair or replace.

(i) Determine location and cause. Repair or replace compressor.

(j) Compressor not unloading for start.

(j) Check oil level and oil pressure. Check capacity control mechanism. Check unloader pistons.

Compressor Short-Cycles

(a) Fans not running on evaporator.

(a) Check all electrical connections, fuses, thermal overloads, and thrown switches.

(b) Coils on evaporator clogged with frost or dirt.
(c) Thermal bulb on expansion

(b) Clean or defrost coils.

(c) Detach thermal bulb from

233

AIR CONDITIONING

Symptom and Possible Cause *Possible Remedy*

valve has lost charge.　　　　　suction line and hold in the palm of one hand, with the other hand gripping the suction line. If flooding through is observed, bulb has lost charge. If no flooding through is noticed, replace expansion valve.

(d) Control set at too close a differential.
(e) Leaky liquid line solenoid valve.
(f) Discharge valve leaks slightly.
(g) Low head pressure.

(d) Check control settings for proper application, clean, and adjust or replace.
(e) Repair or replace.
(f) Replace valve plate.
(g) Check and adjust water-regulating valve. Check refrigerating charge.

Running Cycle Too Long

(a) Control contacts stuck.
(b) Unit too small for application.
(c) Insufficient refrigerant charge.
(d) Excessive head pressure due to overcharge, air in condenser, or dirty condenser.
(e) Leaky valve in compressor.

(a) Check, clean, repair, or replace.
(b) Capacity must be increased by increasing speed (if belt drive), or replace compressor.
(c) Test for leaks and add refrigerant.
(d) Remove excess refrigerant, purge system, clean condenser.
(e) Repair or replace.

Unit Operates Too Long or Continuously

(a) Shortage of gas.
(b) Control contacts frozen.

(a) Repair leak and recharge.
(b) Clean points or replace control.

COMPRESSORS FOR AIR CONDITIONERS

Symptom and Possible Cause   *Possible Remedy*

(c) Dirty condenser.
(d) Location too warm.
(e) Air in system.
(f) Compressor inefficient.
(g) Plugged expansion valve or retainer.
(h) Iced or plugged coil.
(i) Defective insulation.
(j) Service load too great. Unit too small.

(c) Clean condenser.
(d) Change to cooler location.
(e) Purge.
(f) Check valves and pistons.
(g) Clean or replace.
(h) Defrost or clean.
(i) Correct or replace.
(j) Keep doors closed. Add unit or replace.

Fixture Temperature Too High

(a) Refrigerant shortage.
(b) Control set too high.
(c) Expansion valve or strainer plugged.
(d) Compressor inefficient.
(e) Expansion valve set too high.
(f) Iced or dirty coil.
(g) Cooling coils too small.
(h) Unit too small.
(i) Expansion valve too small.
(j) Restricted or small gas lines.

(a) Repair leak and recharge.
(b) Reset control.
(c) Clean or replace.
(d) Check valves and pistons.
(e) Lower setting.
(f) Defrost or clean.
(g) Add surface or replace.
(h) Add unit or replace.
(i) Raise suction pressure with larger valve.
(j) Clear restriction or increase line size.

Compressor Starts, but Motor Will Not Get Off Starting Winding

(a) Low line voltage.
(b) Improperly wired.

(c) Defective relay.

(d) Running capacitor shorted.

(a) Bring up voltage.
(b) Check wiring against diagram.
(c) Check operation manually. Replace relay if defective.
(d) Check by disconnecting running capacitor.

235

AIR CONDITIONING

Symptom and Possible Cause	Possible Remedy
(e) Starting and running windings shorted.	(e) Check resistance. Replace stator if defective.
(f) Starting capacitor weak.	(f) Check capacitance. Replace if low.
(g) High discharge pressure.	(g) Check discharge shutoff valve. Check pressure.
(h) Tight compressor.	(h) Check oil level. Check binding.

Compressor Noisy

(a) Too light a foundation, or loose foundation bolts.	(a) Check size of foundation. Check shim. Tighten all bolts.
(b) Too much oil in circulation, causing hydraulic knocks.	(b) Check oil level. Remove excess, if any. Also check for oil at refrigerant test cock.
(c) Slugging due to flooding back of refrigerant.	(c) Check thermal-bulb location and fastening. Reset expansion valve. Trap suction line so refrigerant will not flood back on the off cycle.
(d) Worn parts such as pistons, piston pins, or connecting rods.	(d) Determine location of cause. Repair or replace compressor.
(e) Restriction in line.	(e) Be sure discharge valve is wide open, or install muffler.

Oil Leaves Compressor

(a) Wet suction causing oil foaming.	(a) Adjust expansion valves and check thermal bulbs for proper mounting. Check refrigerant piping.
(b) High crankcase pressure due to worn rings.	(b) Replace rings and liners if worn.

COMPRESSORS FOR AIR CONDITIONERS

Symptom and Possible Cause　　*Possible Remedy*

(c) Short-cycling.

(c) Check control setting for proper job conditions. Repair or replace faulty control.

Oil Does Not Return to Crankcase

(a) Refrigerant piping laid out incorrectly.
(b) Insufficient charge of refrigerant.
(c) High crankcase pressure.

(d) Separator valve not functioning properly.

(a) Check and correct if necessary.
(b) Test for leaks and add refrigerant.
(c) Check piston ring blowby. Check oil-separator operation.
(d) Check oil-separator operation.

Insufficient Oil Pressure

(a) Excessive oil foaming.

(b) Low oil level.

(c) Defective oil pump.
(d) Broken or loose oil line.
(e) Plug out of crankshaft.

(f) Strainer pickup tube loose, split, or has defective flare.

(a) Adjust expansion valve. Check crankcase heater setting or use of pumpdown cycle.
(b) Check level and add oil if low.
(c) Replace pump.
(d) Check and correct.
(e) Inspect. Install plug if missing.
(f) Check and correct defect if necessary.

Abnormally High Head Pressure

(a) Overcharge of refrigerant.
(b) Air or noncondensable gases in system.
(c) Discharge service valve not fully open.

(a) Remove excess refrigerant.
(b) Purge.

(c) Open full.

Air Conditioning

Symptom and Possible Cause

(d) Restriction in discharge line.
(e) Inadequate air circulation in air-cooled or evaporative condenser.
(f) Dirty condenser.
(g) Fouled water-cooled condenser.
(h) Water flow restricted or too warm on watercooled unit.

Possible Remedy

(d) Determine cause and correct.
(e) Remove obstruction. Clean condenser.
(f) Clean.
(g) Clean tubes.
(h) Check and adjust water-regulating valve.

Low Head Pressure

(a) Low refrigerant charge.
(b) Too much water flowing through condenser.
(c) Liquid refrigerant flooding back from evaporator.

(d) Leaky discharge valve or discharge to suction relief valve leaking.

(a) Check for leak. Repair and add refrigerant.
(b) Adjust water-regulating valve.
(c) Check operation of expansion valve. Also check fastenings and insulation of thermal bulb.
(d) Repair or replace.

High Suction Pressure

(a) Leaky suction valve.
(b) Overfeeding of expansion valve. Leaking or open valve.
(c) Excessive load on evaporator.

(a) Examine and repair or replace.
(b) Regulate expansion valve and check if bulb is attached properly to suction line.
(c) Check air infiltration and insulation.

COMPRESSORS FOR AIR CONDITIONERS

Symptom and Possible Cause   *Possible Remedy*

Low Suction Pressure

(a) Dirty suction screens or suction-stop valve partially closed.
(b) Restriction in liquid line clogged strainer or dehydrator, etc.) causing liquid to flash and restrict flow.
(c) Insufficient refrigerant in system.
(d) Too much oil circulating in system, reducing heat transfer in evaporator.
(e) Improper adjustment or expansion valve too small.
(f) Excessive pressure drop through evaporator, with clogged external equalizer.
(g) Low head pressure.

(h) Light load on evaporator.

(i) Evaporator coil frosted.

(a) Clean screens. Check suction valve and open wide.

(b) Clean strainer. Correct restriction.

(c) Repair leak and charge.

(d) Check for excess oil in system. Remove excess oil.

(e) Adjust valve.

(f) Be sure external equalizer blows clear and not plugged.

(g) Check and adjust water-regulating valve. Check refrigerating charge.
(h) Check capacity against load. Reduce capacity if necessary.
(i) Defrost coil.

SUMMARY

The mechanical components of a gas-compression refrigeration system include the compressor, condenser, receiver, evaporator, and various types of refrigerant controls.

Compressors come in various sizes, depending on the requirements, and are classified according to their method of operation. Methods of operation may be reciprocating, rotary, or centrifugal. The function of a compressor is to provide the necessary pumping action to remove the refrigerant gas from the evaporator, compress the gas, and finally discharge it into the condenser, where it is liquefied.

The condenser causes the heat picked up from the evaporator by the refrigerant to be discharged to the air outside the refrigerated area. Air-cooled condensers are most commonly used for home air conditioners.

Receivers are reservoirs for excess liquid refrigerant not being used in the system. The evaporator is that part of the system where the useful cooling is accomplished by removing the heat from the area to be cooled.

Refrigerant controls are devices that furnish the needed on-off operations for taking the refrigerant from one part of the system to the other at the proper time to produce the removal of heat from a refrigerated space.

The cycle of operation that is the cooling process consists of four separate steps. In the vaporizing process, the low-temperature, low-pressure refrigerant in the evaporator absorbs heat from the material or space being refrigerated and is transformed from a liquid to a vapor in the process. The compression process takes place when the compressor draws the heat-laden vapor from the evaporator through the return piping and compresses this vapor until the temperature is above that of the condensing medium.

The other two parts of the cooling process are the condensing process and the pressure-reducing process. In the compressor the temperature of the refrigerant has been raised above that of the condensing medium (usually 15 to 25°F); the heat of vaporization will flow from the vapor to the condensing medium and so condense the refrigerant to a high-pressure liquid. This high-pressure liquid then flows to the receiver, where it is stored until it is supplied to the cooling unit through the expansion valve.

As the compressor withdraws the refrigerant vapor from the evaporator, the cooling unit must be supplied with more low-temperature, low-pressure refrigerant capable of absorbing heat. A constant flow of liquid is maintained to the evaporator, where it vaporizes and takes the heat of the area back to the compressor again to repeat the cycle.

Compressors can be categorized as either single- or multicylinder with respect to operation. However, according to the

COMPRESSORS FOR AIR CONDITIONERS

method of compression used, they are classified as reciprocating, rotary, or centrifugal. With respect to drive, the compressors are classified as direct drive or belt drive. They may also be classified or grouped with respect to their prime mover. Under this classification they are either independent (belt drive), semihermetic (direct drive; motor and compressor in separate housings), or hermetic (direct drive; motor and compressor in the same housing).

REVIEW QUESTIONS

1. What are the components needed for the operation of a gas-compression refrigeration system?
2. How are compressors classified according to methods of operation?
3. What is the difference between a compressor and a condenser?
4. What is the function of an evaporator?
5. Where is the receiver located in the refrigeration system?
6. Why does the condenser usually need a fan?
7. What does *hermetic* mean?
8. Describe a shaft seal for a compressor.
9. How does the rotary compressor operate?
10. How is the centrifugal compressor different from the rotary compressor?
11. What may be the cause (or causes) of a knock in a compressor?
12. What is the usual reason for a compressor being stuckup?
13. How do you find compressor efficiency?
14. What would you do to repair an air conditioner if the fan on the evaporator doesn't run and the compressor short-cycles?
15. What are the possible causes for the running cycle of an air conditioner being too long?

CHAPTER 8

Condensing Equipment for Air Conditioners

The function of a condenser in a mechanical refrigeration system is to liquefy the refrigerant gas and return it, by means of controls, for a repeat of the refrigeration cycle. When the refrigerant gas leaves the compressor and enters the condenser, it contains the full load of heat, consisting of the heat picked up in the evaporator as net refrigeration, the heat of cooling the liquid from the condenser to the evaporating temperature, the heat picked up in the suction line and cylinder chamber, and the heat of compression.

CLASSIFICATION

Condensers may be classified according to their construction as:

1. Air-cooled
2. Water-cooled
3. Evaporative (combination of air and water)

AIR CONDITIONING

Air-Cooled Condensers—Air-cooled condensers are used universally on smaller refrigerating units but seldom for capacities in excess of 3 tons of refrigeration. They are equipped for natural draft and fan cooling and are termed according to their construction as *plain tubes, finned tubes, plate type,* and *series* or *parallel pass.* The conventional air-cooled condenser consists of an extended-surface coil across which air is blown by a fan. In operation, hot discharge gas enters the coil at the top; and, as it is condensed, it flows to a receiver located below the condenser. Typical air-cooled condensing units are shown in Fig. 8-1.

Air-cooled condensers should always be located in a well-ventilated space so that the heated air may escape and be replaced by cooled air. Because of space requirements, air-cooled condensers are normally constructed with a relatively small face with several rows of tubing in depth. As the air is forced through the condenser, it absorbs heat, and the air temperature rises. Therefore, the efficiency of each succeeding row in the coil decreases, although coils up to eight rows in depth are frequently used.

Water-Cooled Condensers—Water-cooled condensers are used on commercial and larger systems and are classified as *double-pipe, flooded-atmospheric, bleeder, shell-and-tube,* and *shell-and-coil.* In condensers of this type the amount and temperature of the water determine the condensing temperature and pressure, and indirectly the power required for compression. In locations where water must be conserved or where water is restricted, it is usually necessary to install cooling towers or evaporative condensers. When adequate low-cost condensing water is available, water-cooled condensers are often desirable because of the lower condensing pressures and better head-pressure control.

Water, particularly from underground sources, is frequently much colder than daytime temperatures. If evaporative cooling towers are used, the condensing water can be cooled to a point closely approaching the ambient wet-bulb temperature, which allows the continuous recirculation of condensing water and reduces water consumption to a minimum. A water-cooled condensing unit is shown in Fig. 8-2.

A pressure- or temperature-sensitive modulating water-control valve, such as shown in Fig. 8-3, can be used to maintain condens-

CONDENSING EQUIPMENT FOR AIR CONDITIONERS

Courtesy Copeland Refrigeration Corporation

Fig. 8-1A. Typical air-cooled condensing unit.

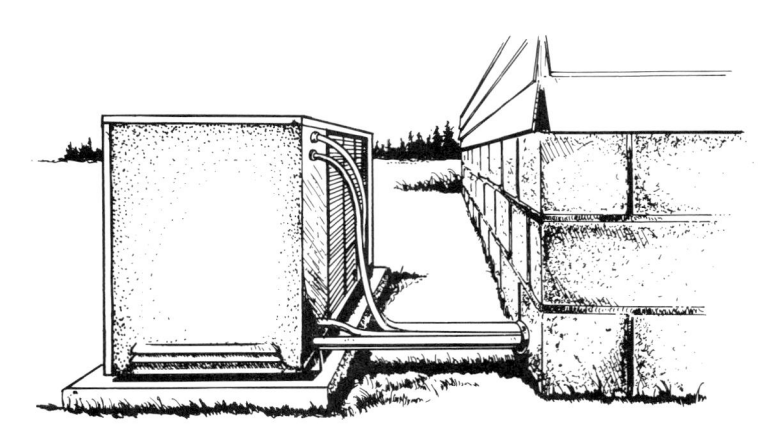

Unit on slab at grade level

Fig. 8-1B. Air-cooled condenser for home use.

245

AIR CONDITIONING

ing pressures within the desired range by increasing or decreasing the rate of water flow as necessary.

Courtesy Copeland Refrigeration Corporation
Fig 8-2. Typical water-cooled condensing unit.

COOLING-WATER CIRCUITS

In compressors with water jackets and in water-cooled condensers, cooling water circuits may be either in series or parallel. The use of parallel circuits results in a lower pressure drop through the circuit and may be necessary when the temperature of the cooling water is such that the water temperature rise must be held to a minimum. Figures 8-4 and 8-5 show how to install water lines for series or parallel flow.

Because of occasional damage to condensers by excessive water velocities, it has been found that the water velocity should not exceed 7 ft/sec. Thus, in order to maintain water velocities at acceptable levels, parallel circuiting of the condenser may be necessary. When a water circulating pump is used, it should be installed so that the condenser is fed from the discharge side of the pump. If the pump is installed on the discharge side of the con-

Fig. 8-3. Temperature-actuated water-regulating valve as employed on water-cooled condensers.

Courtesy Penn Controls, Inc.

denser, the condenser will have a slight vacuum in the water system and the water will be much nearer the boiling point.

DOUBLE-TUBE-TYPE CONDENSERS

There are three types of water-cooled condensers. They are the double-tube, the shell-and-coil and the shell-and-tube types.

The double-tube type consists of two tubes, one inside the other (Fig. 8-6). Water is piped through the inner tube, and refrigerant is piped through the tube that encloses the inner tube. The refrigerant flows in the opposite direction than the water (Fig. 8-7).

This type of coaxial water-cooled condenser is designed for use with refrigeration and air conditioning condensing units where space is limited. These condensers can be mounted vertically, horizontally or at any angle.

These condensers can also be used with cooling towers. They perform at peak heat of rejection with water pressure drop of not

Air Conditioning

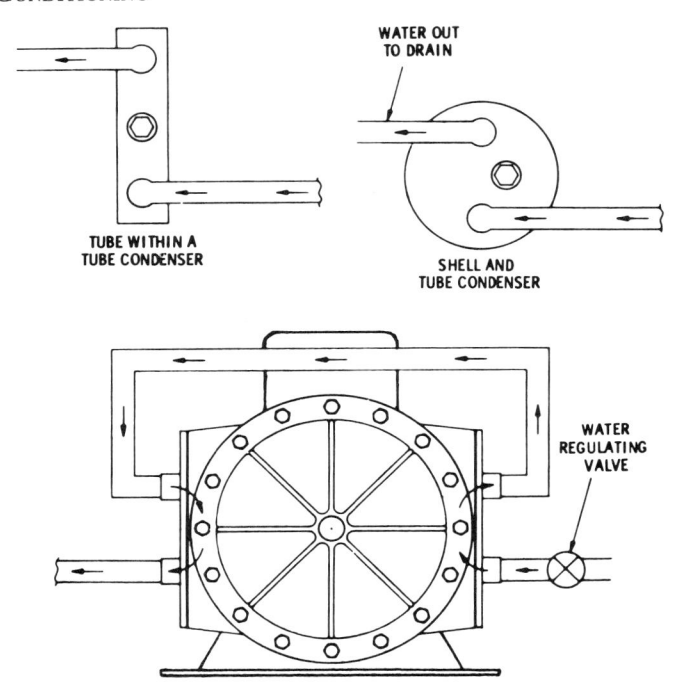

Fig. 8-4. Series-connected condenser water circuit.

more than five pounds per square inch, utilizing flow rates of three gallons per minute per ton.

The typical counter-flow path shows the refrigerant going in at 105°F and the water going in at 85°F and leaving at 95°F. (Fig. 8-8.)

The counter-swirl design shown in Fig. 8-7 gives heat transfer performance of superior quality. The tube construction provides excellent mechanical stability. The water flow path is turbulent. This provides a scrubbing action that maintains cleaner surfaces. The construction method shown also has very high system pressure resistance.

The water-cooled condenser shown in Fig. 8-6 can be obtained in a number of combinations. Some of these combinations are listed in Table 8-1. Copper tubing is suggested for use with fresh

CONDENSING EQUIPMENT FOR AIR CONDITIONERS

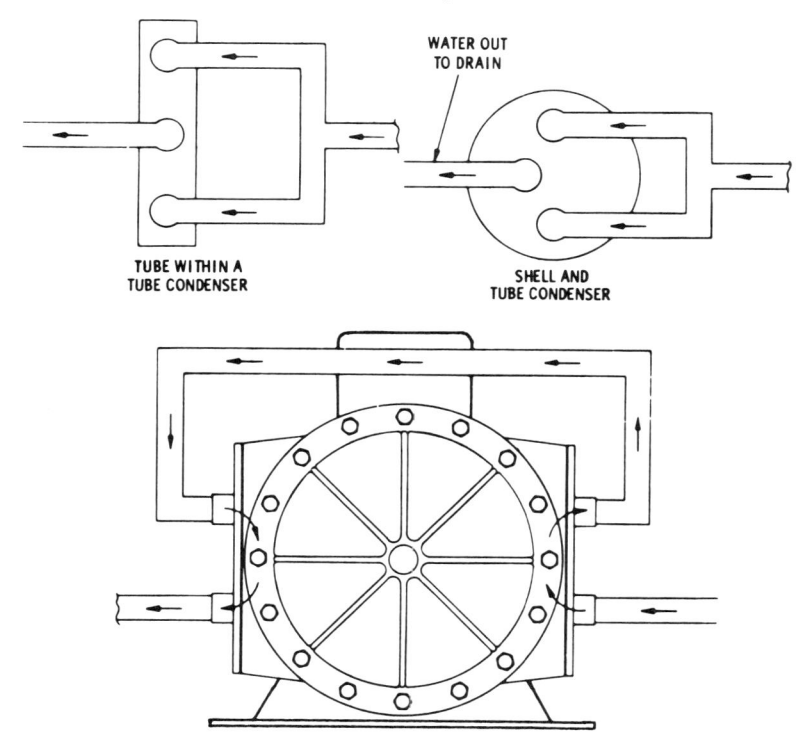

Fig. 8-5. Parallel-connected condenser water circuit.

water and with cooling towers. The use of cupronickel is suggested when salt water is used for cooling purposes.

Convolutions in the water tube result in a spinning, swirling water flow that inhibits the accumulation of deposits on the inside of the tube. This contributes to the antifouling characteristics in this type of condenser. Figure 8-9 shows the various types of construction for the condenser. This type of condenser may be added as a booster to standard air-cooled units.

Figure 8-10 shows some of the configurations of this type of condenser: the spiral, the helix and the trombone. Note the input for the water and input for the refrigerant. The condensers can be further cooled by using a cooling tower to furnish water to contact

AIR CONDITIONING

Fig. 8-6. Coaxial water-cooled condenser. It is used with refrigeration and air-conditioning units where space is limited.

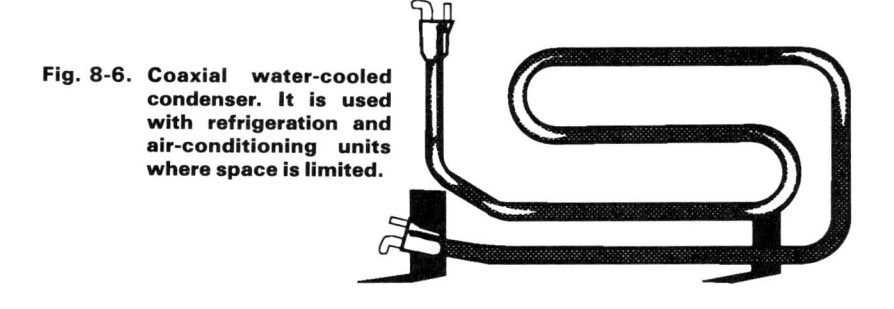

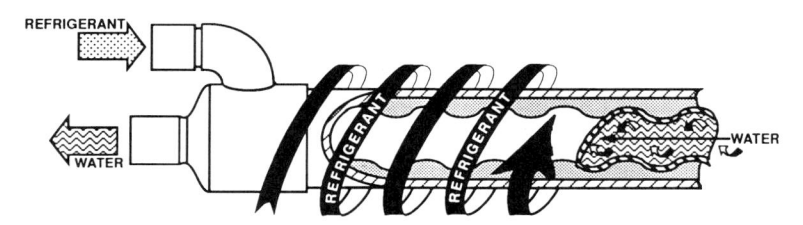

Fig. 8-7. Typical counter-flow path inside a coaxial water-cooled condenser.

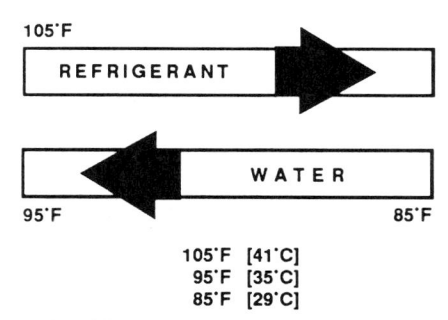

105°F [41°C]
95°F [35°C]
85°F [29°C]

Fig. 8-8. Water and refrigerant temperatures is a counter-flow water-cooled condenser.

the outside tube. Also, a water tower can be used to cool the water sent through the inside tube for cooling purposes. This type of condenser is usable where refrigeration or air conditioning requirements are $1/3$ ton up to 3 tons.

250

Table 8-1. Some Possible Metal Combinations in Water-Cooled Condensers

Shell Metal	Tubing Metal
Steel	Copper
Copper	Copper
Steel	Cupronickel
Copper	Cupronickel
Steel	Stainless Steel
Stainless Steel	Stainless Steel

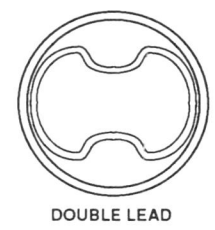

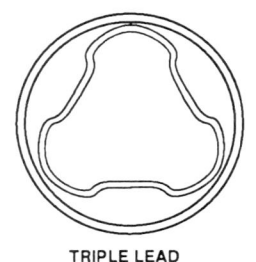

SINGLE LEAD

DOUBLE LEAD

TRIPLE LEAD

Fig. 8-9. Different types of tubing fabrication inside the coaxial water-cooled condenser.

SHELL-AND-COIL CONDENSERS

The shell-and-coil condenser is made by placing a bare tube or a finned tube inside a steel shell (Fig. 8-11). Water circulates through the coils. Refrigerant vapor is injected into the shell. The hot vapor contacts the cooler tubes and condenses. The condensed vapor drains from the coils and drops to the bottom of the tank or shell. From there it is recirculated through the refrigerated areas by way of the evaporator. In most cases, the unit is cleaned by placing chemicals into the water. The chemicals have a tendency to remove deposits that build up on the tubing walls.

EVAPORATIVE CONDENSERS

Evaporative condensers are essentially a combination of a water tower and a water-cooled condenser in which the condenser is

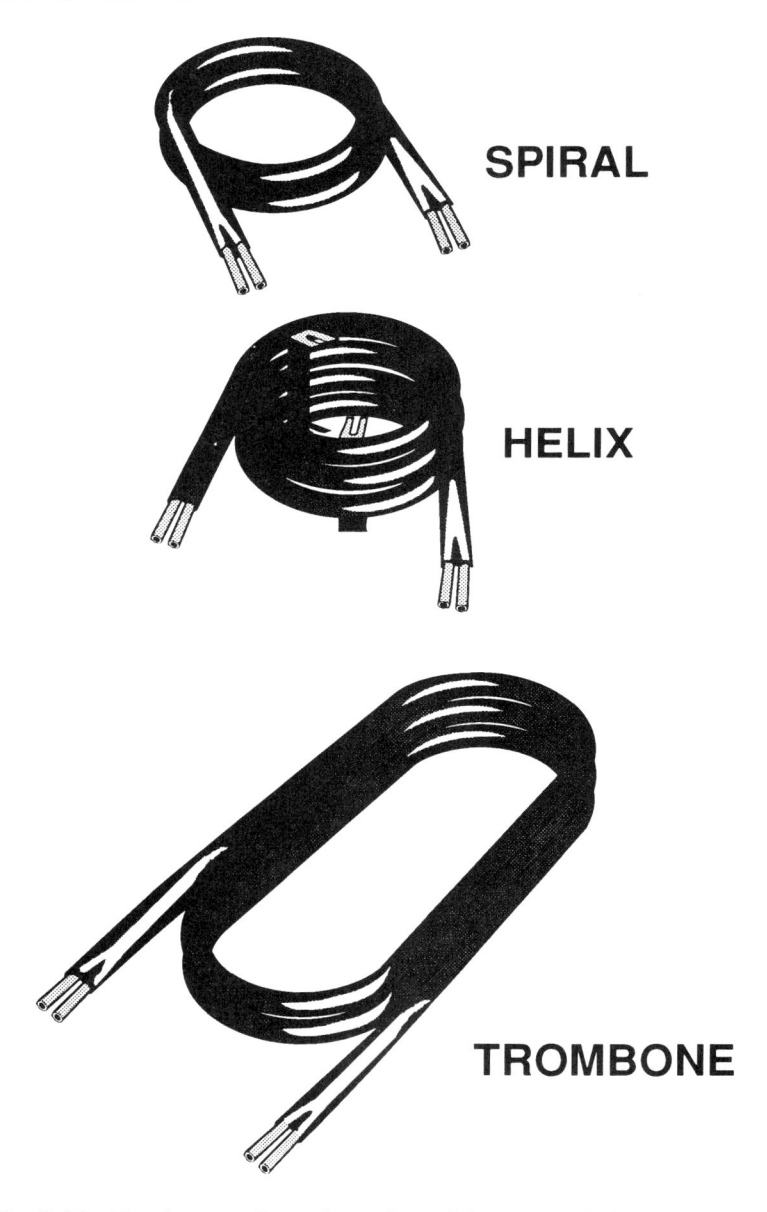

Fig. 8-10. The three configurations of coaxial water-cooled condensers.

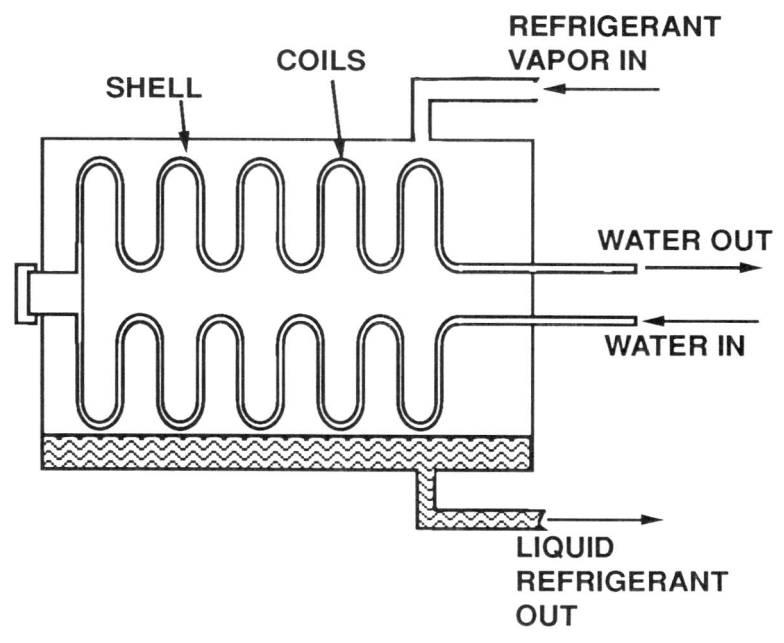

Fig. 8-11. The shell-and-coil condenser. This is a series-type of coil arrangement inside the shell, which the refrigerant enters as a vapor and leaves as a liquid.

placed in intimate contact with the air and water spray. In condensers of this type, the heat in the refrigerant gas is carried away by the air, the circulating water acting merely as an agent in the heat transfer. The effectiveness of any evaporative condenser, however, is dependent on the wet-bulb temperature of the air entering the unit since this determines the minimum temperature to which the spray water can be cooled. Condensers of the evaporative types are manufactured in sizes up to 100 tons or more.

In the operation of a typical evaporative condenser (Fig. 8-12), the circulating pump draws water from the water pan and discharges it above the condensing coils through a series of spray nozzles. The spray nozzles are arranged to distribute the water uniformly over the condensing coil. A large excess of water is circulated by the pump to keep the coil surface clean and maintain the rate of heat transfer. Makeup water is supplied to provide an overflow for the removal of dirt, prevent excessive deposition of

AIR CONDITIONING

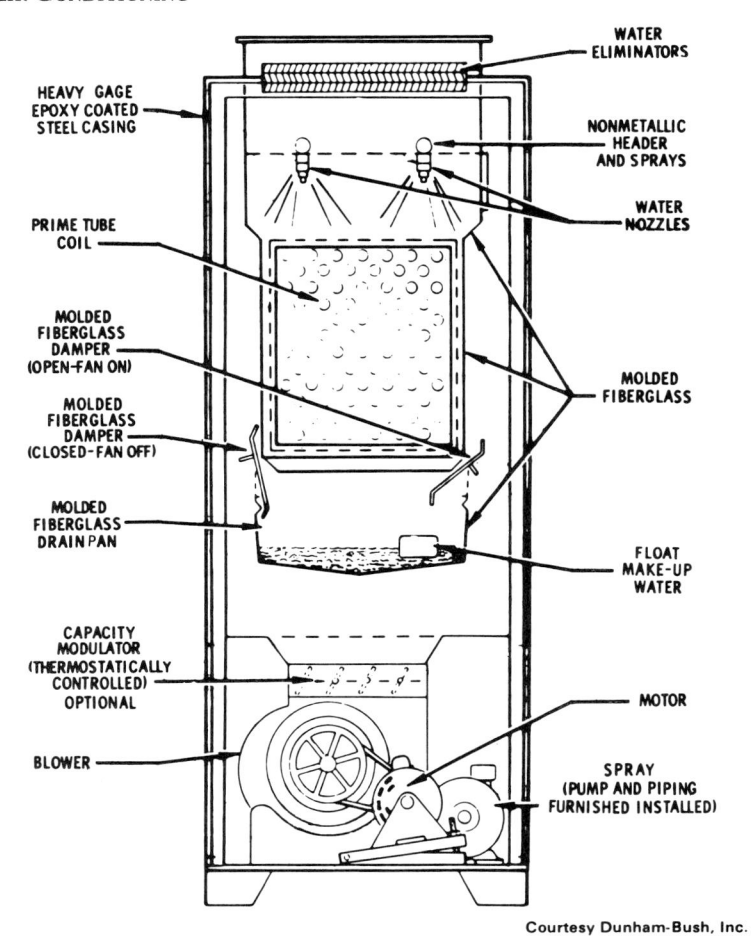

Fig. 8-12. Schematic view of an evaporative condenser.

lime on the condensing coil surfaces, and reduce the acidity of the recirculated water.

Centrifugal fans draw rated quantities of air through the condensing coils and water spray where the relative humidity greatly increases. It is then passed through water-eliminator plates and discharged to the outside in such a manner as to ensure that it will not be recirculated. The hot refrigerant gas enters the condensing coils through a manifold, which ensures a low pressure drop. The gas is condensed to a liquid in the coils, which are arranged to give a counterflow passage of refrigerant and air. The liquid flows from the condenser to a receiver (or receivers) located in the water reservoir, where it is held in storage for use in the evaporator section of the system.

HEAD-PRESSURE CONTROL METHODS

In application of refrigeration condensers, it is well to take advantage of lower-than-design condensing temperatures whenever possible to achieve the effects of more refrigeration capacity from the system, lower brake-horsepower requirements per ton of refrigeration, and lower compressor discharge temperatures. At the same time, it is often necessary to take precautions so that the condensing pressure does not drop down too low during periods of low water temperatures or low outside air temperatures. This is particularly true on installations that have high internal or process loads throughout the year and on systems having compressor capacity control.

The principal problem arising from abnormally low condensing pressures is the reduction of the pressure difference available across the expansion valve to the point where it is unable to feed sufficient refrigerant to the evaporator to handle the cooling load. Liquid feed devices are rated in accordance with the pressure difference across the port. A valve selected at a pressure difference of 60 lb, which is the standard rating point for Freon-12 valves, would have substantially less capacity if the pressure difference available were only 20 lb. Common types of head-pressure control arrangements are used, which are considered in relation to the various types of condensers used.

Water-Cooled Condensers

With water-cooled condensers, head-pressure controls are used for the dual purpose of holding the condensing pressure up and conserving water. On cooling-tower applications, they are used only where it is necessary to hold condensing temperatures up.

The control commonly used with water-cooled condensers is the condenser water-regulating valve. This valve is usually located in the water supply line to the condenser. It is a throttling-type valve and responds to a difference between condensing pressure and a predetermined shutoff pressure. A typical water-cooled condenser is shown in Fig. 8-13. The higher this difference, the more the valve opens up, and vice versa. A pilot connection to the top of the condenser transmits condensing pressure to the valve, which assumes an opening position according to this pressure.

The shutoff pressure of the valve must be set slightly higher than the highest refrigerant pressure expected in the condenser when the system is not in operation. This is to make sure the valve will close and not pass water during off cycles. This pressure will be slightly higher than the saturation pressure of the refrigerant at the highest ambient temperature expected around the condenser.

Condenser water-regulating valves are usually sized to pass the design quantity of water at about a 25- to 30-lb difference between design condensing pressure and valve shutoff pressure. In cooling tower applications, a simple bypass can also be used to maintain condensing temperature by employing a manual valve or an automatic valve responsive to head-pressure change; the valve can be either the two- or three-way type.

Evaporative Condensers

When installed outdoors, this type of unit should be located so that there is a free flow of air. A minimum clearance of twice the width of the unit should be provided at the air inlet and outlet. Care should be taken that the flow of the discharge air is not deflected back to the inlet. Short-circuiting in this fashion can mate-

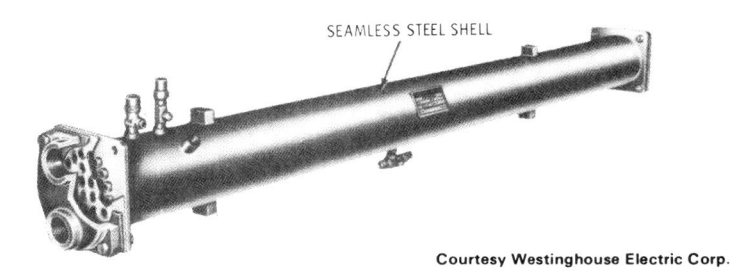

Courtesy Westinghouse Electric Corp.

Fig. 8-13. Exterior view of a shell and tube water-cooled condenser.

rially reduce the cooling capacity. Special consideration should be given to short-circuiting on multiple installations.

It is very desirable to position the unit so that the prevailing summer wind will blow into the inlet. It is also desirable to take precautions against a strong wind blowing into the outlet, which can reduce the capacity and might overload the fan motor. A windbreak erected about 6 feet from the air discharge is often useful in preventing this. In some outdoor locations, it may be desirable to place a coarse screen over the air inlet to exclude large objects, such as birds, paper, and other debris. Although towers and evaporative condensers are exceptionally quiet in operation, it is still good practice to locate the units as far as possible from other air intake hoods where air and water sounds might be transmitted into offices or living spaces.

If the unit is installed indoors, careful consideration must be given to supplying and removing an adequate amount of air from the space. If the discharge air is allowed to recirculate, the wet-bulb temperature will rise and the capacity will be reduced. It is usually possible to omit a duct to the air inlet. The unit can draw air from the room, but provision should be made to supply an adequate amount of replacement fresh air. Blower-type units may be connected to ducts giving an external static pressure of up to $1/4$ inch of water.

The piping to and from the unit should be installed in accordance with good practice, using pipe sizes at least as large as the fittings on the unit. It is desirable to use vibration eliminators on all lines to and from units. The units should be installed so that the top cover can be lifted for maintenance. The makeup water line should be connected to the float valve. If it is desired to use the makeup line to fill the system, it may be desirable to make the pipe size larger than the float-valve-connection size. In some locations, it is permissible to let the overflow drain on the roof; but if this is not desired, the overflow should be piped to the sewer. A bleed or blowdown connection should be incorporated in the piping. This should be installed in the pump discharge pipe at a high enough level so that the bleed flows only when the pump is operating.

It is convenient to have a hose bib available near the unit for use in periodic washing down and cleaning of the case and sump. If

AIR CONDITIONING

the unit is installed outdoors and must operate in winter, the installation should be made using an indoor storage tank in a heated space. Figure 8-14 shows a recommended piping arrangement for evaporative condensers. The pump should be moved indoors and connected to the outlet of the indoor storage tank.

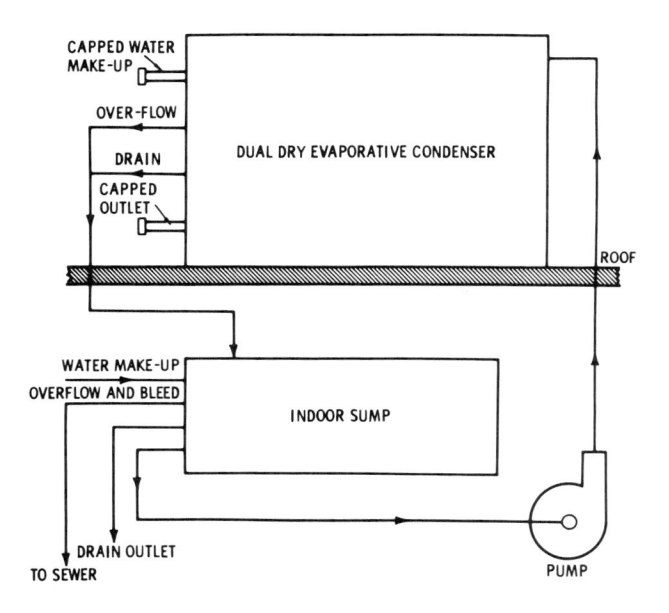

Fig. 8-14. Schematic diagram of piping arrangement for an evaporative condenser.

Air-Cooled Condensers

With air-cooled condensers, various methods employed for condensing-pressure control are cycling fan motor, air-throttling damper, and coil flooding. The first two methods are described under Evaporative Condensers. The third method holds the condensing temperature up by reducing condenser capacity. This is achieved by backing liquid refrigerant up in the coil to cut down on the effective condensing surface.

The basic principle of the pressure stabilizer (Fig. 8-15) consists of a heat-transfer device, which transfers the heat from the hot-gas discharge of the compressor to the subcooled liquid leaving the

CONDENSING EQUIPMENT FOR AIR CONDITIONERS

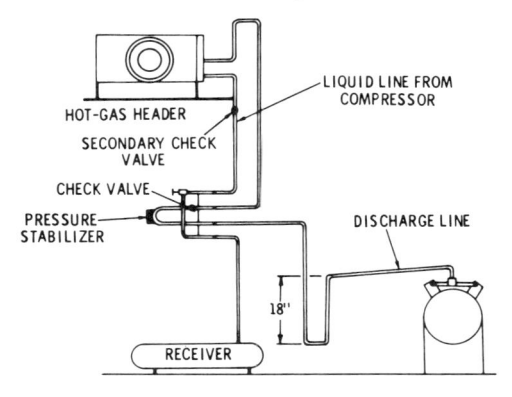

Fig. 8-15. Piping arrangement showing pressure-stabilizing method for an air-cooled condenser.

condenser. This heat exchange is controlled by the regulating valve installed between the condenser and the receiver. This valve is set at the desired operating pressure and throttles from the open position to the closed position as the head pressure drops. The throttling action of the valve forces liquid to flow through the heat-exchanger section of the pressure stabilizer. The heat picked up by the liquid raises the receiver pressure and prevents further flow of liquid from the condenser. This causes flooding of the condenser and, by reducing the effective surface, maintains satisfactory discharge and receiver pressures. This ensures a solid column of liquid at the expansion valve at a pressure that guarantees satisfactory operation. The pressure stabilizer is provided with a spring-loaded check valve in the heat-transfer liquid section which remains closed during warm weather operation to prevent the liquid refrigerant from reheating.

CONDENSER MAINTENANCE

The method of cleaning condenser water tubes will depend on the water characteristics in the locality in which the condensing unit is installed and on the construction of the condenser used. Water has many different impurities, which normally form a hard scale on the walls of the water tubes; this scale is detrimental to condenser performance.

With condensers having removable heads, the scale on the walls of the tubes can be removed by attaching a round steel brush to a rod and working it in and out of the tubes. After the tubes have been cleaned with the brush, flush by running water through them. Some scale deposits are harder to remove than others, and a steel brush may not do the job. There are several different types of tube cleaners on the market for removing hard scale; they may be purchased locally.

When installing the condensing unit, keep in mind that the condensers may need to be cleaned. Allow enough room at the removable head to get a rod long enough to work in and out of the tubes. After cleaning, always use a new condenser head gasket. The simplest method of removing scale and dirt from condenser tubes that are not accessible for mechanical cleaning is with an inhibited acid, which cleans the coils or tubes by chemical action. When the scale deposit is not too great, gravity flow of the acid will provide sufficient cleaning. However, when the deposit is great enough to almost clog the tubes, thus inhibiting gravity flow, forced circulation must be used. The equipment and connections for circulating the inhibited acid through the condenser using the gravity-flow method are shown in Fig. 8-16. The equipment consists of a crock or wooden bucket for the drain, and a 1-inch steel pipe of sufficient length to make the piping connection. The vent pipe should be installed at the higher connection of the condenser.

The equipment and connections used when forced circulation may be necessary are shown in Fig. 8-17. When this method is used, a suitable pump will be required to provide pressure. The inhibited solution is stored in a metal tank (not galvanized) with ordinary bronze or copper fly screening to prevent the large pieces of scale or dirt from getting into the pump intake line. In addition, 1-inch steel piping with connections and globe valves are provided, as shown. The vent pipe should be installed at the higher connection of the condenser.

Cleaning Precautions

When handling the inhibited acid for cleaning condensers, certain precautions will be necessary. The solution will stain the

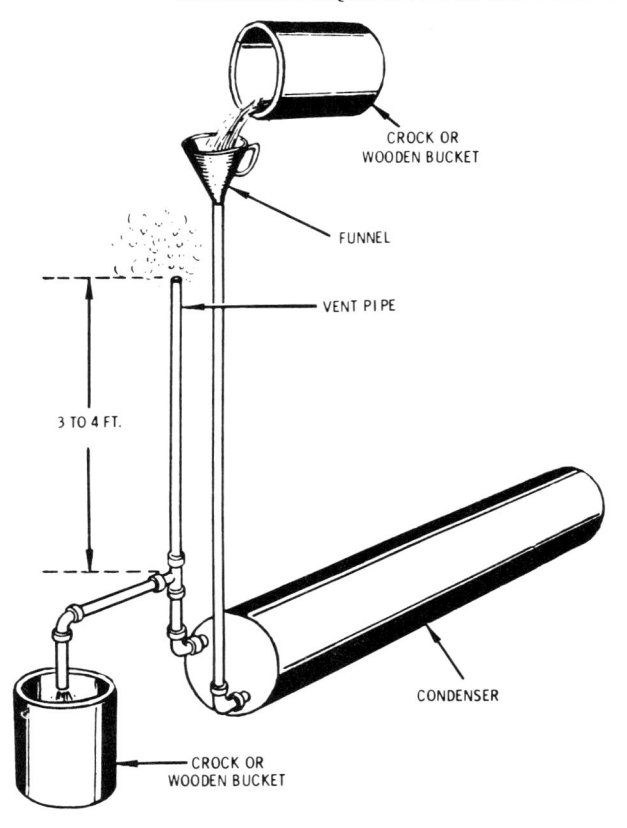

Fig. 8-16. Condenser cleaning method using gravity circulation.

hands and clothing and also attack concrete. Hence, every precaution should be used to prevent spilling or splashing. Where splashing might occur, covering of the surface with burlap or wooden boards is recommended. Gas produced during the cleaning will escape through the vent pipe and is not harmful, but care should be taken to prevent any liquid or spray from being carried through with the gas.

The inhibited acid solution is made up in the following proportions:

AIR CONDITIONING

1. Water: 78 parts by volume
2. Commercial hydrochloric acid (muriatic acid sp. gr. 1.19): 22 parts by volume
3. Inhibitor powder (Grasselli No. 3): 1.7 lb/100 gal of solution, or 0.27 oz/gal of solution

The inhibitor powder (Grasselli No. 3) can be purchased from the Grasselli Chemical Division of E.I. Du Pont de Nemours Company. The properties given in the basic formula should be maintained as closely as possible, but a variation of 5 percent is permissible. To mix the solution, place the calculated amount of water in a tank (not galvanized) or wooden barrel and add the calculated amount of inhibitor powder while stirring the water. Continue stirring until the powder is completely dissolved; then add acid as required.

Charging the System

When gravity flow is used, introduce the inhibited acid as shown in Fig. 8-16. Do not add the solution faster than the vent can exhaust the gases that are generated during the cleaning.

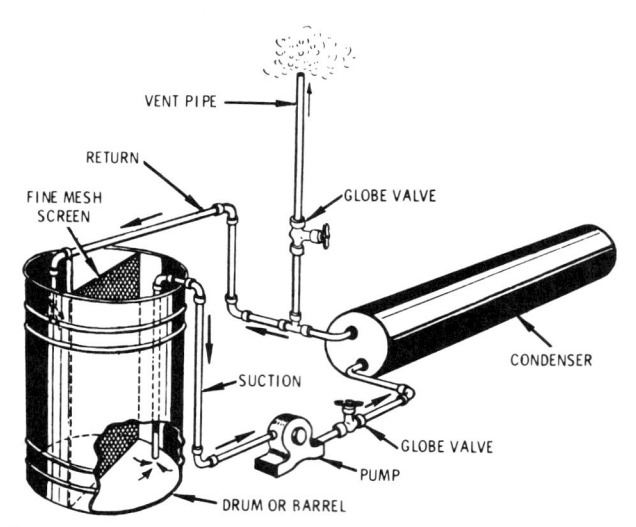

Fig. 8-17. Condenser cleaning method using forced circulation.

When the condenser has been filled, allow the solution to remain there overnight.

When forced circulation is used, the valve in the vent pipe should be fully opened while the solution is being introduced into the condenser but must be closed when the condenser is completely charged and the solution is being circulated by the pump. The valve in the supply line may be fully closed while the pump (if a centrifugal type) is running should the occasion arise to require it.

Cleaning Time

The solution should be allowed to stand or be circulated in the system overnight for cleaning out the average scale deposit. For extremely heavy deposits, forced circulation is recommended, and the time should be increased to 24 hours. The solution acts more rapidly if it is warm, but the cleaning action will be just as thorough with a cold solution if adequate time is allowed.

Flushing the System

After the solution has been allowed to stand or has been circulated through the condenser the required length of time, it should be drained out and the condenser thoroughly flushed with water. If it is desired to clean condensers with removable heads with inhibited acid, the previous procedure can be used without removing the heads; however, extra precaution must be exercised in flushing out such condensers with clear water after the acid has been circulated through the condenser.

Cleaning Evaporative Condensers

The method used in protecting the coils in evaporative condensers from scale and other deposits and removing the deposits after they have formed depends on the characteristics of the water in the particular location. If the water at the installation site is very hard, solids will form between the coil fins. The first step to take for the prevention of this scale formation is to provide ample overflow; this minimizes the concentration of solids in the water

AIR CONDITIONING

and also prevents the formation of slime. If the water composition is such that scale will form readily, the water should be chemically treated. For this treatment, a sample of the water should be sent to a competent water-treatment organization who will make a complete analysis and will recommend the proper chemical solution to use.

Removing Scale Formation

The simplest method of removing scale from condenser coils is with an inhibited acid, which cleans the fins and tubes by chemical action. The method of injecting the chemical into the water and the amount to be used will vary. It is suggested that the recommendations of the firm supplying the chemical be followed. Figure 8-18 shows connections and equipment necessary for cleaning the coils with an inhibited acid: one when the acid is used by the hand or gravity method, the other when forced circulation is used.

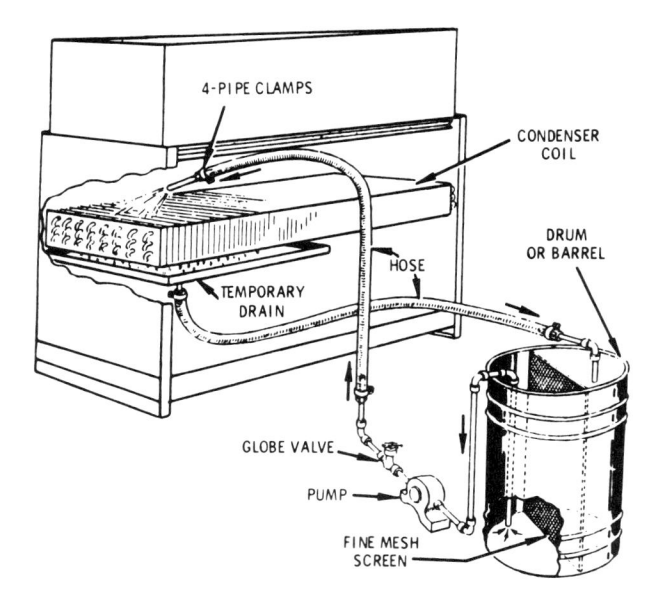

Fig. 8-18. Cleaning evaporative condenser using forced circulation.

CONDENSING EQUIPMENT FOR AIR CONDITIONERS

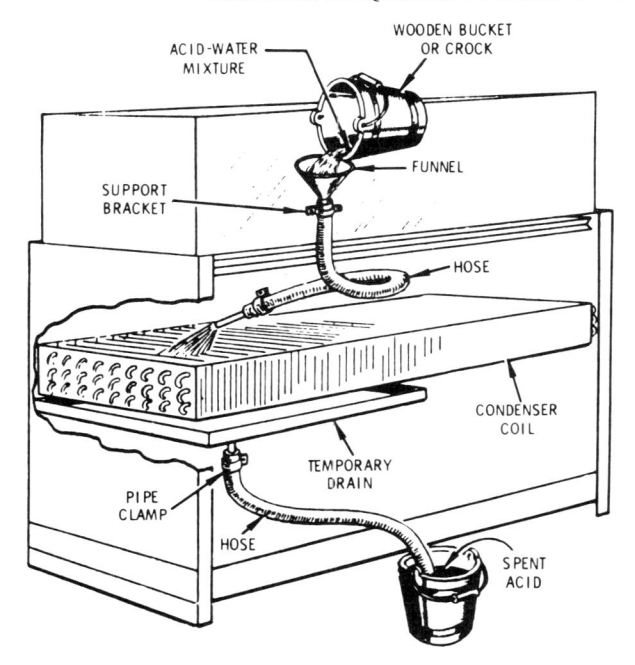

Fig. 8-19. Cleaning evaporative condenser using gravity circulation.

Gravity and Forced Circulation

When the scale deposit is not very great, gravity or hand cleaning with the acid will be sufficient; but when the deposit is heavy enough to block off the space between the fins, then forced circulation should be used. The gravity-flow and forced-circulation methods of removing scale deposits shown in Figs. 8-18 and 8-19 do not differ greatly from that given previously with respect to water-cooled condensers. The basic formula in making up the inhibited-acid solution, the mixing method, and the precautions in handling are identical.

SUMMARY

The function of a condenser in a mechanical refrigeration system is to liquefy the refrigerant gas and return it, by means of

AIR CONDITIONING

controls, for a repeat of the refrigeration cycle. Condensers are classified according to their construction as air-cooled, water-cooled, or evaporative (a combination of air and water).

Evaporative condensers are essentially a combination of water tower and a water-cooled condenser in which the condenser is placed in intimate contact with the air and water spray. Condensers of the evaporative types are manufactured in sizes up to 100 tons or more.

Water-cooled condensers use head-pressure controls for the dual purpose of holding the condensing pressure up and conserving water. On cooling-tower applications, they are used only where it is necessary to hold condensing temperatures up.

Condenser maintenance includes cleaning the condenser water tubes. The method of cleaning used depends upon the locality in which the condensing unit is installed and on the construction of the condenser used. Water has many impurities that can cause scale and other buildup on pipes or tubes. Various chemicals are used to keep the tubes clean of scale and other impurities. Special instructions are given by the manufacturers of chemicals for their safe use.

REVIEW QUESTIONS

1. What is the function of a condenser in a mechanical refrigeration system?
2. How are condensers classified according to their construction?
3. What is the advantage of having a parallel cooling circuit for a condenser?
4. What is the top or largest size of evaporative condensers?
5. What is the principal problem arising from abnormally low condensing pressures?
6. What is the purpose of head-pressure controls on water-cooled condensers?
7. What is the basic principle of the pressure stabilizer?
8. What determines the method used to clean condenser water tubes?

Condensing Equipment for Air Conditioners

9. What chemicals are used for cleaning condenser water tubes?
10. How long does it take to clean condenser water tubes?
11. What determines the method used to protect the coils in evaporative condensers from scale and other deposits?
12. What is the simplest method of removing scale from condenser coils?

CHAPTER 9

Evaporators for Air Conditioners

The evaporator is that part of the system where useful refrigeration is accomplished. This is done by bringing the temperature of the liquid refrigerant below the temperature of the surrounding air. The heat passes into the liquid refrigerant and is absorbed as latent heat, changing the state of the refrigerant from a liquid to a vapor. The vapor is then withdrawn by the action of the compressor.

Since all other parts of a refrigeration system are provided to prepare the refrigerant for this change of state, it follows that the evaporator is an important component of the refrigeration unit. The transfer of heat—from the space or medium surrounding the evaporator to its outside surface, through the metal to the inside surface of the cooling unit, and then to the liquid refrigerant—is made possible because heat always flows from a warmer to a colder object. From the standpoint of temperature, the liquid must be at a lower temperature than the source or medium surrounding it.

AIR CONDITIONING

The cooling medium depends on the refrigeration system used and on the direct expansion system, which consists of the secondary refrigerants such as chilled water or brine. When chilled water is used for space cooling, it is customary to install expansion coils in a spray chamber so that the water sprayed into the air comes into direct contact with the cooling coils.

In the indirect system of refrigeration, brine is cooled by the refrigerant, and the resulting cold brine is used to cool either air or water. This system, however, is rather uneconomical since it is necessary to cool the brine to a temperature between the average brine temperature and that of the substance to be cooled. In any refrigerant evaporator, brine tank, or whatever refrigeration is furnished, heat transfer takes place because of the existence of this difference in temperature.

In modern refrigerating equipment, evaporators may be divided into the following classes, depending on their construction: finned and plain tubing.

Finned Evaporators

Finned evaporators are especially useful for refrigerators, coolers, cases, and counters designed for the production of temperatures above 34°F with gravity circulation and 35°F with forced-air circulation. This covers the general commercial field. For general commercial applications, finned units operate more efficiently and effectively when they defrost automatically during that part of the refrigeration cycle when the condensing unit is not running. When properly applied, the melting of the frost takes place automatically except for an occasional defrosting by manual operation of the controls when unusual conditions are encountered. Finned units with wide-spaced fins are now also used on installations with manual or hot-gas defrosting below 34°F.

Finned cooling units for liquid cooling should be used only with nonfreezing solutions, such as calcium chloride brine and alcohol, where it is impossible for ice to form on the surfaces. This means that in brine-tank installations where the temperatures are below freezing, the chief application for these units will be covered. No attempt should be made to operate finned units in freezing solutions because the ice that forms causes such a reduction in capacity and efficiency that unsatisfactory refrigeration will result.

Plain-Tubing Evaporators

Plain-tubing evaporators, also known as *prime-surface evaporators*, are used successfully for practically any temperature, although they are especially recommended for holding fixtures at 34°F or lower. Prime-surface or bare-tubing evaporators or those with wide fin spacing may be operated in a frosted condition for a longer period of time without materially changing the temperature of the fixture being cooled. It is necessary, however, to defrost such units at regular intervals in order to maintain the required temperatures. The coating of frost will eventually become so thick that the rate of heat transfer is reduced and often air circulation is retarded. The defrosting is usually done manually because generally the units are not permitted to enter the ice-melting zone for a sufficient length of time in the normal refrigeration cycle for automatic defrosting to take place.

Plain-tubing evaporators are recommended for liquid cooling and for freezing solutions. Ice formation does not affect their capacity and operating efficiency as quickly and to such an extent as finned evaporators.

Evaporator-Coil Construction

When refrigeration evaporators are used for cooling air by forced convection, they are usually termed *blast coils* or *unit coolers*. The blast coil is made up of a direct expansion coil mounted in a metal housing, complete with a fan for forced-air circulation. The coil is normally constructed of copper tubing supported in metal-tube sheets, with aluminum fins on the tubing, as shown in Fig. 9-1, to increase heat-transfer efficiency.

If the evaporator is quite small, there may be only one continuous circuit in the coil; but as the size increases, the increasing pressure drop through the longer circuit makes it necessary to divide the evaporator into several individual circuits emptying into a common header. The various circuits are usually fed through a

AIR CONDITIONING

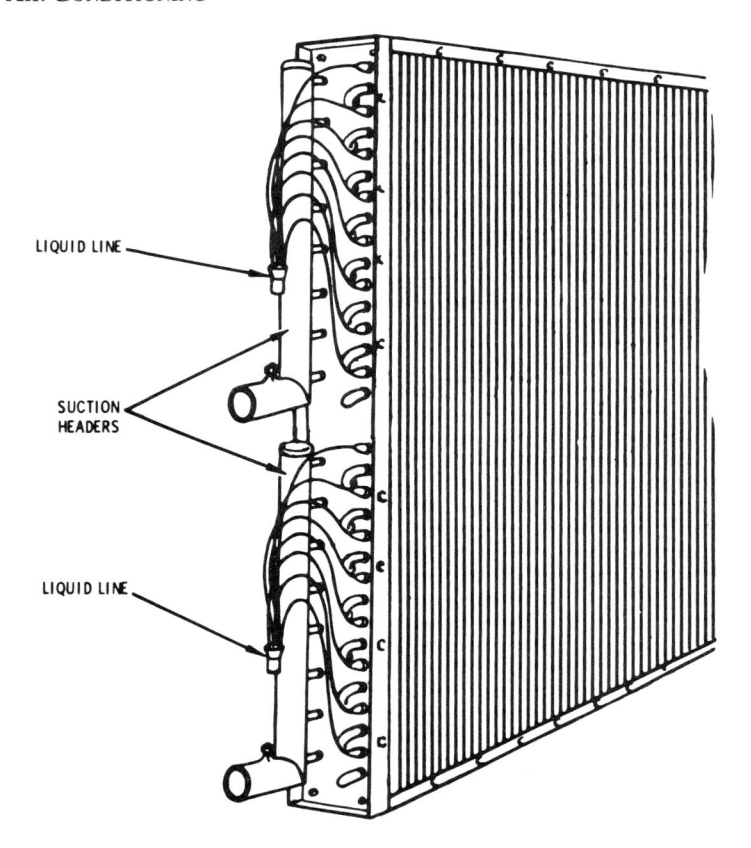

Fig. 9-1A. Direct-expansion evaporator showing coil circulation.

distributor that equalizes the feed in each circuit in order to maintain high evaporator efficiency.

The spacing of fins on the refrigerant tubing will vary depending on the application. Low-temperature coils may have as few as 2 fins per inch, while air-conditioning coils may have up to 12 or more per inch. In general, if the evaporator temperature is to be below 32°F so that frost will accumulate, fin spacings of 4 per inch or less are commonly used, although closer fin spacings are sometimes used if efficient defrost systems are available. In air-conditioning applications, icing of the coil is seldom a problem,

EVAPORATORS FOR AIR CONDITIONERS

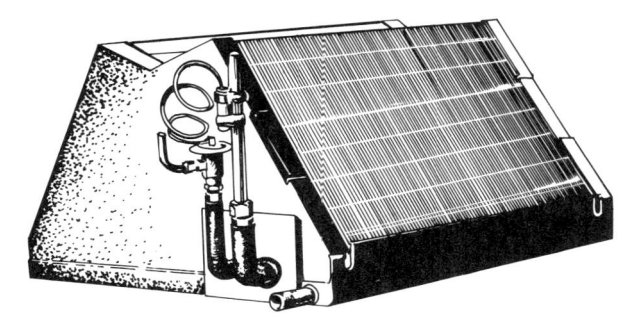

Fig. 9-1B. Evaporator used in home air-conditioning systems. The evaporator fits in the forced air furnace bonnet or plenum. The fan used for heat during the winter is used for circulating air through the evaporator during the summer.

and the limit on fin spacing may be dictated by the coil's resistance to air flow.

Since the heat-transfer efficiency of the coil increases with an increase in the mass flow of air passing through it, high velocities greater than 500 to 600 fpm should not be used. Water collecting on the coil from condensation will be blown off into the airstream if these velocities are exceeded.

REFRIGERANT-CONTROL METHODS

Three types of evaporators or chilling units are in general use: *flooded*, *wet expansion*, and *dry expansion*. The type of evaporator differs with respect to the volume of liquid refrigerant permitted in each and with the refrigerant-control methods.

Although the purpose of each of the evaporator types is the same, each varies with respect to its operation. Thus, the flooded type is operated with approximately 75 percent of its total capacity of liquid refrigerant. Control of the refrigerant level is generally maintained by the use of a high- or low-side float control.

A heat exchanger is generally required in *flooded evaporator* installations to ensure that the suction gas will be superheated. Liquid refrigerant will enter the suction line from the oil-return line. The heat exchanger is necessary to vaporize the liquid refrigerant so that it will not return to the compressor as a slug and make the

273

AIR CONDITIONING

valves inoperative. An oil separator is also desirable to ensure minimum oil circulation through the system. Flooded operation of evaporators using refrigerants such as Freon-12 and Freon-22 can result in problems in oil trapping.

The *wet-expansion* type of evaporator is usually liquid-filled to approximately 50 percent. In this type, the refrigerant mixture flows through the tubes with a rather slow velocity due to large tube diameters, and the possibility of having entrained liquid in the vapor entering the suction line is not pronounced. This permits more of the tail end of the evaporator to be used than would be possible if the vapor velocity were higher; at the same time it ensures positive oil return due to short vapor paths. The temperature difference between the inlet and outlet or superheat of vapor leaving the evaporator should range between 5°F and 7°F, with the latter value on low average-refrigerant-temperature application. The efficiency of an evaporator will be affected 10 to 20 percent by increasing the superheat a few degrees above design conditions, which is attempting, in effect, to refrigerate with vapor.

The *dry-expansion* type of evaporator is of the continuous tube construction. The percentage of evaporator volume filled with liquid is approximately 25 percent. In this type of evaporator, the velocity of the refrigerant mixture is fairly high, causing a turbulent or swirling condition, which throws the liquid refrigerant against the sides of the tube. Therefore, it is necessary to operate with a higher superheat to prevent carrying entrained liquid refrigerant into the suction line. This type of evaporator is usually operated with from 9 to 12°F superheat. The velocity of the vapor in the evaporator increases with the length. In general, the pressure drop through an evaporator is half what it would be if the same amount of pipe or tubing were used as a suction line carrying the same quantity of vapor.

A pipe-coil installation is normally operated at approximately 10°F superheat, depending on the refrigerant temperature, and usually requires a heat exchanger or drier oil due to the low temperature. This type of evaporator is used all the way from ultralow-temperature refrigeration applications up to comfort air-conditioning levels. Evaporators of the foregoing types employ an automatic or thermostatic expansion valve to control the liquid refrigerant. It should be noted, however, that different settings of the expansion valves are required for each type of evaporator.

Superheat

The superheat at any point in a refrigeration or air conditioning system is found by first measuring the actual refrigerant temperature at that point using an electronic thermometer. Then the boiling point temperature of the refrigerant is found by connecting a compound pressure gage to the system and reading the boiling temperature from the center of the pressure gage. The difference between the actual temperature and the boiling point temperature is superheat. If the superheat is zero, the refrigerant must be boiling inside and there is a good chance that some of the refrigerant is still liquid. If the superheat is greater than zero, at least 5°F or better, then the refrigerant is probably past the boiling point and is all vapor.

The method of measuring superheat described here has obvious faults. If there is no attachment for a pressure gage at the point in the system where you are measuring superheat, the hypothetical boiling temperature cannot be found. To determine the superheat at such a point, the following method can be used. This method is particularly useful for measuring the refrigerant superheat in the suction line.

Instead of using a pressure gage, the boiling point of the refrigerant in the evaporator can be determined by measuring the temperature in the line just after the expansion valve where the boiling is vigorous. This can be done by any electronic thermometer. As the refrigerant heats up through the evaporator and the suction line, the actual temperature of the refrigerant can be measured at any point along the suction. Comparison of these two temperatures gives a superheat measurement sufficient for field service unless a distributor metering device is used, or the evaporator is very large with a great amount of pressure drop across the evaporator.

It is possible to read superheat directly using a meter made for the job. This meter works on temperature differential, measuring the temperature at two points and then converting this to superheat by subtracting and presenting the difference.

AIR CONDITIONING

Figure 9-2 illustrates the way superheat works. The bulb "opening" force (F-1) is caused by bulb temperature. This force is balanced against the system back pressure (F-2), and the valve spring force (F-3) to hold the evaporator pressure within a range that will vaporize all of the refrigerant just before it reaches the upper part or end of the evaporator.

The method of checking superheat is shown in Figure 9-3. The procedure is as follows:

1. Measure the suction line temperature at the bulb location. In the example, the temperature is 37°F.
2. Measure the suction line pressure at the bulb location. In the example, the suction line pressure is 27 psi.
3. Convert the suction line pressure to the equivalent saturated (or liquid) evaporator temperature by using a standard temperature pressure chart [27 psi = 28°F.].

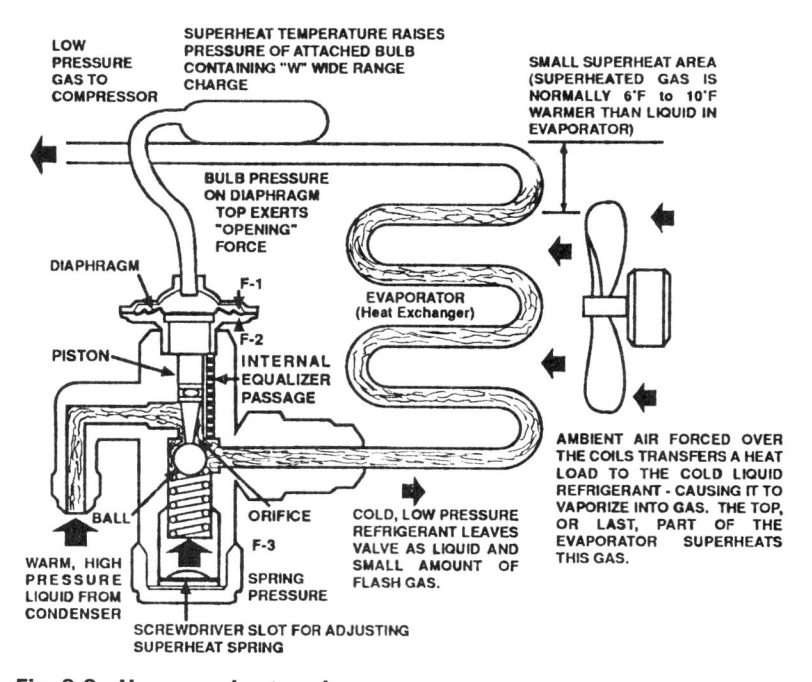

Fig. 9-2. How superheat works.

EVAPORATORS FOR AIR CONDITIONERS

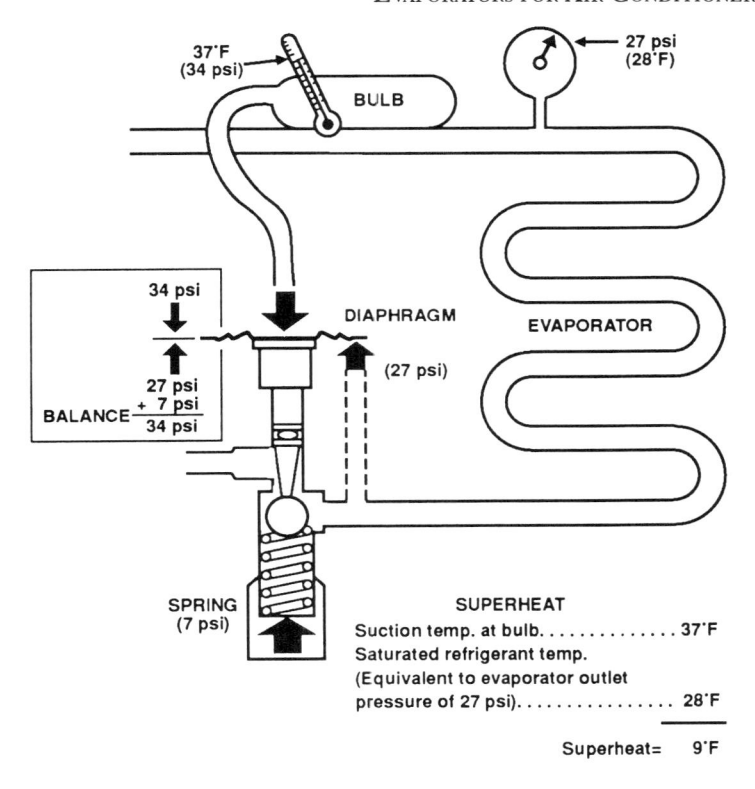

Fig. 9-3. Where and how to check superheat.

4. Subtract the two temperatures. The difference is *superheat.* In this case, superheat is found by the following method:

$$37°F - 28°F = 9°F.$$

Suction pressure at the bulb may be obtained by either of the following methods:

1. If the valve has an external equalizer line, the gage in this line may be read directly.
2. If the valve is internally equalized, take a pressure gage reading at the compressor base valve. Add to this the esti-

277

mated pressure drop between the gage and the bulb location. The sum will approximate the pressure at the bulb.

The system should be operating normally when the superheat is between 6 and 10°F.

Dry Expansion Coil Arrangements

Although there are many ways of arranging dry expansion coils, the following are the most common:

1. By top feed, free-draining piping at a vertical upflow coil, as shown in Fig. 9-4.
2. By a horizontal airflow coil (Fig. 9-5) in which the suction is taken off of the bottom header connection, thus providing free oil draining.
3. By a refrigerant bottom-feed coil with a vertical downflow air arrangement, as shown in Fig. 9-6.

Many coils are supplied with connections at each end of the suction header so that the free-draining connection can be used regardless of which side of the coil is up; the other end is then capped. The preceding arrangements have no oil-trapping problems. With a bottomfeed coil evaporator, such as shown in Fig.

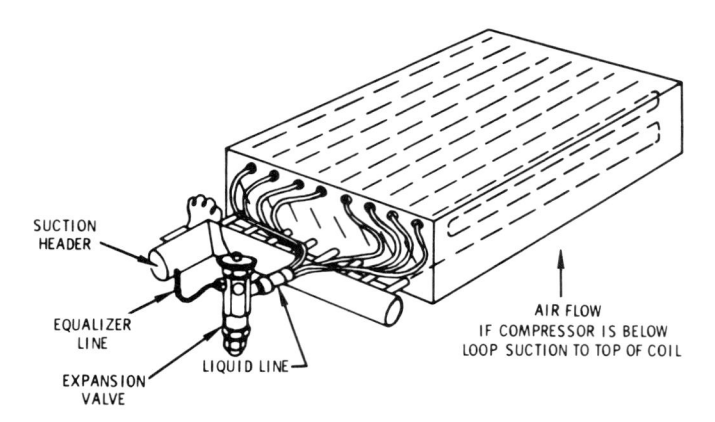

Fig. 9-4. Coils in a top-feed direct-expansion evaporator.

Evaporators for Air Conditioners

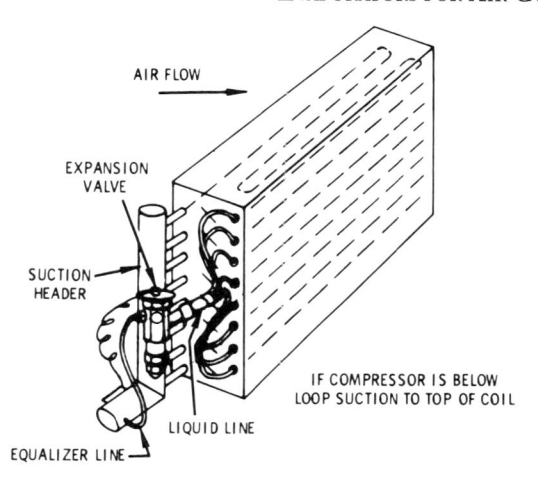

Fig. 9-5. Coil arrangement in a free-drawing direct-expansion evaporator.

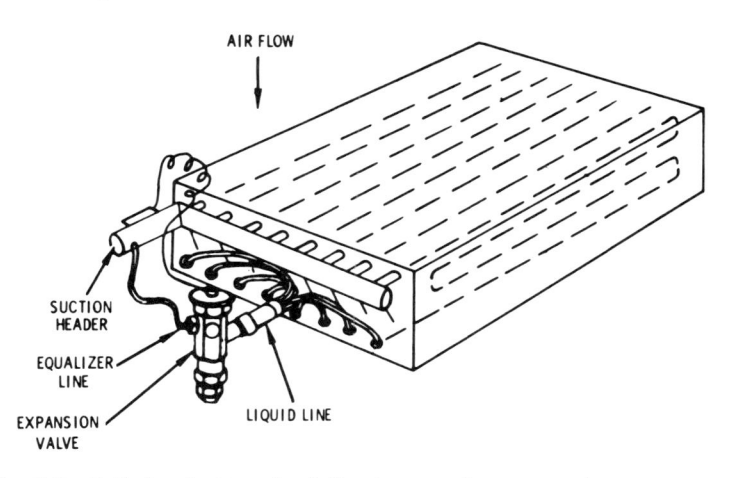

Fig. 9-6. Coils in a bottom-feed direct-expansion evaporator.

9-6, it is necessary that the coil design provide for sufficient gas velocity up through the coil to entrain oil at the lowest partial loadings and carry it into the suction line.

Pumpdown compressor control should be employed on all systems using downfeed evaporators. This is necessary to protect the compressor against a liquid slugback in the case of evaporators where liquid can accumulate in the suction header on the system off cycles.

Capacity Requirements

The factors affecting evaporator capacity are quite similar to those affecting condenser capacity. They are as follows:

1. Surface area or size of the evaporator.
2. Temperature difference between the evaporating refrigerant and medium being cooled.
3. Velocity of gas in the evaporator tubes. In the normal commercial range, the higher the velocity, the greater the heat-transfer rate.
4. The velocity and rate of flow over the evaporator surface of the medium being cooled.
5. Material used in evaporator construction.
6. The bond between the fins and tubing. Without a tight bond, heat transfer will be greatly decreased.
7. Accumulation of frost on evaporator fins. Operation at temperatures below freezing with blower coils will cause the formation of ice and frost on the tubes and fins, which can reduce both the airflow over the evaporator and the heat-transfer rate.
8. The type of medium to be cooled. Heat flows almost five times more effectively from a liquid to the evaporator than from air.
9. Dew point of the entering air. If the evaporator temperature is below the dew point of the entering air, latent as well as sensible cooling will occur.

EVAPORATOR CALCULATIONS

The coefficient of heat transfer (K factor) represents the amount of heat that will pass through a unit area in a unit of time with 1°F surfaces between which the heat is being transferred. This system must be determined initially by experiment. For some types of evaporators, usage and experience have firmly established this value. For other types, the K factor has only been established approximately, yet with sufficient accuracy for practical use.

Evaporators used for air cooling develop a relatively low K factor compared to what the same unit would develop if cooling a liquid such as water, assuming the same unit could be used for both purposes. The speed of movement (velocity) of the cooled medium over the cooling surface affects the K factor. When cooling air, the K factor will vary somewhat with the conditions of the cooling surface.

The amount of heat that can be absorbed by a evaporator, that is, its capacity, follows the fundamental heat-transfer equation:

$$H = KAT_d$$

where K = coefficient of heat transfer
A = total evaporator area, sq ft
T_d = temperature differences between refrigerant and medium being cooled

Therefore, if it is desired to handle a given quantity of heat in a stated period of time, simple mathematics show that a compensating variation must be made in one or both of the remaining factors to hold the capacity H constant.

The following general conditions affect the K factor of any evaporator:

1. Physical construction
2. Conditions on the exterior of the evaporator
3. Conditions on the interior of the evaporator
4. Relation of refrigerant flow to medium being cooled

Some of these factors are difficult to segregate clearly because they interlock with each other. The K factor of finned tubing and surfaces varies greatly and must be determined by tests for each type of construction. With a finned evaporator, the relation of the total surface area to surface area of tubing containing the liquid refrigerant is important. This is sometimes called the R factor and may be written as follows:

$$R = \frac{\text{tube area}}{\text{total area}}$$

In general, the greater the area of fins in proportion to the tubing area, the lower the overall K factor of the evaporator. The physical construction of fins has an effect on the K factor. Finned area extended too far from the tubing becomes ineffective and may be just so much wasted metal. Spacing of fins is important because it affects the K factor and because of its relation to frost formation on cooling units. Closely spaced fins may easily accumulate sufficient frost between them to completely stop air circulation through the unit and are then difficult to defrost. As a general rule, the area of an evaporator is the total surface exposed to the medium being cooled. Area is only of three factors determining the heat-handling capacity of an evaporator. Its importance is often overemphasized.

Exterior Conditions

Conditions on the exterior of the evaporator that affect the coefficient of heat transfer are:

1. Medium being cooled
2. Velocity of cooled medium over cooling unit surface
3. Condition of exterior surface
4. Temperature of medium being cooled

Surface-Area Determination

In determining surfaces, it should be noted that the plain-tubing type will haven a higher K than the finned construction. The finned-surface evaporator, however, will often permit the use of more heat-absorbing surface in the space allowed for the cooling unit and has been used more in recent years than the plain-tubing evaporator. The evaporator surface-area determination is illustrated in the following example.

Example — If it is assumed that a tube evaporator has a K of 2 and a finned-tube evaporator has a K value of 0.5, calculate the

Evaporators for Air Conditioners

required tube surface of each when it takes a T_d of 20°F to produce 10,000 Btu/hr.

Solution — Substitution of values in the heat-transfer formula for the plain-surface evaporator gives:

$$A = \frac{H}{K \times T_d}$$

$$= \frac{10,000}{2 \times 20}$$

$$= 250 \text{ sq ft of tube surface}$$

For the finned-surface evaporator we obtain similarly:

$$A = \frac{10,000}{0.5 \times 20}$$

$$= 1000 \text{ sq ft of tube surface}$$

Example — If 10,000 Btu/hr are to be absorbed in an evaporator having a K factor of 10 and an A of 100, what is the temperature difference between the refrigerant and the medium being cooled?

Solution — With reference to the heat-transfer formula, we obtain:

$$H = KAT_d$$

or

$$10,000 = 10 \times 100 \times T_d$$

$$T_d = \frac{10,000}{10 \times 100}$$

$$= 10°$$

If the K factor has been reduced for some reason to 2, the product of A and T_d would have to be raised to 5000 if H is to remain at 10,000:

$$H = KAT_d$$

AIR CONDITIONING

or
$$10{,}000 = 2 \times 100 \times T_d$$
$$A \times T_d = \frac{10{,}000}{2}$$
$$= 5000$$

So, if A remains at 100, T_d would have to be 50:

$$H = KAT_d$$

or
$$10{,}000 = 2 \times A \times T_d$$
$$T_d = \frac{10{,}000}{2 \times 100}$$
$$= 50°$$

If K and A are *fixed* quantities (for example, $K = 2$ and $A = 200$), and T_d is varied, H will be found to vary directly as T_d varies. For example, if $T_d = 20$, $H = 8000$; if $T_d = 10$, $H = 4000$.

The preceding example illustrates the change necessary in T_d should some of the factors affecting K be violated.

EVAPORATOR MAINTENANCE

For low-temperature applications with dry-coil-type units, defrosting can be obtained by providing arrangements for the intake of outside air and the exhaust of this air to the outside during the defrosting period. It should be noted, however, that the outside temperature must be above 35°F when using this defrosting method. At very low temperatures, hot-gas defrosting may be employed only when there is more than one evaporator connected to a condensing unit, so that the other unit can continue to provide refrigeration to the fixture.

Where operating conditions (design-fixture temperature, operating refrigerant temperature, and condensing-unit running time) will permit, the removal of frost may be accomplished automatically by use of controls normally supplied with condensing units by most manufacturers. By proper adjustment of these controls, the condensing unit is cycled to suit the job at hand.

Another variety of automatic defrosting can be obtained on some installations by the use of a time clock that will shut the system down at suitable and convenient intervals and return it to normal operation thereafter. This system is usable only as a general rule when fixture temperatures of around 32°F or higher are employed or when the fixture can be quickly warmed up to approximately 35°F in a reasonable period of time.

The first method described (cycling the condensing unit) is decidedly preferable when it can be used. It does not require the purchase and installation of special equipment and permits the use of controls normally furnished with a condensing unit by most manufacturers.

In certain types of refrigeration requiring frosted evaporators, it is impossible to defrost automatically by normal cycling of the condensing unit. Therefore, it is necessary to defrost by mechanical methods or by shutting down the system. The mechanical method requires the use of picks and scrapers to remove the ice and frost accumulated. This calls for the use of steel-pipe or plate-type cooling units, which are sturdy enough to withstand the action of these defrosting instruments. Finned or coppertubing evaporators should not be specified where mechanical defrosting is used because they would be damaged and possibly made inoperative by the use of picks and scrapers.

A second method of manual defrosting is usable when design-fixture temperatures are not below 35°F or where it is permissible to allow the room to warm up to 35°F or more at suitable intervals to dispose of the frost accumulated on the cooling units. This may be accomplished simply by opening the main switch to the condensing unit or units operating in the room and allowing them to stand idle until defrosting is completed. If forced-air evaporators are used, the fans should be kept moving to speed the defrosting operation. Close the main switch to resume normal operation.

Time Clocks

Time clocks are used frequently in defrosting operations where it is permissible to stop the compressor for a period of time. In or-

AIR CONDITIONING

der to ensure that the defrosting is done regularly and at convenient times, a time clock is used to connect or disconnect the compressor motor from its source at preset time intervals. Time clocks, such as shown in Fig. 9-7, are available for both 24-hour and 7-day cycles. The defrost interval, time of starting, and termination of the defrost cycle can be adjusted to suit various conditions. There are several types of defrost-control circuits operating as a function of time:

1. Time-initiated and time-terminated circuits
2. Time-initiated and pressure-terminated circuits
3. Time-initiated and temperature-terminated circuits

In this connection it should be noted that on circuits with pressure or temperature termination, an overriding time termination is normally provided in the event that the defrost cycle is unusually prolonged for some reason.

ACCESSORY EQUIPMENT

Accessory equipment required in air-cooling applications usually consists of the following:

Courtesy Paragon Electric Company

Fig. 9-7. Typical defrost time clock.

Evaporators for Air Conditioners

1. Heat exchangers
2. Surge drums
3. Oil separators
4. Dryers
5. Dehydrators
6. Strainers
7. Mufflers
8. Subcoolers
9. Check valves

Heat Exchangers

The function of a heat exchanger is to remove the heat from the liquid refrigerant and transfer it to the suction gas. A typical heat exchanger is shown in Fig. 9-8. Suction gas flows through the large center tube, while liquid is piped through the smaller tube wrapped around the suction tubing. The cold suction vapor absorbs heat from the warm high-pressure liquid through the tube-to-tube metal contact. Internal fins are often provided in the suction-gas section to increase the heat transfer between the suction gas and liquid refrigerant.

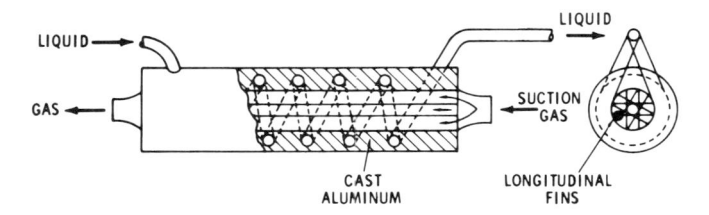

Fig. 9-8. Typical heat-exchanger for capacities up to 10 tons.

A heat exchanger is generally required in flooded-evaporator installations to ensure that the suction gas will be superheated. Liquid refrigerant will enter the suction line from the oil return line. The heat exchanger is necessary to vaporize this liquid refrigerant gas so that it will not return to the compressor as a slug and damage the valves. The theoretical amount of cooling through the superheating of the suction gas within the heat exchanger is

AIR CONDITIONING

determined by the increase in the temperature of the gas as it leaves the exchanger, multiplied by the specific heat of the suction gas and the amount of gas pumped. The amount of subcooling actually taking place in the liquid refrigerant as it leaves the heat exchanger is determined by the number of degrees to which it is cooled, multiplied by its specific heat and the pounds of refrigerant circulated.

The pounds of refrigerant circulated per hour at the temperature and pressure conditions in question can be determined by dividing the actual Btu-per-hour capacity by the refrigerating effect (Btu/lb) of the refrigerant in circulation. By the use of a heat exchanger, the net refrigerating effect (Btu/lb) is increased in direct proportion to the amount in degrees the liquid refrigerant is subcooled. In running a calorimeter test, the suction gas is generally superheated within the calorimeter to a temperature corresponding to that of the test room. Thus, the maximum refrigerating effect is obtained within the calorimeter, and the use of an exchanger would be of no advantage.

In an ordinary installation the suction gas can only be superheated to approximately the temperature within the unit, which is usually around 40°F. In an installation of this kind, a heat exchanger will increase the net refrigeration (Btu/lb) by taking advantage of this cold-suction gas to remove some of the heat from the liquid refrigerant before it enters the evaporator.

Heat exchangers may be constructed in a number of different ways. Where it is desired to make a low-priced exchanger, the suction and liquid line may be soldered together, starting at the point of emergence from the fixture and extending as far as practical towards the condensing unit. No definite recommendations as to the length of this type of exchanger can be made; but, in general, low-temperature fixtures require a greater length of exchanger, and the longer the exchanger, the more effective it will be. When lines are soldered together inside the fixture, the result will be to increase the effectiveness of the suction line as a drier.

Another method of constructing a heat exchanger is to use two concentric copper tubes, as shown in Fig. 9-9. Here the inner tube carries the suction gas from the evaporator and is connected in the suction line. The outer tube carries the liquid refrigerant from the receiver on the condensing unit to the expansion valve and is connected to the liquid line. The suction gas travels counterflow to the liquid refrigerant. Before the exchanger is assembled, the inner surface of both tubes and the outside of the suction-line tube should be cleaned to make sure no dirt or moisture is on the surface since these surfaces are exposed to the refrigerant.

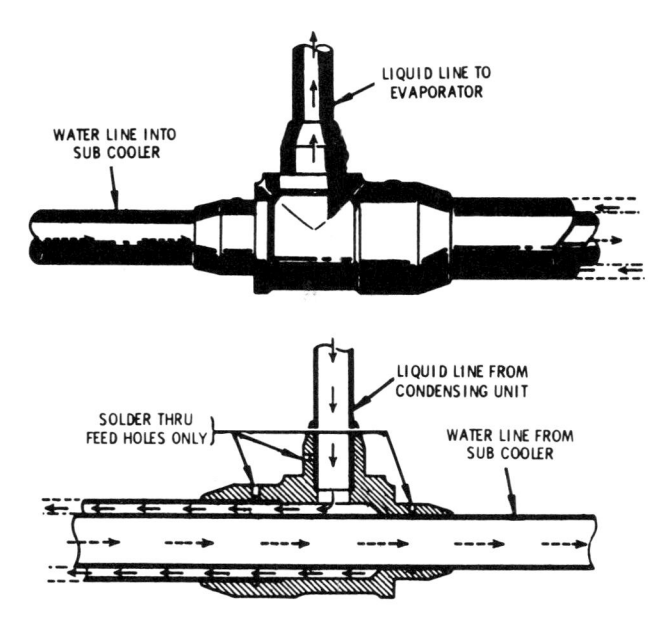

Fig. 9-9. Sectional view of typical two-pipe heat exchangers.

As the liquid section of the exchanger must be full in order to be most effective, it will be necessary to add refrigerant to the system to ensure this condition. When installing the exchanger, it should be mounted as near to the fixture as possible. It should be located on the outside of the fixture when space is available. If mounted inside the fixture, it should be insulated. The pressure drop through the exchanger should be added to the pressure drop through the rest of the suction line in order to determine the entire line drop.

Certain types of installations *spill* large quantities of liquid refrigerant when starting or may be subject to temporary surges. Low-temperature and pipe-coil applications are most prominent

AIR CONDITIONING

in this regard. Such liquid surges may pass through an otherwise satisfactory exchanger. To prevent this, an accumulator effect may be provided in the exchanger. Excess liquid will be collected in the accumulator exchanger until its refrigerating effect has been absorbed by the warm liquid. Such an operation will prevent the loss of refrigerating effect of "spillover" liquid as well as any possible damage to the compressor from this condition.

Surge Drums

Surge drums, also termed *suction accumulators* (Fig. 9-10), are required on the suction side of almost all flooded-type evaporators to prevent liquid slopover to the compressor. The exceptions are shell-and-tube coolers and similar shell-type evaporators, which are designed to provide ample surge space above the liquid level, or which contain eliminators for the separation of gas and liquid.

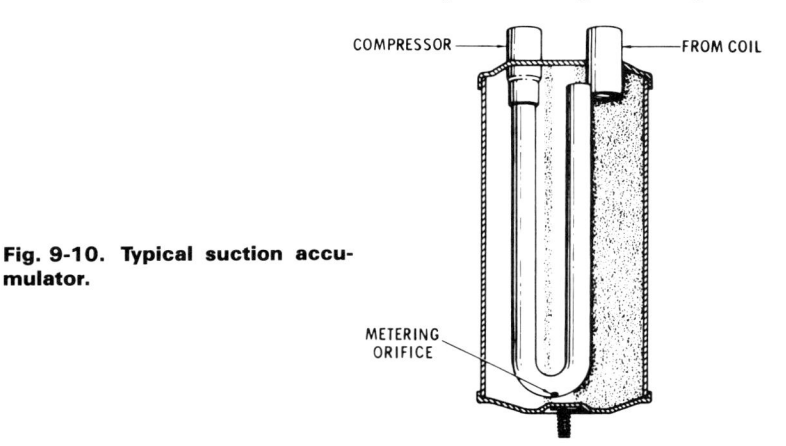

Fig. 9-10. Typical suction accumulator.

Courtesy Refrigeration Research, Inc.

A horizontal surge drum is sometimes used where head room is limited. The drum may be designed with baffles or eliminators to separate the liquid from the suction gas returning from the top of the shell to the compressor. More often, there is space allowed for sufficient separation above the liquid level. Such a design is usually of the vertical type with a separation height above the liquid level of from 24 to 30 inches and with the shell diameter sized to keep the suction-gas velocity at a low enough value to allow the liquid droplets to separate and not be entrained with the returning suction gas from the top of the shell.

Oil Separators

Oil separators, when used, are installed in the discharge line between the compressor and receiver. As shown in Fig. 9-11, oil separators are used because refrigerants and oil are mixed in a refrigerant system. Unless some device is used to prevent it, oil will be carried along with the refrigerant into the low side. Different refrigerants are soluble in oil to a greater or lesser degree. Freon-12 and Freon-22 are more soluble in oil than sulfur dioxide (SO_2) under similar conditions. Thus, no hard-and-fast rule can be made regarding the use of oil separators. Some of the applications where discharge-line oil separators can be useful are:

1. In systems where it is impossible to prevent substantial absorption of refrigerant in the crankcase oil during shutdown periods.
2. In systems using flooded evaporators where refrigerant bleedoff is necessary for oil removal from the evaporator.
3. In direct-expansion systems using coils or tube bundles that require bottom feed for good liquid distribution and where refrigerant slopover is essential from the top of the evaporator for proper oil removal.
4. In low-temperature systems where it is advantageous to have as little oil as possible going through the low side, especially Freon-13 systems, where the oil is not miscible with the refrigerant.

In applying oil separators in refrigeration systems, certain potential hazards must be fully recognized and properly dealt with:

1. The main hazard to guard against in the use of oil separators is their tendency to condense-out liquid refrigerant during compressor off cycles and compressor startup. This is true if the condenser is in a warm location, such as an evaporative condenser and receiver on a roof. During the off cycle, the

AIR CONDITIONING

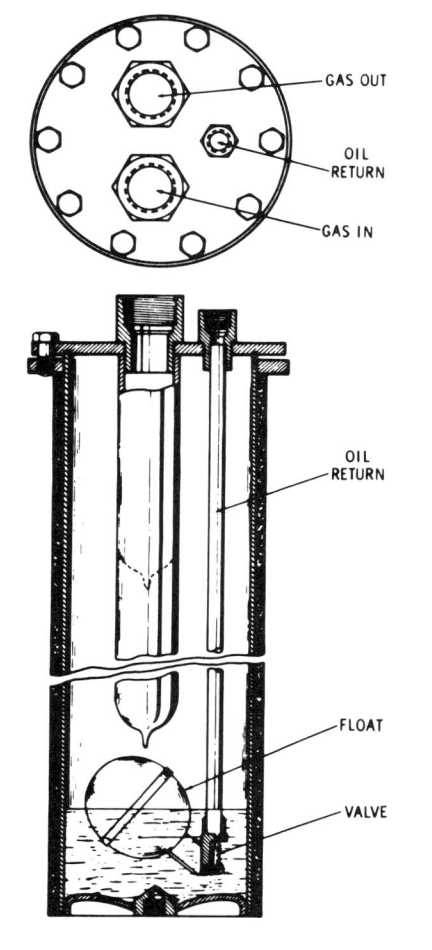

Fig. 9-11. Sectional view of an oil separator.

oil separator cools down and acts as a condenser for liquid refrigerant that evaporates in warmer parts of the system. Thus, a cool oil separator will act as a liquid condenser during off cycles and also on compressor startup until the separator has warmed up; it will automatically drain this condensed liquid into the compressed crankcase. On startup, there is excessive boiling in the crankcase because of the presence of a large amount of liquid refrigerant. This will

result in poor lubrication and wear on the compressor and may even end up completely draining the oil out of the crankcase as a result of this violent boiling action. If not protected by an oil-safety switch, compressor failure may result.
2. Oil separators are not 100 percent efficient, and therefore it is still necessary to design the complete system for oil return to the compressor.
3. The float valve is a mechanical device that may stick open or closed. If it sticks open, hot gas will be continuously bypassed to the compressor crankcase, which will cause the compressor to operate at an elevated temperature and, of course, reduce capacity. If the valve sticks closed, then no oil is returned to the compressor.

Where oil separators are used, it is recommended that precautions be taken to prevent draining condensed refrigerant into the crankcase. To minimize this possibility, the drain connection from the oil separator can be connected into the suction line entering the compressor. In this way, any liquid refrigerant returning from the separator will go into the suction manifold rather than directly into the crankcase. This drain line should be equipped with a shutoff valve, hand throttling valve, solenoid valve, and sight glass. The throttling valve should be adjusted so that the flow through this line is very small, in fact, only a little greater than would be normally expected for the return of oil through the suction line. The use of the sight glass will help in adjusting this flow. The solenoid valve should be wired so that it is open only when the compressor is running and is closed when the compressor is off.

The preceding arrangement will prevent liquid from draining down into the compressor during an off cycle. When the compressor starts up with liquid refrigerant in the oil separator, it will also prevent this liquid from dumping directly into the crankcase. It will allow it to be bled slowly into the suction line where it can be taken care of in the normal manner, similar to oil returning from the system.

The hazard of draining condensed refrigerant into the crankcase can also be minimized by insulating the oil separator and installing it ahead of a hot-gas loop to the floor that there will be a trap between it and any evaporative condenser installed above. Insulating the oil separator keeps it warm for a longer period after shutdown of the compressor and cuts down on the amount of condensation in the shell during shutdown unless it is for a prolonged period.

One manufacturer returns oil from an oil separator through an oil reservoir, which is heated, boiling off the refrigerant into the suction line. Return of liquid refrigerant from an oil separator to the compressor crankcase is a serious hazard.

Refrigerant Driers

The function of a drier in a refrigeration system is to remove moisture. The use of a permanent refrigerant drier is recommended on most systems using Freon-12 and Freon-22. It is a must on all low-temperature systems, including those using Freon-13. The decision of whether or not to use a permanent drier on higher-temperature systems is one of judgment. In the case of package-type air conditioners and water chillers it may not be necessary if proper dehydration techniques have been followed at the factory, no refrigerant connections are broken in the field, and any refrigerant added is at a sufficiently low moisture content. If there is any doubt that the system is initially dry in operation, then a drier is indicated for protection of the compressor and to prevent freezing at the liquid-feed device.

In the case of hermetic compressors, there is an additional reason for keeping moisture out of the system. The motor windings are exposed to the refrigerant gas and excessive moisture can cause breakdown of the motor-winding insulation, which can result in motor burnout and distribution of the products of decomposition throughout the refrigeration system. A full-flow drier is usually recommended in hermetic compressor systems to keep the system dry and prevent the products of decomposition from getting into the evaporator in the event of a motor burnout.

Side-outlet driers are preferred since the drying element can be replaced without breaking any refrigerant connections. The drier is best located in a bypass of the liquid line near the liquid receiver. The drier may be mounted horizontally, as shown in Fig. 9-12, but should never be mounted vertically with the flange on

EVAPORATORS FOR AIR CONDITIONERS

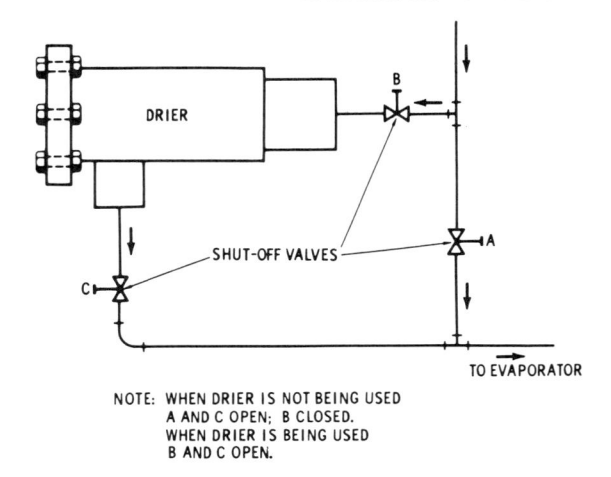

Fig. 9-12. Drier-piping connection.

NOTE: WHEN DRIER IS NOT BEING USED
A AND C OPEN; B CLOSED.
WHEN DRIER IS BEING USED
B AND C OPEN.

top since any loose material would then fall into the line when the drying element is removed.

Dehydrators

A dehydrator is a device that removes moisture from the refrigerant in a system. Usually it consists of a shell filled with a chemical moisture-absorbing agent. Its permanent installation in any field-connected installation is extremely desirable for both economy and operation. Its first cost is negligible compared to the service parts and labor it will save and also the improvement in expansion valve operation by the elimination of moisture.

In systems that are field-connected, there is always a possibility of taking in moisture, especially where connections are hastily or carelessly made. Even carefully made connections may be loosened by continued vibration or possible abuse after installation. Obviously, these leaks should be repaired as soon as they are detected. Before detection, however, considerable moisture may be absorbed by the system, especially on low-temperature installations. Damage from such sources may be avoided if a dehydrator has been installed as a permanent fixture.

Temporary dehydrator installation as a service procedure to

Air Conditioning

overcome moisture is necessary, of course, but two things must be kept in mind: First, the condition probably would not have developed had a permanent dehydrator been installed; and second, when the temporary dehydrator is removed, the system is necessarily opened up and moisture is given a chance to reenter and nullify all the good that had been accomplished. These points further emphasize the advisability of installing permanent dehydrators at the time of original installation.

The dehydrating agent used is capable of absorbing only a certain amount of moisture. Should a greater quantity of moisture get into the system, the dehydrator will become saturated and inactive and then must be replaced or refilled. Should dehydrators be opened or not tightly sealed in the stockroom, they will absorb moisture from the air, becoming saturated and inactive if put on an installation. Therefore, be sure the dehydrating agent is still active in any dehydrator that is to be installed.

Strainers

Strainers, or *filters* as they are sometimes called, are used to filter or strain out undesirable material (Fig. 9-13). Their function is to protect all automatic valves and the compressor from foreign material in the system, such as pipe scale, rust, and metal chips. Refrigeration systems usually have strainers at several points, for example:

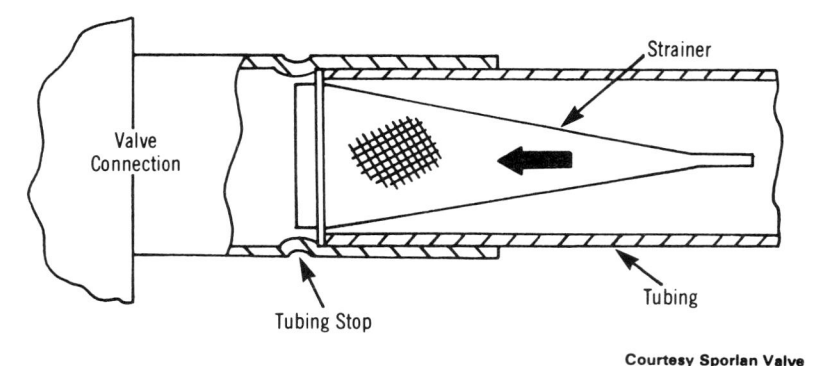

Courtesy Sporlan Valve

Fig. 9-13. Strainer inserted in a refrigeration system's tubing. Note direction of flow.

1. At the liquid-line outlet for the receiver
2. At the expansion-valve inlet
3. Ahead of the solenoid valves
4. At the inlet side of the water valve

Installation of a liquid-line strainer iii front of each automatic valve is recommended to prevent particles from lodging on the valve seats. Where multiple expansion valves with internal strainers are used at one location, a single, main liquid-line strainer will be sufficient to protect the system. Care should be taken to pipe the suction line at the compressor so that the built-in strainer is accessible for servicing.

Both the liquid and suction-line strainers should be properly sized to ensure adequate foreign-material storage capacity without excessive pressure drop. In steel-piping systems, it is recommended that an external suction-line strainer be used in addition to the compressor strainer.

Mufflers

Noise in hot-gas lines between interconnected compressors and evaporative condensers may be eliminated by the installation of mufflers. This noise may be particularly noticeable in large installations and is usually caused by the pulsations in gas flow due to the reciprocating action of the compressors and velocity of gas through the hot-gas line from the compressor. The proper location of a muffler is in a horizontal or down portion of the hot-gas line, immediately after leaving the compressor. It should never be installed in a riser.

Subcoolers

The problem off decreased system capacity due to excessive vertical lifts in the liquid line is usually solved by installation of subcoolers. The decrease in system capacity is due to two factors:

1. Operating head pressures must be increased to offset the static pressure exerted by the long vertical column of liquid refrigerant. Thus, condensing-unit capacity is decreased.

AIR CONDITIONING

2. Since liquid refrigerant follows definite pressure-temperature relations, the large fall in pressure due to long vertical lifts should be accompanied by a decrease in temperature if a solid column of liquid is to be maintained up to the expansion valve. Pressure and temperature drop will be accompanied by the evaporation of liquid refrigerant. This forms gas in the liquid line, and since gas has a larger volume per pound than liquid refrigerant, an expansion valve will handle a smaller number of pounds of liquid and gas than it will liquid in a unit of time. Thus, expansion valve capacity is decreased.

A subcooler installed in the liquid line, as shown in Fig. 9-14, will eliminate this decrease in capacity by supplying an external cooling source to reduce the temperature of the liquid refrigerant. Thus, a solid column of liquid is maintained at the valve. The valve capacity, therefore, is unaffected. The subcooling of the liquid also improves the quality of the refrigerant enough so that, even though condensing-unit-head pressures are increased, the evaporator performance is not decreased.

Check Valves

Check valves are used in the suction line of low temperature fixtures when they are multiplied with high temperature fixtures. Their use is most important when the condensing unit is controlled by low-pressure control. Their application is illustrated in Fig. 9-15.

During idle periods of the condensing unit, if there is no check valve in the suction line from the evaporator C, some of the refrigerant vapor from evaporator A will back up into evaporator C. This refrigerant from evaporator A will be condensed in evaporator C, causing at least two difficulties: first, heat is added into the cold evaporator C from the condensing process, causing a decrease in the refrigerating effect; second, the pressure that should have started the condensing unit to supply refrigeration to evaporator A has been reduced and the condensing unit will remain idle.

The foregoing outlined performance will give erratic temperature control. A check valve is important for accurate temperature

EVAPORATORS FOR AIR CONDITIONERS

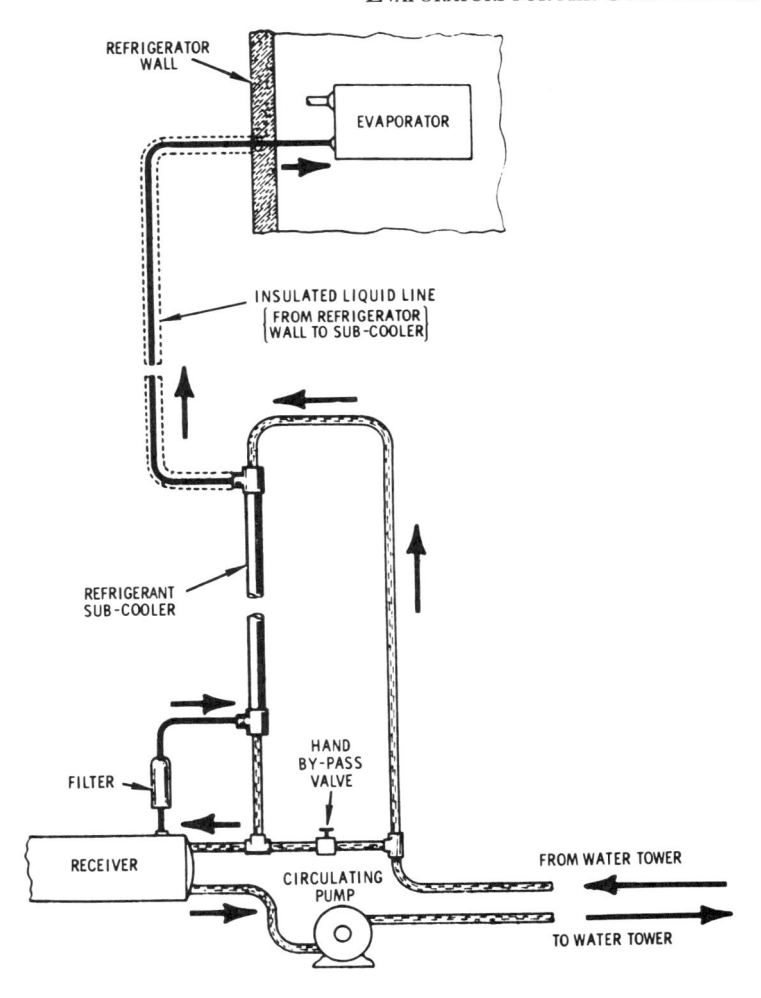

Fig. 9-14. Installation of subcooler on city water pressure.

control with low-pressure switches on installations involving two or more fixtures where there is a wide variation in temperature. This is particularly true if the low-temperature fixture has a large evaporator in proportion to the warm evaporator and also if it has long holdover periods.

AIR CONDITIONING

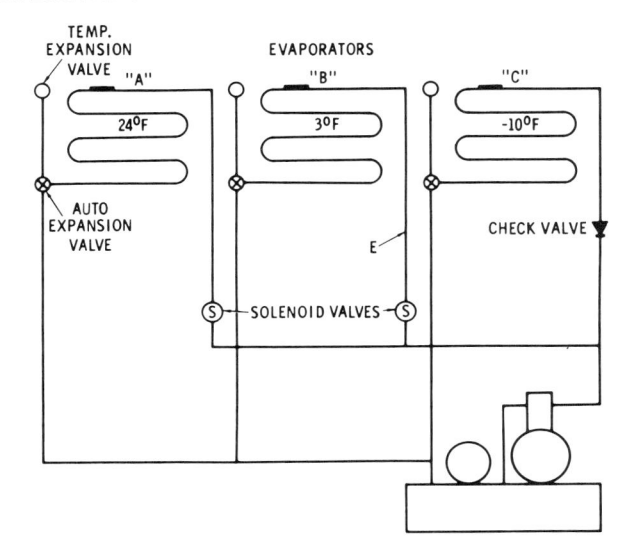

Fig. 9-15. Installation of suction-line check valve.

SUMMARY

The principle that heat always flows from a warmer to a colder surface makes the evaporator operate. In modern refrigerating equipment, evaporators may be divided into the following classes, depending on their construction: finned or plain tubing.

The finned class of evaporator is ordinarily useful for refrigerators, coolers, cases, and counters designed for the production of temperatures above 34°F with gravity circulation and 35°F with forced-air circulation.

Plain-tubing evaporators are also known as prime-surface evaporators. They are used successfully for practically any temperature, although they are especially recommended for holding fixtures 34°F or lower.

Three types of evaporators or chilling units are in general use: flooded, wet expansion, and dry expansion. The type of evaporator differs with respect to the volume of liquid refrigerant permitted in each and with the refrigerant-control methods. Factors affecting evaporator capacity are quite similar to those affecting condenser capacity.

The amount of heat that can be absorbed by an evaporator, that is, its capacity, follows the fundamental heat-transfer equation:

$$H = KAT_d$$

where, K is the coefficient of heat transfer, A is the total evaporator area in square feet, and T_d is the temperature differences between the refrigerant and the medium being cooled.

Time clocks are used frequently in defrosting systems where it is permissible to stop the compressor for a period of time. There are several types of defrost-control circuits operating as a function of time: time-initiated and time-terminated, time-initiated and pressure-terminated, and time-initiated and temperature-terminated.

The function of a heat exchanger is to remove the heat from the liquid refrigerant and transfer it to the suction gas. A heat exchanger is generally required in flooded-evaporator installations to ensure that the suction gas will be superheated.

Surge drums are required on the suction side of almost all flooded-type evaporators to prevent liquid slopover onto the compressor.

Oil separators, when used, are installed in the discharge line between the compressor and receiver. They are used because refrigerants and oil are mixed in a refrigerant system. Unless some device is used to prevent it, oil will be carried along with the refrigerant into the low side. Where oil separators are used, it is recommended that precautions be taken to prevent draining condensed refrigerant into the crankcase.

Refrigerant driers in a refrigeration system are used to remove moisture. They are a must on most all low-temperature systems.

A dehydrator is a device that removes moisture from the refrigerant in a system. It usually consists of a shell filled with a chemical moisture-absorbing agent. In systems that are field connected, there is always a possibility of taking in moisture.

Strainers are sometimes called *filters* for they filter, or strain out, undesirable material. Their function is to protect all automatic valves and the compressor from foreign material in the system. Such foreign material may be rust, scale, metal chips, or any number of other things that can get into a system during installation and manufacture.

Noise in hot-gas lines between interconnected compressors and evaporator condensers may be eliminated by the installation of mufflers. The proper location of a muffler is in a horizontal or down portion of the hot-gas line, immediately after leaving the compressor. It should never be installed in a riser.

The problem of decreased system capacity due to excessive vertical lifts in the liquid line is usually solved by the installation of subcoolers.

Check valves are used in the suction line of low-temperature fixtures when they are multiplied with high-temperature fixtures. Their use is most important when the condensing unit is regulated by low-pressure control.

REVIEW QUESTIONS

1. Define an evaporator.
2. What is the indirect system of refrigeration?
3. How are evaporators classified according to construction?
4. What are the three types of evaporators, or chilling units, in use today?
5. What is the coefficient of heat?
6. What four general conditions affect the K factor of an evaporator?
7. What is the heat-transfer formula?
8. How are time clocks used in defrosting systems?
9. What is the function of a heat exchanger?
10. Where are heat exchangers needed?
11. What is a surge drum?
12. Why are oil separators used in refrigeration systems?
13. Where are refrigerant driers used?
14. What is a dehydrator?
15. What is the purpose of a strainer or filter?
16. Why are mufflers needed?

17. What is a subcooler?
18. Where are subcoolers placed in the refrigeration system?
19. What is a check valve?
20. What do check valves do in a refrigeration system?

CHAPTER 10

Water-Cooling Systems

Water-cooled condensing units require an unfailing supply of suitable water. Condenser water may be obtained from city mains, rivers, lakes, wells, or by means of recirculating water towers. In large-tonnage installations, water from city mains is seldom used because of excessive cost.

During the winter months, the low ambient temperature introduces problems. The running time of the condensing unit is materially reduced due to the refrigeration-load decrease, which, with the decrease of ambient temperatures, creates undesirable fixture conditions. Since water-cooled condensing units usually have considerably more capacity than air-cooled models using the same body size or motor horsepower, the use of water-cooled units may further aggravate winter-operating conditions because they tend to reduce the running time still further.

One of the main problems is to vary the unit capacity with the change in load in such a manner that suitable running time is obtained in the winter as well as in the summer. Experience has shown that one method of improving winter operating difficulties is by the use of a combination air- and water-cooled condensing unit. By using the proper combination unit, good summer operating characteristics may be obtained with reduced water consumption during the periods of peak temperatures. To obtain satisfactory operation, the selection of the unit should be based on the following:

1. Moderate ambient temperatures
2. Use of air-cooled condensing-unit speeds and capacities
3. Balancing the unit with proper-size evaporators to produce the desired temperature difference

The combination unit is primarily an air-cooled unit. During periods of moderate or low ambient temperatures, the refrigerant is condensed by the air-cooled condenser, and the water-cooled condenser acts only as a receiver. The water valve remains closed, and no water is used. As a rule, these units should be chosen to match the load, calculated in the usual manner, using 80°F air as the condensing medium. Advantage may be taken of the speed changes possible with the various pulley combinations available for the condensing units selected. Such a selection will usually take care of the system with economical use of water to 100°F and somewhat beyond.

Atmospheric cooling is usually employed in the larger condensing units. Artificial methods of water cooling are used in cases where the water temperature is high due to exposure to the sun or where it is desired to use the water over again. This water-cooling method is usually obtained by means of water-cooling towers, cooling ponds, or spray-cooling ponds.

WATER TOWERS

Water towers are usually applicable in localities having excessive water rates or where sufficient quantities of cold water are not available and evaporative condensers are not suitable. Tower construction is usually of steel, cypress, or heart redwood, and all metal parts such as piping, nails, etc., should be rustproof. The water basin should be approximately 12 inches deep and louvers and nozzles properly spaced to secure efficient operation of the tower. A typical tower construction is shown in Fig. 10-1.

The principal operation in cooling water by means of a tower is by spraying the water from a series of nozzles directly into a stream of air and thereby dissipating the heat carried in the water, as shown in Fig. 10-2. This is accomplished by thoroughly mixing the air and water. The greater amount of cooling accomplished is due to the latent heat of vaporization of the water itself rather

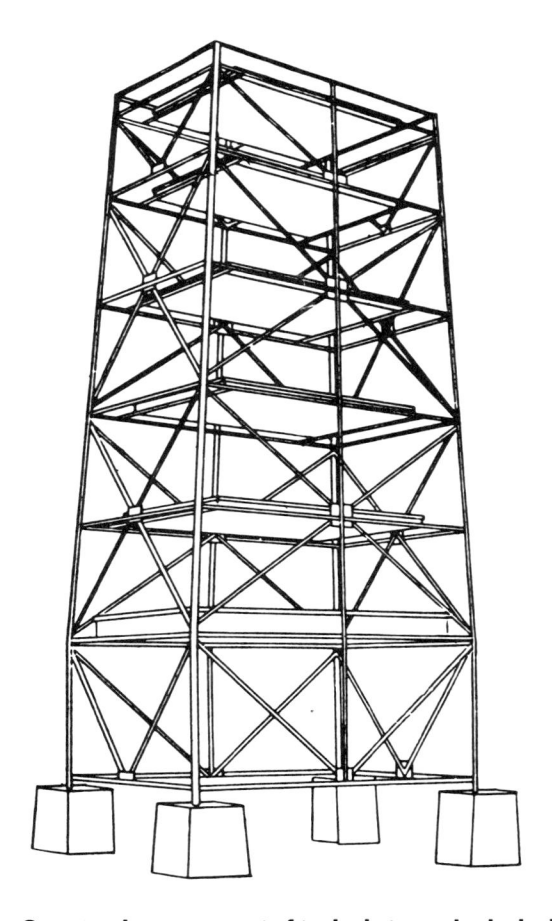

Fig. 10-1. Structural arrangement of typical atmospheric deck tower.

AIR CONDITIONING

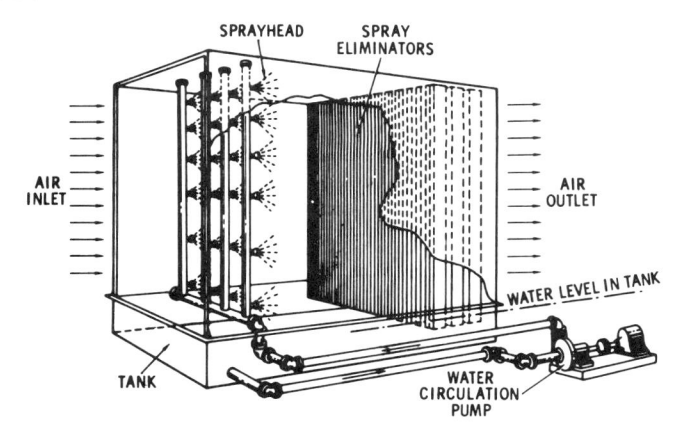

Fig. 10-2. Spray arrangement in a typical water-cooling tank.

than by sensible heat transfer. In this process of cooling, the water is circulated by means of a pump through the condenser tubes, picking up the heat and passing over the tower, and dissipating the heat to the atmosphere. The water can be used over and over again by this method. It is only necessary to add a small amount of makeup water to compensate for evaporative cooling and winddrift losses. A forced-draft water-cooling tower is shown in Fig. 10-3.

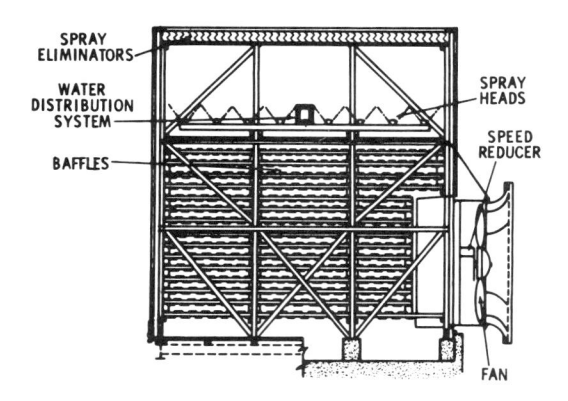

Fig. 10-3. A typical forced-draft water-cooling tower.

WATER-COOLING SYSTEMS

The wet-bulb temperature of the air passing through the water and the water spray determines the lowest temperature the water can be cooled. It is impossible to cool below the wet-bulb temperature and commercially impractical to cool the water down to this point. The water will usually cool to within 6 to 8° of the wet-bulb temperature if properly designed. The air velocity through a forced-air tower should be between 250 to 600 ft/min. The efficiency of a tower will range from 35 to 90 percent, the usual design being around 67 percent.

The circulating water pump should have a capacity to handle at least 5 gal/min/ton of cooling. Approximately 1 gallon of every 100 gallons of water circulated is lost by drift; approximately 1 gallon of every 1000 gallons circulated per degree of cooling is lost by evaporation. Care should be exercised in selecting a location for the tower in order to obtain efficient operation and yet not conflict with buildings, cornices, etc., for drift.

If the tower is not to be used in winter, it should be drained completely and the lines and basin cleaned to prevent corrosion. Water towers need occasional treatment of inhibitors against impurities contained in almost all water. This requires an analysis of the water and the amounts used, as well as the working temperatures, and should be done by an experienced chemist who will prescribe the proper treatment necessary in each individual instance.

Algae moss formation is a common growth on towers and can be treated with a concentrated solution of potassium permanganate in hot water. Add a sufficient quantity to the tower water to hold a pinkish color for 15 to 20 minutes.

Pressure-actuated water-regulating valves are designed for condensing units cooled either by atmospheric or forced-draft cooling towers. A typical water-regulating valve is shown in Fig. 10-4, and the piping arrangement is shown in Fig. 10-5. Valves of this type may be used on single- or multiple-condenser hookups to the tower to provide the most economical use of the tower.

In operation, low refrigerant head pressure may be the result of low tower water temperature, which causes the cooling ability of the refrigeration system to fall off rapidly. The valve senses the compressor head pressure and allows cooling water to flow to the condenser, to bypass the condenser, or to allow water flow to both

AIR CONDITIONING

Courtesy Penn Controls, Inc.

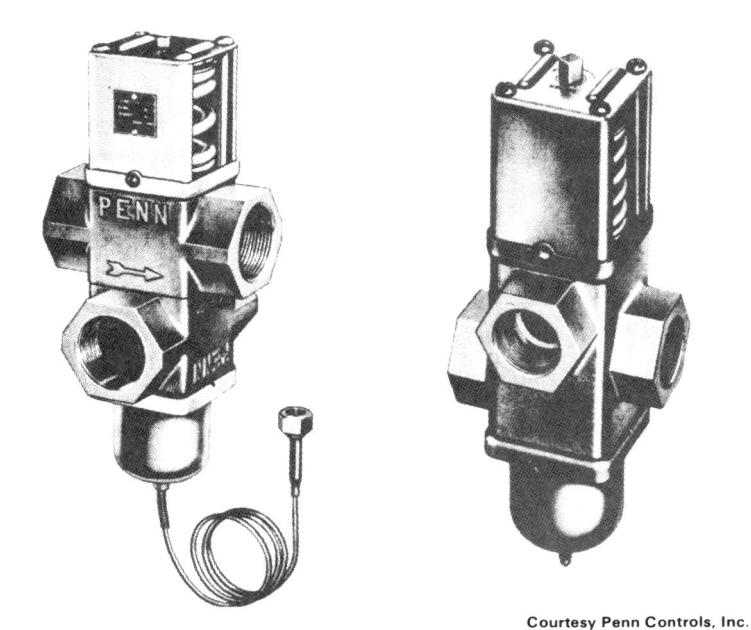

Fig. 10-4. Three-way pressure-regulating valve used in condensing units cooled by either an atmospheric or forced-draft cooling tower.

condenser and bypass line in order to maintain correct refrigerant-head pressures. With the correct valve size and adequate pump capacity, the three-way valve will maintain refrigerant condensing temperatures between 90 and 105°F with cooling tower water temperatures of 85 to 40°F. A water-temperature safety-control unit with time-delay switch is shown in Fig. 10-6.

The three-way valve permits water flow to the tower through the bypass line even though the condenser does not require cooling. This provides an adequate head of water at the tower at all times so that the tower can operate efficiently with a minimum of maintenance on nozzles and wet surfaces.

COOLING PONDS

The cooling pond is usually resorted to when location of the plant is such that a natural body of water is not available nearby.

WATER-COOLING SYSTEMS

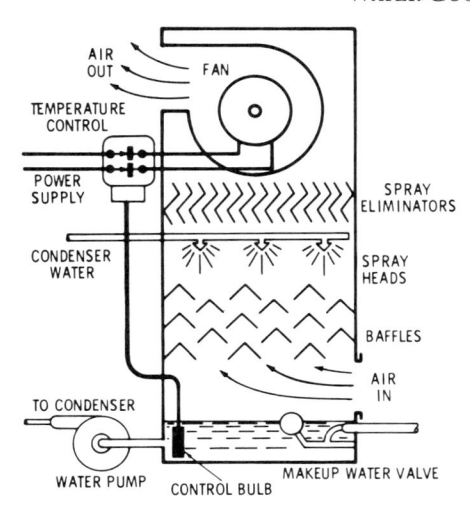

Fig. 10-5. Typical cooling-tower control equipped with forced draft.

Courtesy Penn Controls, Inc.

Fig. 10-6. Exterior view of water-temperature safety control with built-in time-delay switch.

Here, the water is cooled by being forced into surface contact with the air, and the amount of cooling will depend on the relative temperature of the air and the humidity.

Any available pond or lake may be used as a means of cooling recirculated water without the use of sprays provided it is large enough for the purpose. The heated water should be delivered as far from the suction connections as possible; and since the cooled water tends to sink due to its greater density, the best results will be achieved where the suction is applied at the bottom of the pond.

SPRAY-COOLING PONDS

The spray-cooling pond differs from the cooling pond discussed previously in that a means is introduced to accelerate the cooling effect. It consists essentially of properly designed sprays or nozzles, which break the water into fine drops. The drops must fall back into the pond and not be carried away. The number of nozzles required depends on the size of the pond and the water requirement of the plant. In average practice, the nozzles are located from 3 to 8 feet above the water level and in horizontal rows of from 10 to 16 inches apart.

The pressure at the inlet to the spray system may be from 5 to 12 psi. As in the cooling pond, the final temperature of the water depends on air temperature and humidity. In most cases, however, a reduction of water temperature of from 10 to 15° may be obtained with a single spraying.

NEW DEVELOPMENTS

All-metal towers with housing, fans, fill, piping and structural members made of galvanized or stainless steel are now being built. One factor in this change is that some local building codes are becoming more restrictive with respect to fire safety; another is low maintenance.

Engineers are beginning to specify towers, such as all-steel or all-metal, less subject to deterioration due to environmental conditions. Already, galvanized steel towers have made inroads into the air conditioning and refrigeration market. Stainless-steel towers are being specified in New York City, northern New Jersey and Los Angeles, due primarily to a polluted atmosphere which can

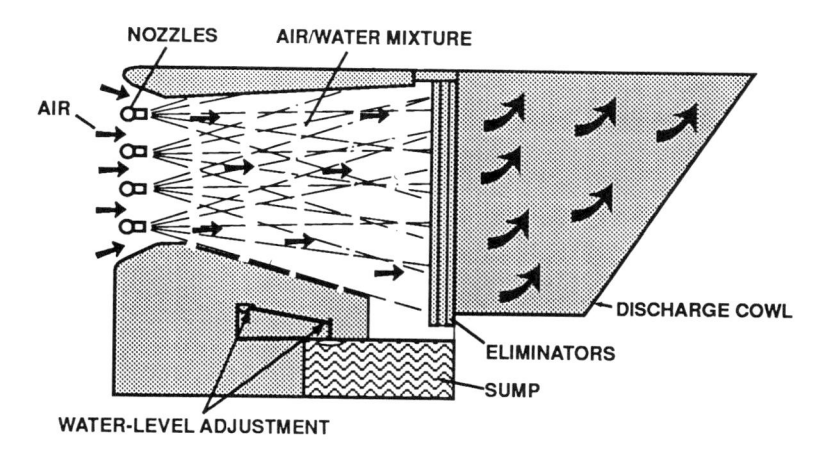

Fig. 10-7. Newer no-fans tower design.

lead to early deterioration of nonmetallic towers and, in some cases, metals.

Figure 10-7 shows a no-fans design for a cooling tower. Large quantities of air are drawn into the tower by cooling water as it is injected through spray nozzles at one end of a venturi plenum. No fans are needed. Effective mixing of air and water in the plenum permits evaporative heat transfer to take place without the fill required in conventional towers. The cooled water falls into the sump and is pumped through a cooling-water circuit to return for another cycle. The name applied to this design is Baltimore Aircoil. In 1981, towers rated at 10 to 640 tons with 30 to 1920 gallons per minute were standard. The nozzle clogging problems have been minimized by using prestrainers in the high pressure flow. There are no moving parts in the tower. This results in very low maintenance costs.

Air-cooled condensers are reaching 1000 tons in capacity. Air coolers and air condensers are quite attractive for use in refineries and natural gas compressor stations. They are also used for cooling in industry, as well as for commercial air conditioning purposes. Figure 10-7 shows how the air cooled condensers are used in a circuit system that is completely closed. These are very popular where there is little or no water supply.

COOLING-UNIT PIPING

During the passage of refrigerant through the piping system, a pressure drop takes place similar to that which occurs in the suction and discharge valves of the compressor. The pressure drop through the piping between the compressor and condenser requires a higher pressure inside the compressor during discharge than in the condenser. Unless the liquid is subcooled, pressure losses in the liquid line between condenser, receiver, and expansion-valve may cause flashing of the liquid refrigerant with consequent faulty expansion-valve operation.

In selecting refrigerant piping, frictional losses should be kept to a minimum. Piping should be selected that will give the smallest possible losses. It is necessary to consider the optimum sizes with respect to economics, friction losses, and oil return.

Pressure-drop consideration in liquid lines, however, is not as critical as with suction and discharge lines. The important things to remember are that the pressure drop should not be so excessive as to cause gas formation in the liquid line or insufficient liquid pressure at the liquid-feed device. A system should normally be designed so that the pressure drop in the liquid line, due to friction, is not greater than that corresponding to about a 1 to 2° change in saturation temperature. Friction pressure drops in the liquid line include accessories, such as solenoid valves, strainer-driers, and hand valves, as well as the actual pipe and fittings from the receiver outlet to the refrigerant-feed device at the evaporator.

Pressure drop in the suction line means a loss in system capacity because it forces the compressor to operate at a lower suction pressure to maintain the desired evaporating temperature in the coil. It is usually standard practice to size the suction line to have a pressure drop due to friction not any greater than the equivalent of about a 1 to 2° change in saturation temperature.

Where a reduction in pipe size is necessary to provide sufficient gas velocity to entrain oil up vertical risers at partial loads, greater pressure drops will be imposed at full load. These can usually be compensated for by oversizing the horizontal and downcomer lines to keep the total pressure drop within the desired limit. It is important to minimize the pressure loss in discharge lines because losses in these lines increase the required compressor horsepower per ton of refrigeration and decrease the compressor capacity.

Piping Principles

From the foregoing discussion, it follows that a refrigerant piping system should be designed and operated to accomplish the following:

1. Ensure proper refrigerant feed to evaporators.
2. Provide practical refrigerant-line sizes without excessive pressure drop.
3. Prevent excessive amounts of lubricating oil from being trapped in any part of the system.
4. Protect the compressor from loss of lubricating oil at all times.
5. Prevent liquid refrigerant from entering the compressor during either operating or idle time.
6. Maintain a clean and dry system.

LIQUID LINES

Liquid lines present no particular design problem. Refrigeration oil is sufficiently miscible with Freon-12 and Freon-22 in the liquid form to ensure adequate mixing and positive oil return. Low liquid velocities and traps in the line do not pose oil-return problems. In designing a liquid line, however, it must be recognized that it is desirable to have slightly subcooled liquid reach the liquid-feed device at a sufficiently high pressure for proper operation of the device.

Where liquid subcooling is required, this is usually accomplished in one or both of the following methods: use of a liquid-suction heat interchanger or use of liquid subcooling coils in evaporative condensers. The determination of how much liquid subcooling is required to offset friction and static head losses is quite simple. At normal liquid temperatures, the static pressure loss due to elevation at the top of a liquid lift is equal to 1 psi for every 1.8 feet for Freon-12, and 2 feet for Freon-22.

SUCTION LINES

Suction lines are the most critical from a design and construction standpoint (Fig. 10-8). Several important aspects must be considered in the design of refrigerant suction lines:

1. The suction line should be sized for a practical pressure drop at full load.
2. The suction line should be designed to return oil from the evaporator to the compressor under minimum load conditions.
3. The suction lines should be designed to prevent liquid from draining into the compressor during shutdown.
4. The suction line should be designed so that oil from an active evaporator does not drain into an idle evaporator.

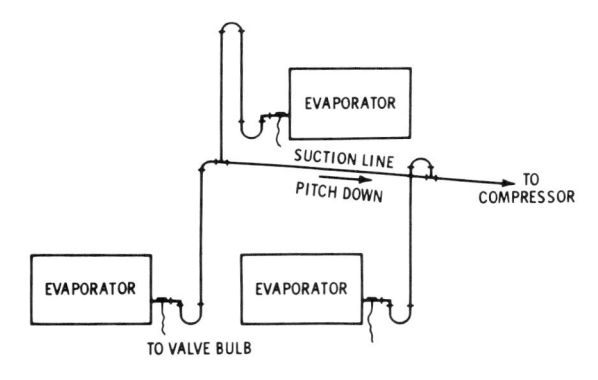

Fig. 10-8. Typical piping from evaporators located above and below common suction line.

Oil Circulation

Lubricating oil is lost from reciprocating compressors during normal operation because these compressors are designed so that the pistons and piston rings ride on a film of oil while traversing the cylinder. A small quantity of this oil is continuously pushed on through the cylinders and out with the discharge gas. Since it is inevitable that oil will leave the compressor with the refrigerant, means must be provided for in systems using Freon-12 and Freon-22 to return this oil at the same rate at which it leaves.

Oil that leaves the compressor reaches the condenser and there becomes dissolved with the liquid refrigerant. In this condition, it readily passes through the liquid supply lines to the evaporators. In the evaporator, however, a distillation process occurs, and there is almost complete separation of the oil and refrigerant. Very little oil can remain in the relatively cold and less dense refrigerant vapor at the temperature and pressure corresponding to its evaporating condition. Therefore, oil that is separated in the evaporator can be returned to the compressor only by gravity or entrainment with the returning gas.

Suction risers must be sized for minimum system capacity. It is extremely important that oil be returned to the compressor at the operating condition corresponding to the minimum displacement and minimum suction temperature at which the compressor will operate. For compressors with capacity control, the minimum capacity is represented by the lowest capacity at which the compressor can operate. In the case of multiple compressors with capacity control, it is important that the minimum capacity is the lowest at which the last operating compressor can run.

DISCHARGE LINES

The design of discharge (hot gas) lines with self-contained condensing units does not present any problem because these are usually factory-assembled. In the case of remotely located condensers, the essential element of hot-gas-line design should include the following:

1. The hot-gas line should be sized for a practical pressure drop.
2. The hot-gas line should be designed to avoid trapping oil at partial-load operation.
3. The hot-gas line should be designed to prevent refrigerant from condensing and draining back to the head of the compressor.
4. Connections of multiple compressors to a common hot-gas line should be carefully chosen and designed.

AIR CONDITIONING

5. The lines should not develop excessive noise and vibration as the result of hot-gas pulsations or compressor vibration (Fig. 10-9).

Even though a low loss is desired in hot-gas lines, they should not be oversized to the extent that gas velocities are reduced to a point where the refrigerant will not be able to carry along any entrained oil. In the usual application, this will not be a problem in hot-gas lines. In the case of multiple compressors with capacity control, it is wise to make sure any hot-gas risers will carry oil along at all possible loadings, as shown in Fig. 10-10.

REFRIGERANT-PIPING ARRANGEMENT

The refrigerant piping should be arranged so that normal inspection and servicing of the compressor and other equipment will not be hindered. In order to minimize tubing, refrigerant and

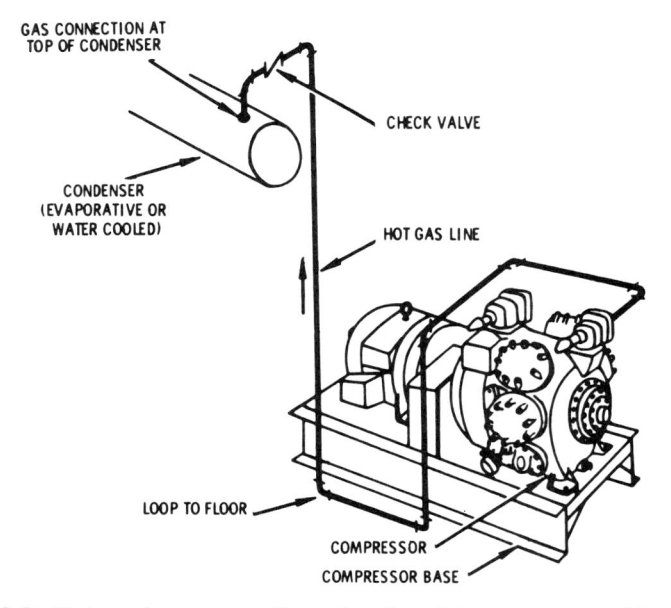

Fig. 10-9. Hot-gas loop connections showing piping arrangement to prevent liquid and oil from draining back to the compressor head (single compressor).

WATER-COOLING SYSTEMS

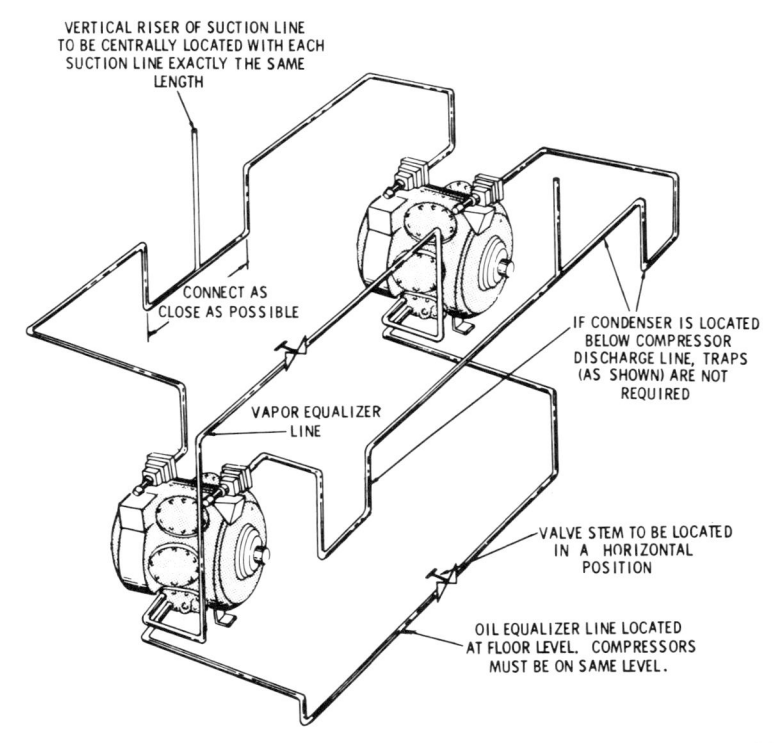

Fig. 10-10. Piping connections for suction and hot-gas headers (multiple compressors).

pressure drop lines should be as short and run as directly as possible. The number of piping joints, fittings, and elbows should be as small as possible. Sufficient flexibility, however, should be provided in order to absorb compressor vibration. In locations where copper tubing will be exposed to mechanical injury, the tubing should be enclosed in rigid or flexible conduit. In addition, piping should be run so as not to interfere with passages, obstruct headroom, or affect the opening of windows or doors.

Piping Supports

Rod-type hangers, perforated metal strips, or wall brackets that are commercially available are satisfactory and provide a simple means of supporting horizontal pipe runs. On vertical pipe runs, the piping weight may be supported with riser clamps bearing on structural members of the building, or by means of a platform at the bottom of the riser.

Supports should be strong enough to support any load caused by thermal expansion or contraction of the pipe so that stresses will not be placed on the equipment connected to the piping. In some cases, it may be necessary to provide some form of vibration isolation between pipe and support, thereby preventing objectionable vibrations that may be transmitted through the structure.

Piping Insulation

Insulation of refrigeration pipes serves a two-fold purpose: to prevent *moisture condensation*, and to prevent *heat gain* from the surrounding air. The desirable properties of pipe insulation are as follows:

1. It should have a low efficiency of heat transmission.
2. It should be easy to apply and have a high degree of permanency, and it should be readily protected against air and moisture infiltration.
3. It should have a reasonable installation cost.

All suction lines should be insulated to prevent moisture condensation and heat gain. Ordinarily, liquid lines do not require insulation on installations where the suction and liquid lines are clamped together, but the two lines may be insulated as a unit. It is also advisable to insulate the liquid line when it passes through an area of higher temperature to minimize heat gain. Normally hot-gas lines are not insulated. They should be, however, if the heat dissipated would be objectionable or where there is a possibility of people being burned by the pipe. In the latter case, it is not essential to provide an insulation with a tight vapor seal since there is no problem of moisture condensation. Fig. 10-11 shows various insulated pipe fittings.

Insulation covering lines on which moisture can condense or lines subjected to outside conditions must be vapor sealed to prevent any moisture travel through the insulation or condensation in

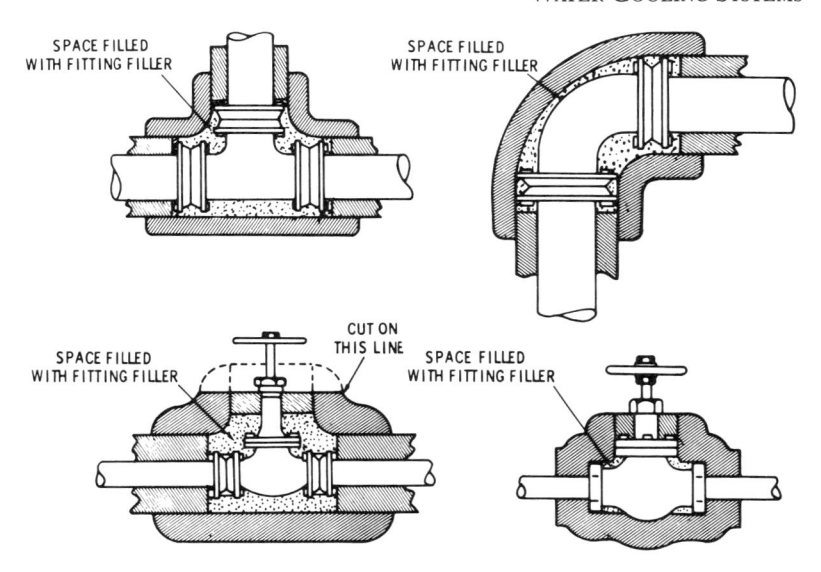

Fig. 10-11. Various fittings insulated with cork covering.

the insulation. Many commercially available types are provided with an integral waterproof jacket. On these and other types it is essential that the manufacturer's recommendations regarding sealing be followed carefully. Although all joints and fittings should be covered, it is not advisable to do so until the system has been thoroughly leak-tested and operated for a period of time.

SUMMARY

Water-cooled condensing units require an unfailing supply of suitable water. Condenser water may be obtained from city mains, rivers, lakes, wells, or by means of recirculating water towers.

Water towers are usually applicable in localities having excessive water rates or where sufficient quantities of cold water are not available and evaporative condensers are not suitable. In a cooling tower approximately 1 gal/100 gal of water circulated is lost by drift. Approximately 1 gal/1000 gal circuited per degree of cooling is lost by evaporation.

Algae moss formation is common growth on towers and can be treated with a concentrated solution of potassium permanganate in hot water.

Cooling ponds are usually resorted to when location of the plant is such that a natural body of water is not available nearby.

Pressure drop in the suction line means a loss in the system capacity because it forces the compressor to operate at a lower suction pressure to maintain the desired evaporating temperature in the coil.

Suction lines are the most critical from a design and construction standpoint. There are several important aspects that must be considered in the design of refrigerant lines.

Lubricating oil is lost from reciprocating compressors during normal operation because these compressors are designed so that the pistons and piston rings ride on a film of oil while traversing the cylinder. Oil that leaves the compressor reaches the condenser and there becomes dissolved with the liquid refrigerant.

The design of discharge (hot-gas) lines with self-contained condensing units does not present any problem since these are usually factory-assembled.

The refrigerant piping should be arranged so that normal inspection and servicing of the compressor and other equipment will not be hindered.

REVIEW QUESTIONS

1. What are some of the problems associated with water-cooling systems during the winter months?
2. Where are water towers used?
3. How does a water tower cool water or refrigerant?
4. What should you do with a water tower if it is not used during the winter?
5. What is the most common growth on towers that needs treatment?
6. Where are pressure-actuated water-regulating valves used?
7. Why are cooling ponds used?

WATER-COOLING SYSTEMS

8. How does the spray-cooling pond differ from the regular cooling pond?
9. What does a pressure drop in the suction line mean in the way of capacity of a system?
10. How is lubricating oil lost during normal operation of a compressor?
11. How must suction risers be designed?
12. What is another name for the discharge line from a compressor?
13. How should the piping be arranged in the refrigeration system?
14. What are the two purposes of piping insulation?
15. Why should all suction lines be insulated?

CHAPTER 11

Air-Conditioning Control Methods

Controls are an essential part of every air-conditioning installation. Some means must be provided for regulating the temperature and relative humidity in order to maintain conditions that are comfortable to human beings or that are required for industrial applications. Outside temperatures and relative humidities are seldom constant for any length of time, and throughout the seasons these atmospheric conditions fluctuate over a wide range. Thus, without controls, the temperatures would be too low at times and too high at other times, while humidities would vary considerably. With such variable conditions, it is quite obvious that any installation would be unsatisfactory to both the customer and the manufacturer.

Economy of operation is also a very important factor since it will not be necessary to operate the equipment continually throughout the season. Therefore, controls should be furnished to actuate the various operating devices in order to keep the cost of

AIR CONDITIONING

operation to a reasonable figure. In this connection it should be clearly understood that air-conditioning equipment is generally specified to cover the average maximum demand made upon it, and that this maximum does not exist except for a relatively few days during the year.

There are three general types of air-conditioning controls, namely: manual, automatic, and semiautomatic.

Manual Control

Manual control requires personal attention and regulation at frequent intervals, especially when the load varies continually. Even if the unit is in operation only at certain times of the day, inside and outside conditions may vary so much that constant changes in adjustment are necessary. A manual control may be an ordinary snap or knife switch or similar switches, which must be operated by hand when it is desired to start or stop the unit.

This type of control, while giving temperature and humidity regulation for a given set of conditions, will not provide the most desirable regulation due to fluctuating demands. Constant attention is required by the operator to maintain the predetermined conditions as closely as humanly possible, with the result that cost of operation is usually found to be excessive. If a manual control is provided, the user will permit the unit to operate until the room is too cold or will forget to start the machine until the room has reached an uncomfortable temperature. In either case, the user will not be satisfied. It is therefore wise always to install an automatic control.

Automatic Control

A completely automatic installation will meet a varying load almost instantly. This type of control involves the use of a large number of instruments, such as time clocks, thermostats, humidity controllers, and automatic dampers operated either by electricity or compressed air. Automatic controls provide all the features that are considered most desirable for temperature and humidity regulation as well as for economy of operation. The operator or owner is relieved of all responsibility for maintaining the

AIR-CONDITIONING CONTROL METHODS

predetermined conditions, and operating costs will be kept at a minimum.

As an example, consider an office being cooled by a unit conditioner. The unit is specified to maintain comfortable conditions during the hottest days of the summer, and it also must furnish comfortable conditions on mild days in the spring and fall. Unless an automatic control is used, the office will get too cold on mild days, and a complaint of unsatisfactory operation will be registered.

Semiautomatic Control

The semiautomatic control system differs from the manual and the automatic systems mainly in the number of control devices required. Thus, for example, a differential thermostat will be entirely satisfactory for temperature regulation in most cases. Time clocks are also very desirable on installations where definite shutdown periods are required. The fresh-air intake from outdoors should always be equipped with adjustable louvers so that the maximum amount of air can be regulated for both summer cooling and winter heating.

It is desirable on any job below 25 tons to keep the control system as simple and free from service liabilities as possible. On large theater or building applications, the expenditure of a considerable amount for control equipment for thermostatic or manual control should constitute the extent of the control system. This may include provisions for manually controlled ventilation and time-clock operation. If the additional expense is warranted, the job can be equipped with automatic humidity controls.

BASIC CONTROL TYPES

With respect to the preceding discussion, the various controls as applied to air-conditioning systems consist of different devices for maintaining *temperature, humidity,* and *pressure* within certain predetermined limits. These devices may be classified broadly as: temperature-control devices, humidity-control devices, and pressure-control devices.

AIR CONDITIONING

Residence-type control equipment may range from automatic regulation of heating plants to control of complete year-round air conditioning. Various actuating devices sensitive to changes in condition are: thermostats, hygrostats, and pressure regulators.

The various residence heating controls employed depend on the fuel supply used for the appliance, which may be classified as coal-, oil-, gas-, or electrically-heated.

All the foregoing include a room thermostat as part of their control. Various kinds of thermostats in common use are the *room*, *duct*, and *immersion* types.

Room Thermostats

By definition, a room thermostat is an automatic device for regulating temperature by opening or closing a damper or other device of a heating appliance, such as a heating furnace or electric heat element. Thermostats are based on the principles of expansion and contraction of metals due to temperature changes. They consist essentially of a loop or coil made up of a bimetal strip, such as brass and steel, securely fastened together, having one end fixed and the other attached to a pointer as shown in Fig. 11-1. As room temperature falls below a certain value, the pointer bends toward one of the contacts until it closes an electric circuit, which, in turn, energizes a furnace-control device. This operation will supply additional heat to the room. When the temperature rises above that desired, the pointer moves toward the other contact, which, due to its electrical circuit device, acts to reduce the heat produced.

Thermostats utilize either the differential expansion of solids, liquids, or gases subject to heat, or the thermopile principle. A fixed-temperature thermostat operates at predetermined fixed temperatures, whereas a rate-of-rise thermostat operates when the rate of increase in temperature exceeds a predetermined amount. Room thermostats are usually mounted on the wall of the room or space where the temperature is to be controlled. In locating the room thermostat, the following rules should be observed:

1. The thermostat should always be on an inside partition, never on an outside wall.

AIR-CONDITIONING CONTROL METHODS

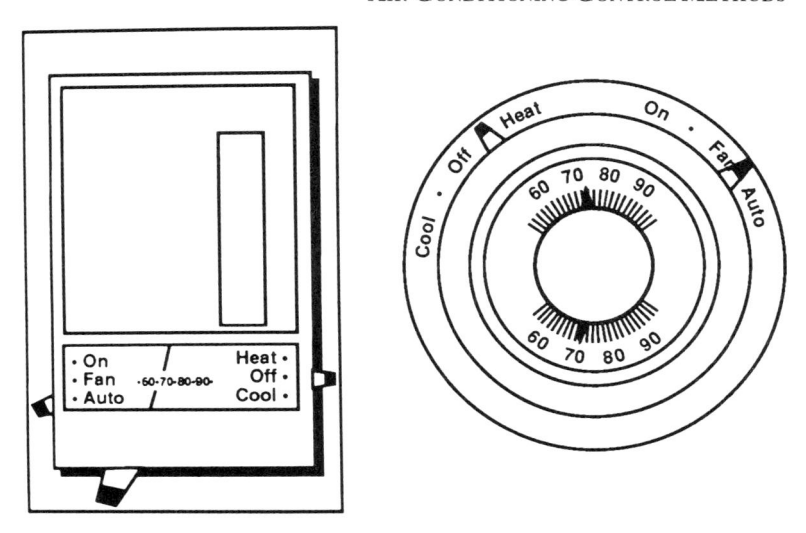

Fig. 11-1A. Two types of wall mounted room thermostats.

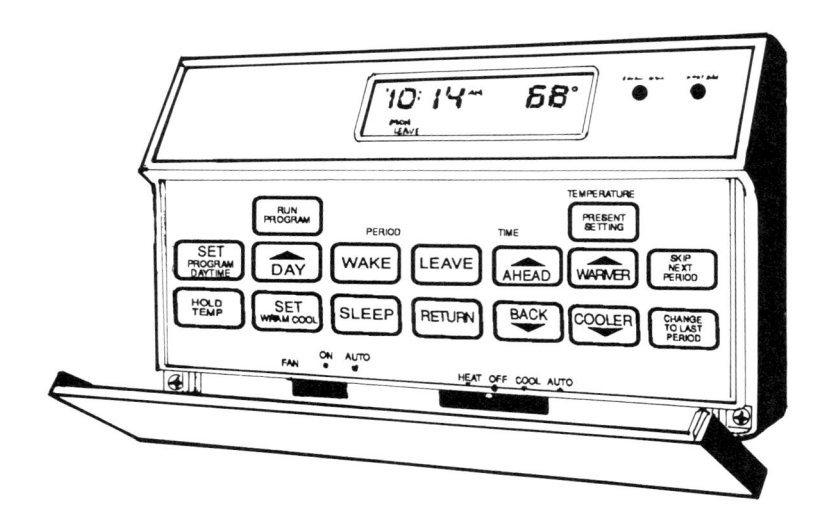

Fig. 11-1B. Microprocessor used for air-conditioning and furnace control in a home.

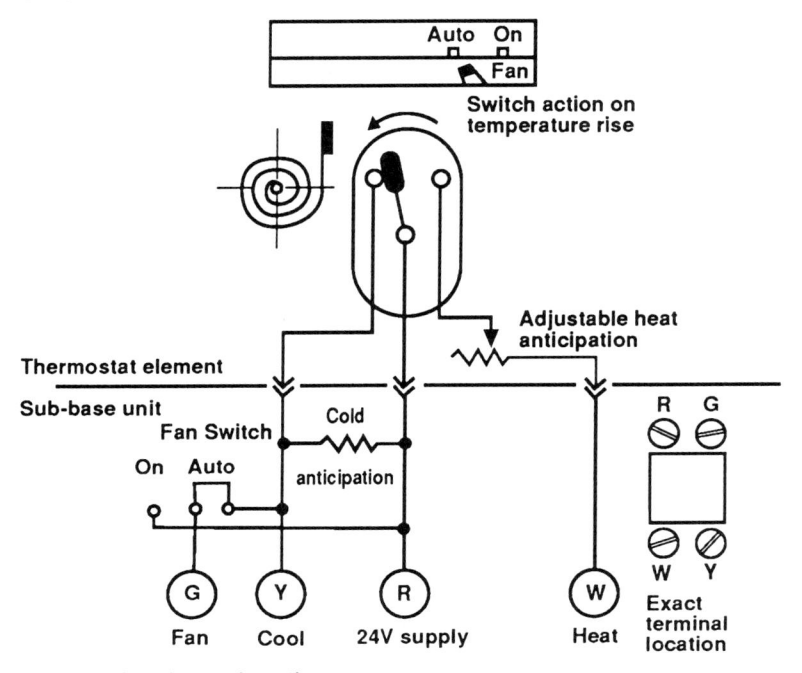

AIR CONDITIONING

Fig. 11-1C. Electrical circuitry for a home heat-cool thermostat.

2. Do not mount the instrument on a part of the wall that has steam or hot water pipes or warm air ducts behind it.
3. The location should be such that direct sunshine or fireplace radiation cannot strike the thermostat.
4. Be careful that the spot selected is not likely to have a floor lamp near it or a table lamp under it.
5. Do not locate it near an outside door or a radiator, or where warm air from a supply grille can strike it.
6. Do not locate it where heat from kitchen appliances can affect it or on a wall that has a cold unused room on the other side.
7. After a thermostat has been mounted, it is wise to fill the stud space behind the instrument with insulating material to prevent any circulation of cold air. Furthermore, the hole

AIR-CONDITIONING CONTROL METHODS

behind the thermostat for the wires should be sealed so that air cannot emerge from the stud space and blow across the thermostat element. It is quite common to find considerable air motion through this hole caused by a chimney effect in the stud space.

Remember that the thermostat location has nothing to do with the heat balance of the various rooms. For example, if one room runs cold, it does not help to move the thermostat to that room. True, putting the thermostat will make the room warm, but at the same time the other rooms will be overheated. The same result could be achieved at the original location by setting the thermostat to a higher temperature. All instances of certain rooms being too hot or too cold must be corrected by changing the heat input to that room or the heat loss from it, not by moving the thermostat.

Twin-Type Thermostats

The so-called *night-and-day thermostat* controls the heating or cooling source to maintain either of two selected temperatures. In principle, this is an assembly of two thermostats mounted on a single base with one cover. The electric clock can be set to throw the temperature control from one thermostat for the daytime temperatures to the other for nighttime temperatures at a predetermined time setting on the clock. This conveniently permits a lower temperature at night and normal temperature during the day.

Modern heating thermostats are built to incorporate a *heat leveling*, or heat-anticipation device. A heating coil consisting of a high-resistance unit drawing very low current is automatically connected in the thermostat circuit when the thermostat contacts close as it calls for heat (Fig. 11-2). The heat-anticipation feature may be described as follows: After the burner (or stoker, as the case may be) has been in operation a short time, the heating coil warms up slightly, which, in turn, has a warming effect on the thermostat mechanism. This causes the contacts to open in anticipation of the heat that has been generated in the boiler or furnace but has not yet reached the room.

AIR CONDITIONING

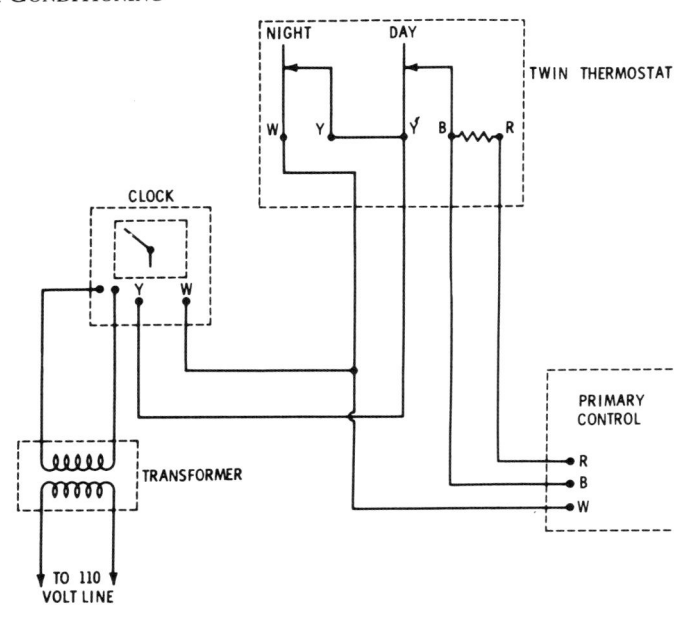

Fig. 11-2. Wiring diagram of a twin-type thermostat showing connections between the clock and primary control.

Duct Thermostats

A duct thermostat is provided with suitable fittings for installation in heating or air-conditioning ducts. The insertion-type is provided with a rigid bulb and is arranged so that the temperature-responsive element or bulb extends through the wall of the duct. The remote bulb-type thermostat is arranged so that the bulb and instrument head are connected by means of a flexible tube of the desired length. In this arrangement, the bulb is inserted in the duct, and the head is located where it is readily accessible for adjustment and inspection.

Immersion Thermostats

Immersion-type thermostats are used in automatic gas water heaters where a liquid-tight connection is required. This kind of

AIR-CONDITIONING CONTROL METHODS

actuating device is of the vapor- or liquid-filled type and operates a bellows, which, in turn, operates the gas valve. Changing the thermostat setting is accomplished by altering the relation of these various elements so that the gas valve closes at a lower or higher temperature (Fig. 11-3).

Contraction of the thermal element immersed in the stored hot water, through mechanical linkage, serves to open the main gas valve in the unit on a drop in the temperature of the water in the tank. Expansion of this element serves to close the gas supply to the main burner when the tank water has reached a predetermined temperature.

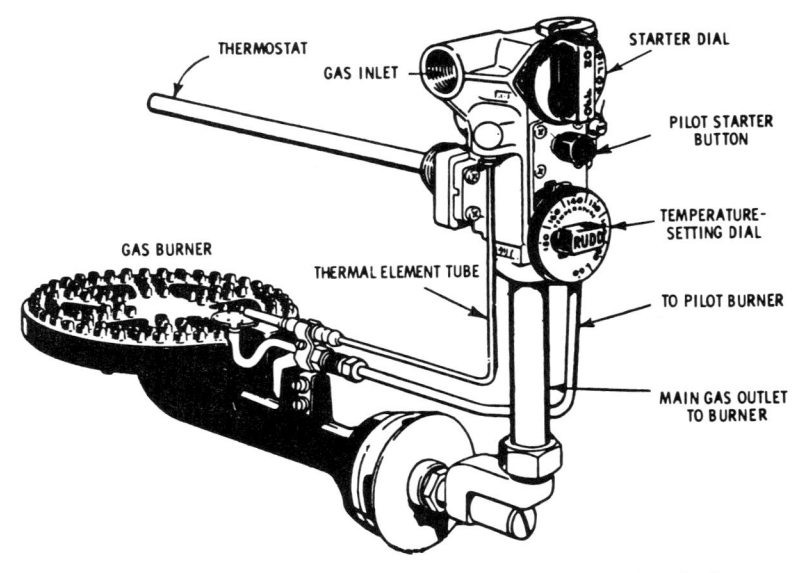

Courtesy Rudd Manufacturing Company

Fig. 11-3. Twin-dial thermostat used as a temperature control in an automatic gas water heater.

Hygrostats

The hygrostat, or *humidistat*, as it is sometimes called, is an instrument for regulating or maintaining the degree of humidity within a room or enclosed space. The hygrostat is actuated by changes in humidity and is used for automatic control of relative

333

AIR CONDITIONING

humidity. The control element in the instrument is some hygroscopic material, such as human hair, paper, wood, or similar material (Fig. 11-4). When the material absorbs water from the air, expansion takes place; when the material gives up moisture to the air, contraction takes place. It is this action which is used to make and break the circuit controlling the humidifying or dehumidifying apparatus.

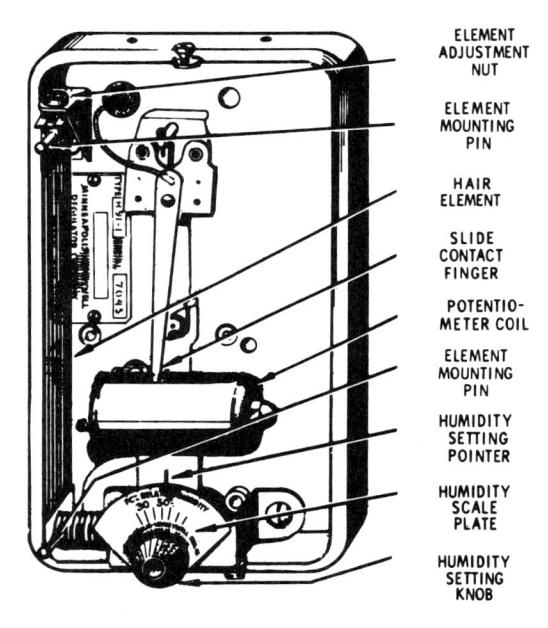

Courtesy Honeywell, Inc.

Fig. 11-4. Typical hygrostat showing component parts with cover removed.

As with thermostats, the hygrostat generally makes or breaks an electric circuit controlling the operation of an electric motor or valve, or opens or closes a compressed-air circuit from which air is obtained for operating an air motor or valve. Both electric- and air-actuated equipment are supplied for various types of duty; but, as a rule, electrical apparatus is used in most air-conditioning installations.

AIR-CONDITIONING CONTROL METHODS

In locating a room hygrostat, the following rules should be observed:

1. The hygrostat should be located on an inside wall about 5 feet above the floor. A hallway or dining room is preferred. Do not install in kitchen or bathroom.
2. Locate the instrument where natural air circulation is unrestricted.
3. Do not install the instrument where its operation might be affected by lamps, sunlight, fireplace, registers, radiant heaters, radios, television, concealed air ducts, pipes, or other heat producing appliances.

Pressure Regulators

Pressure regulators are manufactured in a number of different types and sizes and are designed to regulate the pressure of fluids, air, and gases. When designed for the regulation of a single pressure, they are classified as *single-pressure regulators;* and when designed to maintain a predetermined difference between two pressures, they are termed *dual-pressure regulators.* A cutaway view is shown in Fig. 11-5.

LIMIT CONTROLS

A limit control is a device responsive to changes in pressure, temperature, or liquid level and is used to turn on, shut off, or throttle the fuel supply to a heating appliance. In steam-boiler operation, the limit control consists of a bellows arrangement, which, being responsive to boiler pressure, breaks an electric contact when the pressure exceeds a predetermined value, thus preventing additional heat from being delivered to the boiler.

The limit control for a hot water boiler consists of an immersion thermostat inserted in a well of the boiler. The limit control stops the fuel supply when a predetermined water temperature has been reached. In a warm-air heating system a thermostat is inserted in the furnace bonnet. The thermostat will cut off the heating source when a predetermined temperature has been attained.

AIR CONDITIONING

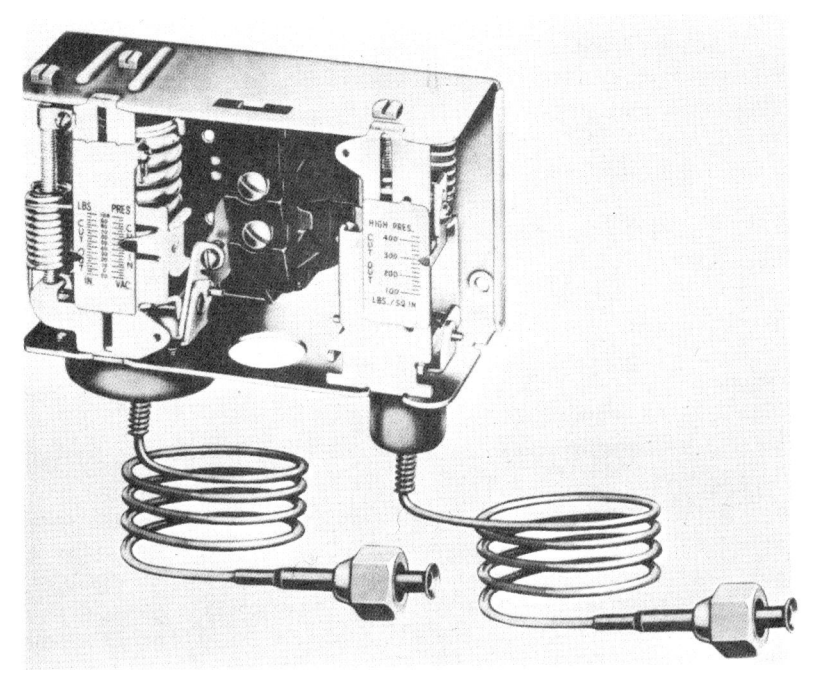

Courtesy Penn Controls, Inc

Fig. 11-5. Interior view of a dual-function pressure control.

CONTROL VALVES, DAMPERS, AND RELAYS

Control valves are designed to control the flow of fluids and may be actuated in various ways, such as by thermostat or other controlling device. Depending on their construction, they may be classified as *normally open* or *normally closed*. Depending on their application, they may be termed *single-seated, double-seated*, or *three-way*.

Single-seated valves are designed for tight shutoff. *Double-seated valves* are designed for applications where tight shutoff is not required. *Three-way valves* are designed to operate between two ports and to close one port as the other is opened.

AIR-CONDITIONING CONTROL METHODS

Dampers

Dampers are designed to control the flow of air or gases and may be operated manually, by electricity, or by compressed air. Depending on service conditions, they are known as the *single* or *multiblade* type. In large installations, dampers are motor operated, the motor being actuated by a controlling device and connected to the blades as required to give the desired movement.

Motors for operating dampers are provided when automatic regulation of airflow is required in heating systems, ventilating ducts, air-conditioning systems, and air-conditioning supply ducts for zone control. The motors are powered by electricity and controlled by the action of a thermostatic element or by compressed air. They can be connected to single- or multiple-louvered dampers. The dampers can be positioned by the motors as required and in accordance with indications from the governing device.

Two general classes of electrically operated motors are manufactured: positive and reversing. Each class is divided into several types, the positive motor being known as the *two-position positive*, *four-position positive*, and *five-position positive*, all unidirectional in operation and with or without speed adjustment. The reversing motor is known as the *two-position, four-position*, or *five-position* motor, which may be stopped, restarted in the original direction, or reversed at any point. These motors can be supplied with or without adjustable speed.

Standard three-wire thermostats or proportioning thermostats are generally supplied for controlling the operation of the motors, although hygrostats and other control instruments can be used when other types of regulation are required. Dampers can also be operated by thermostatic elements, the expansion and contraction of a liquid or gas supplying the power required to actuate the dampers.

Relays

Relays are a necessary part of many control and pilot circuits. By definition, a relay is a device that employs an auxiliary source of energy to amplify or convert the force of a controller into available energy at a valve or damper motor. Depending on their function in an air-conditioning system, relays may be classified as *pneumatic, switching,* and *positioning.*

A *pneumatic relay,* usually electrically operated, starts or stops the flow of air as required. A *switching relay* may be used to switch the operation of a controlled device from one controller to another. A *positioning relay* may be directly connected to a damper motor lever or valve and is affected by both valve or damper position.

REFRIGERANT CONTROL DEVICES

In order to control the flow of liquid refrigerant between the low and high side of a refrigeration system, some form of expansion device must be used. The various expansion and control devices necessary in order to obtain automatic control are as follows:

1. Automatic expansion valves
2. Thermostatic expansion valves
3. Low-side float valves
4. High-side float valves
5. Capillary tubes

Automatic Expansion Valves

The function of an automatic expansion valve in a mechanical refrigeration system is that of maintaining a constant pressure in the evaporator. Automatic expansion valves are usually employed on direct-expansion refrigeration systems but are not satisfactory in air conditioning due to load fluctuations. Fig. 11-6 shows a typical expansion valve.

Thermostatic Expansion Valves

The thermostatic expansion valve is designed to meter the flow of refrigerant into the evaporator in exact proportion to the rate of evaporation of the refrigerant in the evaporator, thus preventing the return of liquid refrigerant to the compressor. By being re-

AIR-CONDITIONING CONTROL METHODS

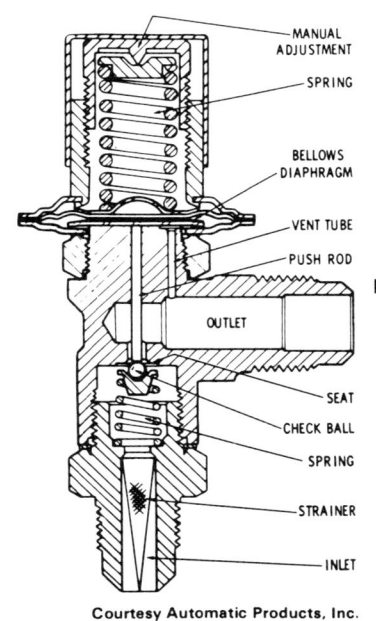

Fig. 11-6. Sectional view of an automatic expansion valve.

Courtesy Automatic Products, Inc.

sponsive to the temperature of the refrigerant gas leaving the evaporator and the pressure in the evaporator, the thermostatic expansion valve can control the refrigerant gas leaving the evaporator at a predetermined superheat.

The construction of a thermostatic expansion valve is shown in Fig. 11-7. In construction, it is similar to the automatic expansion valve except that it has, in addition, a thermostatic element. This thermostatic element is charged with a volatile substance (usually a refrigerant, the same as that used in the refrigeration system in which the valve is connected). It functions according to the pressure-temperature relationship and is connected (clamped) to the evaporator suction line of the evaporator and to the bellows operating diaphragm of the valve by a small sealed tube.

Its operation is governed by temperature. Thus, a rise in the evaporator temperature will increase the temperature of the evaporated gas passing through the suction line to which the thermostatic expansion-valve bulb is clamped. The bulb absorbs heat; and since its charge reacts in accordance with pressure-

339

AIR CONDITIONING

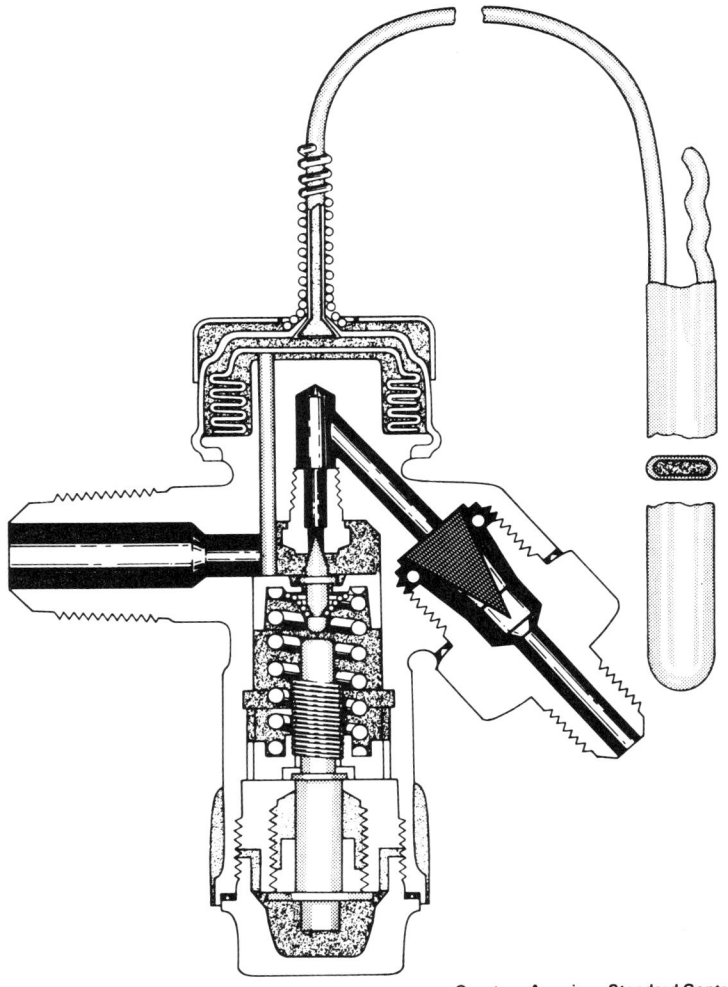

Fig. 11-7. Sectional view of a typical thermostatic expansion valve.
Courtesy American Standard Controls

temperature relations, the pressure tending to open the valve needle is increased, and it opens proportionately.

Briefly, the greater the evaporator gas temperature rise in the evaporator, the wider the valve opens, and vice versa. The wider the valve opens, the greater is the percentage of coil flooding,

AIR-CONDITIONING CONTROL METHODS

which improves heat transfer and also causes the compressor to operate at a higher average suction pressure. Hence, the compressor capacity may be somewhat increased, which increases the overall system capacity.

Sizing of Thermostatic Expansion Valves

The sizing of thermostatic expansion valves should be done carefully to avoid the evils of being undersize at full load and excessively oversize at partial load. The refrigerant pressure drops through the system (distributor, coil condenser, and refrigerant lines including liquid lifts) must be properly evaluated to arrive at the correct pressure drop available across the valve on which to base the selection. Variations in condensing pressure during the period of refrigeration requirements have a very important bearing on the pressure available across the valve and hence on its capacity.

Occasionally, valves turn out to be undersized because of either unexpected flash gas in the liquid line, liquid lifts that were not taken into consideration, or condensing temperatures below design. More often, however, they turn out to be oversized due to one or more of the following reasons:

1. Over-conservative selections on the high side.
2. Some manufacturers' valves are underrated.
3. Presence of much higher pressure drop available across the valve than was considered in the selection, or (more likely) it was not considered at all. This frequently happens on air-cooled condensing and low-temperature applications.
4. Sufficient increments in sizes not available for a close selection in all cases.
5. Grossly overestimating the refrigeration load.

Oversized thermostatic expansion valves do not control as well at full-system capacity as properly sized valves and become progressively worse as the coil load decreases. This results in a cycling condition: alternate flooding and starving the coil because the valve is attempting to throttle at a capacity below its capabilities, causing difficulties such as periodic flooding of liquid back to the compressor and wide variations in air temperature leaving the coil.

Capacity reduction available in most compressors further aggravates this problem and necessitates closer selection of expansion valves to match realistic loads. In systems having multiple coils, a solenoid valve can be located in the liquid line feeding each evaporator or group of evaporators, to be individually closed off as compressor capacity is reduced.

Low-Side Float Valves

Liquid-refrigerant control by means of low-side float valves consists essentially of a ball float located in either the receiver or the evaporator fastened to the low-pressure side of the refrigeration system. The ball float is fastened to a lever arm, which is pivoted at a given point and connected to a needle that seats at the valve opening. If there is no liquid in the evaporator, the ball-lever arm rests on a stop and the needle is not seated, thus leaving the valve open. When liquid refrigerant under pressure from the compressor again enters the float chamber, the float rises with the liquid until a predetermined level is reached and the needle closes the valve opening. Figure 11-8 shows a typical float valve.

In refrigeration work, if the boiling is proceeding at a slow rate, the float will tend to hold the needle off its seat at a point so that only a thin trickle of refrigerant enters the float chamber, the quantity entering as a liquid being identical to the quantity represented by the vapor leaving. If boiling goes on a rapid rate, the needle is held considerably off its seat and a considerable quantity of liquid will be admitted, the amount still being the same as that being removed as a vapor by the compressor. Since low-side float valves can be used only with a flooded evaporator, they have been used extensively on household refrigerators and to some extent in commercial refrigeration.

High-Side Float Valves

The high-side float valve gets its name because the float valve and its housing are subject to high-side pressure. The high-side float operates in reverse of a low side in that it only opens and ad-

AIR-CONDITIONING CONTROL METHODS

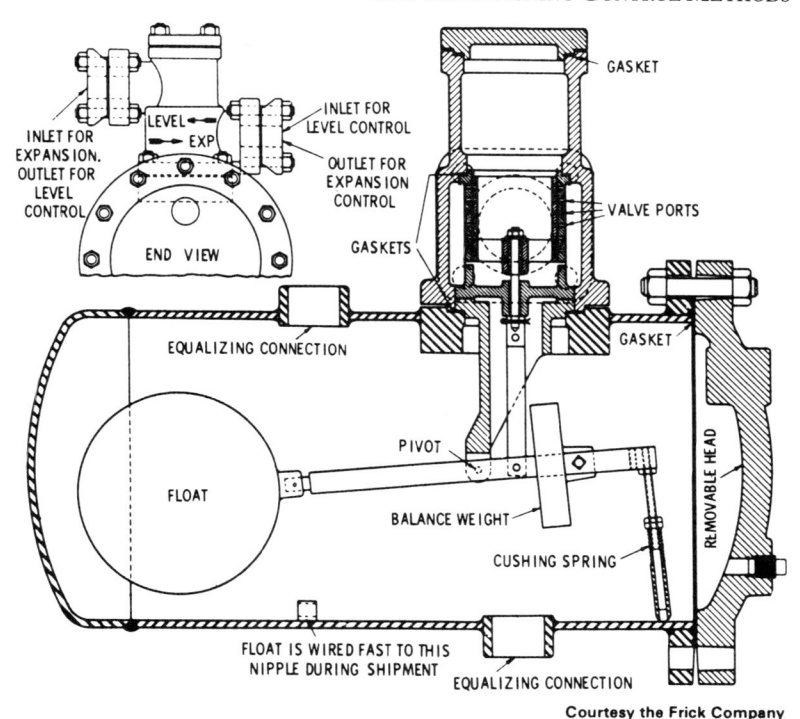

Fig. 11-8. Construction of a typical float valve.

mits liquid to the low side upon a rise in the liquid level in the float housing.

In the low-side float system, a receiver can be used to hold a surplus of liquid. As soon as the compressor shuts down, a little liquid will trickle past the needle, raise the liquid levels, and the float and needle will be forced tightly on its seat. Regardless of the quantity of liquid held in the receiver, no more will be admitted to the low-side float chamber until the machine has been started and some of the liquid vaporized. A little surplus is rather an excellent thing to have for there is bound to be a very tiny leak somewhere. If a leak develops, one large enough to smell or hear, the serviceman should be called to tighten the leaking joint.

Capillary Tubes

A capillary-tube, or restrictor, system may also be used as a liquid-refrigerant expanding device. It consists essentially of a tube with an extremely small bore having a length of from 5 to 20 feet. In this connection it should be noted that the bore diameter and the length of the tube are critical, and for these reasons its application has been largely limited to completely assembled factory refrigeration units.

The restrictor, or capillary-tube system accomplishes reduction in pressure from the condenser to the evaporator, not with a needle valve or pressure reducing valve, but by using the pressure drop or friction loss through a long small opening. With this system, there is no valve to separate the high-pressure zone of the condensing unit from the low-pressure zone of the unit and evaporator. Therefore, pressures through the system tend to equalize during the off cycle, being retarded only by the time required for the gas to pass through the small passage of the restrictor. The expansion system is sometimes referred to as being the *dry system*, while the low-side float system is the *flooded system*. This is sometimes true, but not necessarily so, for the expansion valve may be used on a flooded system where the evaporator is so designed as to take the low-pressure vapor out of the top with the evaporator practically filled with liquid refrigerant.

Crankcase Pressure-Regulating Valve

The crankcase pressure-regulating valve is designed to limit the compressor suction pressure below a preset limit to prevent compressor overloading. The valve should be located in the suction line between the evaporator and compressor. The valve setting is determined by a pressure spring, and the valve modulates from a fully open to a fully closed position in response to the outlet pressure, closing on a rise in the outlet pressure.

Use of the crankcase pressure-regulating valve permits the application of a larger displacement compressor without motor overloading. The pressure drop through the valve may result in an unacceptable loss in system capacity unless the valve is properly sized. A typical crankcase pressure-regulating valve is shown in Fig. 11-9.

AIR-CONDITIONING CONTROL METHODS

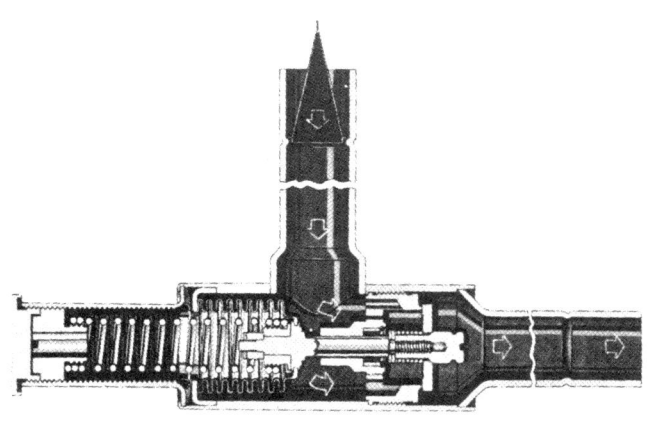

Courtesy Controls Company of America

Fig. 11-9. Sectional view of a crankcase pressure-regulating valve.

Evaporator Pressure-Regulating Valve

Evaporator pressure-regulating valves are used in multiple evaporators operating at different temperatures or on refrigerating systems where the evaporating temperature must not be allowed to fall below a certain predetermined value. This valve operates in a manner similar to that of the crankcase pressure-regulating valve except that it is responsive to inlet pressure. It should be installed in the suction line at the evaporator outlet (Fig. 11-10).

Reversing Valves

Reversing valves are used in certain air-conditioning applications operating on the *heat-pump* principle. Because of its flexibility in that it can supply heating and cooling alternately by means of a reversing valve, this method of year-round air conditioning has become increasingly popular.

In operation, the four-way reversing valve (Fig. 11-11) will enable the evaporator and condenser function to be switched by a change in refrigerant flow, as desired, so that the indoor coil becomes the evaporator for cooling purposes and the condenser for that of heating. In this manner, the outdoor coil functions as a condenser during the cooling cycle and as an evaporator during

345

AIR CONDITIONING

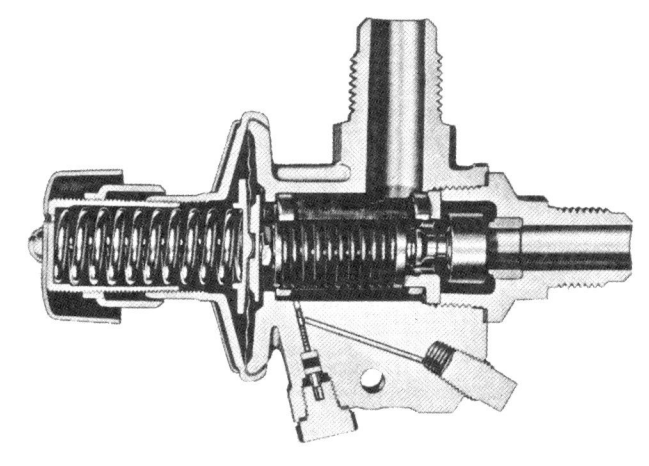

Courtesy Controls Company of America

Fig. 11-10. Sectional view of a typical evaporator pressure-regulating valve.

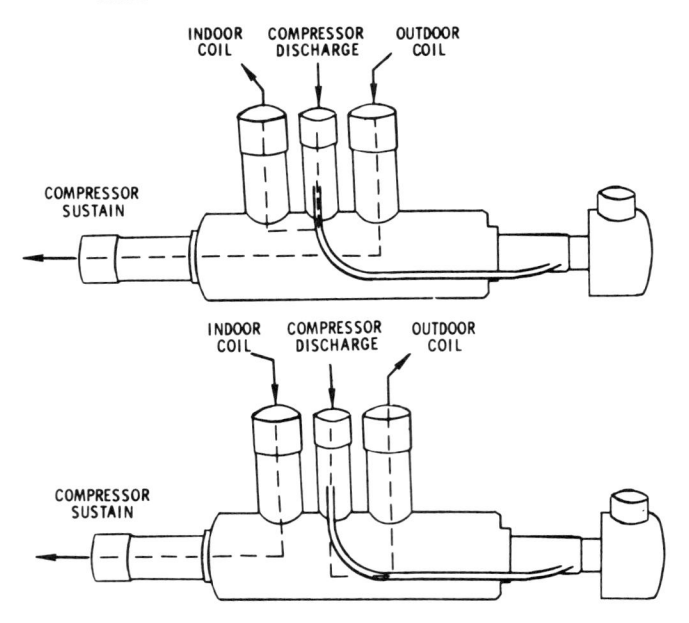

Courtesy Sporlan Valve Company

Fig. 11-11. Typical four-way reversing valve when actuated for cooling and heating, respectively.

AIR-CONDITIONING CONTROL METHODS

the heating cycle. The control of the valve is by means of a slide action actuated by a solenoid. Thus, the connection from the compressor suction and the discharge ports to the evaporator and condenser can be reversed to obtain either cooling or heating, as conditions require.

SUMMARY

Controls are essential for air-conditioning systems. Some means must be provided for regulating the temperature and relative humidity in order to maintain conditions that are (a) comfortable to human beings or (b) required for industrial applications.

There are three general types of air-conditioning controls: manual, automatic, and semiautomatic.

Devices used to control temperature, humidity, and pressure within predetermined limits are classified broadly as temperature control, humidity control, and pressure control. Sensing devices, such as thermostats, hygrostats, and pressure regulators work to control the valves that control the refrigeration systems.

A limit control is a device responsive to changes in pressure, temperature, or liquid level and is used to turn on, or throttle, the fuel supply to a heating appliance.

Control valves are designed to control the flow of fluids and may be actuated in various ways. They may be controlled by thermostats or some other device. Valves are used to control the flow of refrigerant. Single-seated valves are designed for tight shutoff. Double-seated valves are designed for applications where tight shutoff is not required. Three-way valves are designed to operate between two ports and to close one port as the other is opened.

Dampers are designed to control the flow of air or gases and may be operated manually, by electricity, or by compressed air.

Relays are a necessary part of many control and pilot circuits. A relay is a device that uses an auxiliary source of energy to amplify or convert the force of a controller into available energy at a valve or damper motor.

An automatic expansion valve in a mechanical refrigeration system maintains a constant pressure in the evaporator. Automatic expansion valves are usually employed on direct-expansion refrigeration systems. They are not suitable for use on air-conditioning systems due to load fluctuations.

The thermostatic expansion valve is designed to meter the flow of refrigerant into the evaporator in exact proportion to the rate of evaporation of the refrigerant in the evaporator. Thermostatic expansion valves should be sized carefully in order to avoid the evils of being undersize at full load and excessively oversize at partial load.

A capillary tube, or restrictor system may also be used as a liquid-refrigerant expanding device. The restrictor, or capillary-tube system accomplishes the reduction in pressure from the condenser to the evaporator, not with a needle valve or pressure-reducing valve, but by using the pressure drop or friction loss through a long small opening.

The crankcase pressure-regulating valve is designed to limit the suction pressure below a preset limit to prevent compressor overloading. Evaporator pressure-regulating valves are used in multiple evaporators operating at different temperatures or on refrigerating systems where the evaporating temperature must not be allowed to fall below a certain predetermined value. Reversing valves are used in certain air-conditioning applications operating on the heat-pump principle.

REVIEW QUESTIONS

1. Why are controls needed in an air-conditioning system?
2. What is the term for the general classification of air-conditioning controls?
3. What are the three basic control types?
4. What is a hygrostat?
5. What are three types of thermostats in common use?
6. Where are immersion thermostats used?
7. What is a humidistat?
8. Where do you use a limit control?
9. Why are dampers necessary?
10. What is a relay?
11. List five refrigerant-control devices.
12. What is the purpose of a thermostatic expansion valve?

AIR-CONDITIONING CONTROL METHODS

13. What is a capillary tube?
14. Where are crankcase pressure-regulating valves used?
15. Where are evaporator pressure-regulating valves used?
16. What is a reversing valve?
17. Where is the reversing valve used?
18. How does the four-way reversing valve operate?

CHAPTER 12

Weather Data and Design Conditions

This chapter includes recommended design conditions, indoor and outdoor, for summer and winter. Data on outdoor design conditions are presented for over 150 weather stations in the United States.

INDOOR DESIGN CONDITIONS

The indoor conditions to be maintained within a building are the dry-bulb temperature and relative humidity of the air at a breathing line 3 to 5 feet above the floor in an area that would indicate average conditions at that level and that is not affected by abnormal or unusual heat gains or losses from the interior or exterior.

Because it is difficult to maintain the mean radiant temperature at the dry-bulb temperature in actual practice, the effect of warmer walls normally encountered during the cooling season

Air Conditioning

will require a design condition lower than stated for optimum comfort. For this reason, the recommended 75°F design dry-bulb temperature, slightly lower than the optimum range of 76.5 to 77.6°F (with walls at room temperature), will partially compensate for the higher mean radiant temperature normally found in comfort-cooling applications.

During the heating season, the mean radiant temperature is normally lower than air temperature (in some panel-heating installations the mean radiant temperature could be higher than air temperature), thereby suggesting a higher dry-bulb temperature than 75°F for comfort conditions. Because the type of clothing an individual wears has an effect on the optimum indoor design temperature, and because the recent ASHRAE comfort tests were performed with lightly clothed subjects, the overall effect is a recommended 75°F design dry-bulb for both winter and summer. Where unusual application conditions exist, such as large exposures of glass and inside partitions exposed to adjacent roads of widely different temperatures, consideration should be given to adjustment of the design dry-bulb temperature to maintain comfort conditions. In no case should the correction be made for mean radiant temperatures outside the range of 70 to 80°F.

When the recommended design condition of 75°F is applicable, the dry-bulb temperature may vary from 73 to 77°F within the occupied zone (the region within a space between a level 3 inches above the floor and the 72-inch level and more than 2 feet from walls or fixed air-conditioning equipment).

The air temperature at the ceiling may vary beyond the comfort range and should be considered in the overall heat transmission to the outdoors. A nominal 0.75°F increase in air temperature per foot elevation above the breathing level would be expected in normal applications, with approximately 75°F difference between indoor and outdoor temperatures.

For summer cooling, a relative humidity of 50 percent may be chosen for the average job in the United States. In arid climates, a design relative humidity of 40 percent is more realistic. During winter conditions, the relative humidity to be maintained in the structure is dependent upon the severity of the outdoor conditions. The relative humidity should not exceed 60 percent at any point in the occupied space (to prevent material deterioration) and normally should not fall below 20 percent to prevent human nostrils from becoming dry and furniture from drying out.

OUTDOOR DESIGN CONDITIONS

Winter

Recommended design temperatures, together with the range of wind velocities associated with severe cold, are presented in Columns 4 and 5 of Table 12-1. All data are based on detailed records from official weather stations of the U.S. Weather Bureau, U.S. Air Force, and U.S. Navy.

With few exceptions, the data were from airport stations, where hourly observations have been made by trained observers for a sufficient number of years to permit a reasonably accurate definition of the climatic range for each station.

Three temperature levels are offered for each station. The *median of extremes* is the middle value of the coldest temperature recorded each year for periods up to 25 or 30 years and is approximately equal numerically to the *average annual minimum*, published in some previous editions. The 99 and 97 percent values represent temperatures that equalled or exceeded those proportions of the total hours (2160) in December, January, and February. In a normal winter there would be approximately 22 hours at or below the 99 percent value and approximately 54 hours at or below the 97 percent design value.

The wind data shown in Column 5 pertains specifically to velocities that occur coincident with periods of extreme cold. The four classes of velocity (very light, light, moderate, and high) are defined in footnote *e* of Table 12-1.

Summer

Recommended design dry- and wet-bulb temperatures and outdoor daily temperature ranges are presented in Columns 6 to 8 of Table 12-1. All data are based on detailed records from official weather stations of the U.S. Weather Bureau, U.S. Air Force, and U.S. Navy. With few exceptions, the data was from airport sta-

Table 12-1. Climatic Conditions for the United States*,[a]

			Winter					Summer						
			Col. 4			Col. 5	Col. 6			Col. 7	Col. 8			
Col. 1	Col. 2	Col. 3	Median of Annual Ex-			Coinci- dent Wind Ve-	Design Dry-Bulb			Out- door Daily	Design Wet-Bulb			
State and Station[b]	Latitude[c] ° '	Elev.,[d] ft	tremes	99%	97½%	locity[e]	1%	2½%	5%	Range[f]	1%	2½%	5%	
ALABAMA														
Birmingham AP	33 3	610	14	19	22	L	97	94	93	21	79	78	77	
Huntsville AP	34 4	619	−6	13	17	L	97	95	94	23	78	77	76	
Mobile CO	30 4	119	24	28	32	M	96	94	93	16	80	79	79	
ALASKA														
Anchorage AP	61 1	90	−29	−25	−20	VL	73	70	67	15	63	61	59	
Fairbanks AP	64 5	436	−59	−53	−50	VL	82	78	75	24	64	63	61	
Kodiak	57 3	21	4	8	12	M	71	66	63	10	62	60	58	
ARIZONA														
Flagstaff AP	35 1	6973	−10	0	5	VL	84	82	80	31	61	60	59	
Nogales	31 2	3800	15	20	24	VL	100	98	96	31	72	71	70	
Yuma AP	32 4	199	32	37	40	VL	111	109	107	27	79	78	77	
ARKANSAS														
Fayetteville AP	36 0	1253	3	9	13	M	97	95	93	23	77	76	75	
Hot Springs Nat. Pk.	34 3	710	12	18	22	M	99	97	96	22	79	78	77	
Texarkana AP	33 3	361	16	22	26	M	99	97	96	21	80	79	78	
CALIFORNIA														
San Fernando	34 1	977	29	34	37	VL	100	97	94	38	73	72	71	
San Francisco CO	37 5	52	38	42	44	VL	80	77	73	14	64	62	61	
Yreka	41 4	2625	7	13	17	VL	96	94	91	38	68	66	65	

354

Table 12-1. Climatic Conditions for the United States (Cont'd)*,a

Col. 1 State and Station[b]	Col. 2 Latitude[c] ° '	Col. 3 Elev.[d] ft	Winter Col. 4 Median of Annual Extremes	Winter Col. 4 99%	Winter Col. 4 97½%	Col. 5 Coincident Wind Velocity[e]	Summer Col. 6 Design Dry-Bulb 1%	Col. 6 2½%	Col. 6 5%	Col. 7 Outdoor Daily Range[f]	Col. 8 Design Wet-Bulb 1%	Col. 8 2½%	Col. 8 5%
COLORADO													
Alamosa AP	37 3	7536	−26	−17	−13	VL	84	82	79	35	62	61	60
Durango	37 1	6550	−10	0	4	M	88	86	83	30	64	63	62
Grand Junction AP	39 1	4849	−2	8	11	M	96	94	92	29	64	63	62
CONNECTICUT													
Hartford, Brainard Field	41 5	15	−4	1	5	M	90	88	85	22	77	76	74
New London	41 2	60	0	4	8	M	89	86	83	16	77	75	74
Windsor Locks, Bradley Field	42 0	169	−7	−2	2	M	90	88	85	22	76	75	73
DELAWARE													
Dover AFB	39 0	38	8	13	15	M	93	90	88	18	79	78	77
Wilmington AP	39 4	78	6	12	15	M	93	90	87	20	79	77	76
DISTRICT OF COLUMBIA													
Andrews AFB	38 5	279	9	13	16	L	94	91	88	18	79	77	76
Washington National AP	38 5	14	12	16	19	VL	94	92	90	18	78	77	76
FLORIDA													
Key West AP	24 3	6	50	55	58	VL	90	89	88	9	80	79	79
Sarasota	27 2	30	31	35	39	M	93	91	90	17	80	80	79
Tallahassee AP	30 2	58	21	25	29	H	96	94	93	19	80	79	79

355

Table 12-1. Climatic Conditions for the United States (Cont'd)*,[a]

Col. 1 State and Station[b]	Col. 2 Latitude[c] ° '	Col. 3 Elev.[d] ft	Winter Col. 4 Median of Annual Ex- tremes	99%	97½%	Col. 5 Coinci- dent Wind Ve- locity[e]	Summer Col. 6 Design Dry-Bulb 1%	2½%	5%	Col. 7 Out- door Daily Range[f]	Col. 8 Design Wet-Bulb 1%	2½%	5%
GEORGIA													
Americus	32 0	476	18	22	25	L	98	96	93	20	80	79	78
Dalton	34 5	702	10	15	19	L	97	95	92	22	78	77	76
Voldosta-Moody AFB	31 0	239	24	28	31	L	96	94	92	20	80	79	78
HAWAII													
Hilo AP	19 4	31	56	59	61	L	85	83	82	15	74	73	72
Honolulu AP	21 2	7	58	60	62	L	87	85	84	12	75	74	73
Wahiawa	21 3	215	57	59	61	L	86	84	83	14	75	74	73
IDAHO													
Burley	42 3	4180	−5	4	8	VL	95	93	89	35	68	66	64
Idaho Falls AP	43 3	4730r	−17	−12	−6	VL	91	88	85	38	65	64	62
Lewiston AP	46 2	1413	1	6	12	VL	98	96	93	32	67	66	65
ILLINOIS													
Carbondale	37 5	380	1	7	11	M	98	96	94	21	80	79	78
Freeport	42 2	780	−16	−10	−6	M	92	90	87	24	78	77	75
Peoria AP	40 4	652	−8	−2	2	M	94	92	89	22	78	77	76
INDIANA													
Jeffersonville	38 2	455	3	9	13	M	96	94	91	23	79	78	77
South Bend AP	41 4	773	−6	−2	3	M	92	89	87	22	77	76	74
Valpariso	41 2	801	−12	−6	−2	M	92	90	87	22	78	76	75

Table 12-1. Climatic Conditions for the United States (Cont'd)*,a

Col. 1 State and Station[b]	Col. 2 Latitude[c] ° '	Col. 3 Elev.,[d] ft	Winter Col. 4 Median of Annual Extremes	Winter Col. 4 99%	Winter Col. 4 97½%	Col. 5 Coincident Wind Velocity[e]	Summer Col. 6 Design Dry-Bulb 1%	Col. 6 2½%	Col. 6 5%	Col. 7 Outdoor Daily Range[f]	Col. 8 Design Wet-Bulb 1%	Col. 8 2½%	Col. 8 5%
IOWA Mason City AP	43 1	1194	−20	−13	−9	M	91	88	85	24	77	75	74
KANSAS													
Garden City AP	38 0	2882	−10	−1	3	M	100	98	96	28	74	73	72
Hutchinson AP	38 0	1524	−5	2	6	H	101	99	96	28	77	76	75
Wichita AP	37 4	1321	−1	5	9	H	102	99	96	23	77	76	75
KENTUCKY													
Ashland	38 3	551	1	6	10	L	94	92	89	22	77	76	75
Covington AP	39 0	869	−3	3	8	L	93	90	88	22	77	76	75
Hopkinsville, Campbell AFB	36 4	540	4	10	14	L	97	95	92	21	79	78	77
LOUISIANA													
Alexandria AP	31 2	92	20	25	29	L	97	95	94	20	80	80	79
Natchitoches	31 5	120	17	22	26	L	99	97	96	20	81	80	79
New Orleans AP	30 0	3	29	32	35	M	93	91	90	16	81	80	79
MAINE													
Augusta AP	44 2	350	−13	−7	−3	M	88	86	83	22	74	73	71
Caribou AP	46 5	624	−24	−18	−14	L	85	81	78	21	72	70	68
Waterville	44 3	89	−15	−9	−5	M	88	86	82	22	74	73	71
MARYLAND Baltimore AP	39 1	146	8	12	15	M	94	91	89	21	79	78	77

Table 12-1. Climatic Conditions for the United States (Cont'd)*,[a]

Col. 1 State and Station[b]	Col. 2 Latitude[c] °	Col. 3 Elev.,[d] ft	Winter Col. 4 Median of Annual Ex-tremes	99%	97½%	Col. 5 Coincident Wind Ve-locity[e]	Summer Col. 6 Design Dry-Bulb 1%	2½%	5%	Col. 7 Outdoor Daily Range[f]	Col. 8 Design Wet-Bulb 1%	2½%	5%
MARYLAND (continued)													
Baltimore CO	39 2	14	12	16	20	M	94	92	89	17	79	78	77
Cumberland	39 4	945	0	5	9	L	94	92	89	22	76	75	74
MASSACHUSETTS													
Framingham	42 2	170	−7	−1	3	M	91	89	86	17	76	74	73
Greenfield	42 3	205	−12	−6	−2	M	89	87	84	23	75	74	73
New Bedford	41 4	70	3	9	13	H	86	84	81	19	75	73	72
MICHIGAN													
Detroit Met. CAP	42 2	633	0	4	8	M	92	88	85	20	76	75	74
Flint AP	43 0	766	−7	−1	3	M	89	87	84	25	76	75	74
Sault Ste. Marie AP	46 3	721	−18	−12	−8	L	83	81	78	23	73	71	69
MINNESOTA													
Alexandria AP	45 5	1421	−26	−19	−15	L	90	88	85	24	76	74	72
Bemidji AP	47 3	1392	−38	−32	−28	L	87	84	81	24	73	72	71
Minneapolis/St. Paul AP	44 5	822	−19	−14	−10	L	92	89	86	22	77	75	74
MISSISSIPPI													
Biloxi− Keesler AFB	30 2	25	26	30	32	M	93	92	90	16	82	81	80
Columbus AFB	33 4	224	13	18	22	L	97	95	93	22	79	79	78
Jackson AP	32 2	330	17	21	24	L	98	96	94	21	79	78	78

Table 12-1. Climatic Conditions for the United States (Cont'd)*,[a]

Col. 1 State and Station[b]	Col. 2 Latitude[c] ° '	Col. 3 Elev.[d] ft	Winter Col. 4 Median of Annual Extremes	99%	97½%	Col. 5 Coincident Wind Velocity[e]	Summer Col. 6 Design Dry-Bulb 1%	2½%	5%	Col. 7 Outdoor Daily Range[f]	Col. 8 Design Wet-Bulb 1%	2½%	5%
MISSOURI													
Kirksville AP	40 1	966	-13	-7	-3	M	96	94	91	24	79	78	77
Rolla	38 0	1202	-3	3	7	M	97	95	93	22	79	78	77
Sikeston	36 5	318	4	10	14	L	98	96	94	21	80	79	78
MONTANA													
Butte AP	46 0	5526r	-34	-24	-16	VL	86	83	80	35	60	59	57
Helena AP	46 4	3893	-27	-17	-13	L	90	87	84	32	65	63	61
Missoula AP	46 5	3200	-16	-7	-3	VL	92	89	86	36	65	63	61
NEBRASKA													
Chadron AP	42 5	3300	-21	-13	-9	M	97	95	92	30	72	70	69
Hastings	40 4	1932	-11	-3	1	M	98	96	94	27	77	75	74
Kearney	40 4	2146	-14	-6	-2	M	97	95	92	28	76	75	74
NEVADA													
Elko AP	40 5	5075	-21	-13	-7	VL	94	92	90	42	64	62	61
Las Vegas AP	36 1	2162	18	23	26	VL	108	106	104	30	72	71	70
Reno AP	39 3	4404	-2	2	7	VL	95	92	90	45	64	62	61
NEW HAMPSHIRE													
Berlin	44 3	1110	-25	-19	-15	L	87	85	82	22	73	71	70
Keene	43 0	490	-17	-12	-8	M	90	88	85	24	75	73	72
Portsmouth, Pease AFB	43 1	127	-8	-2	3	M	88	86	83	22	75	73	72

359

Table 12-1. Climatic Conditions for the United States (Cont'd)*,a

Col. 1 State and Station[b]	Col. 2 Latitude[c] ° '	Col. 3 Elev.,[d] ft	Winter Col. 4 Median of Annual Extremes	99%	97½%	Col. 5 Coincident Wind Velocity[e]	Summer Col. 6 Design Dry-Bulb 1%	2½%	5%	Col. 7 Outdoor Daily Range[f]	Col. 8 Design Wet-Bulb 1%	2½%	5%
NEW JERSEY													
Atlantic City CO	39 3	11	10	14	18	H	91	88	85	18	78	77	76
Newark AP	40 4	11	6	11	15	M	94	91	88	20	77	76	75
Phillipsburg	40 4	180	1	6	10	L	93	91	88	21	77	76	75
NEW MEXICO													
Clovis AP	34 3	4279	2	14	17	L	99	97	95	28	70	69	68
Grants	35 1	6520	-15	-7	-3	VL	91	89	86	32	64	63	62
Las Cruces	32 2	3900	13	19	23	L	102	100	97	30	70	69	68
NEW YORK													
Geneva	42 5	590	-8	-2	2	M	91	89	86	22	75	73	72
Massena AP	45 0	202r	-22	-16	-12	M	86	84	81	20	75	74	72
NYC-Kennedy AP	40 4	16	12	17	21	H	91	87	84	16	77	76	75
NORTH CAROLINA													
Henderson	36 2	510	8	12	16	L	94	92	89	20	79	78	77
Rocky Mount	36 0	81	12	16	20	L	95	93	90	19	80	79	78
Wilmington AP	34 2	30	19	23	27	L	93	91	89	18	82	81	80
NORTH DAKOTA													
Bismarck AP	46 5	1647	-31	-24	-19	VL	95	91	88	27	74	72	70
Devil's Lake	48 1	1471	-30	-23	-19	M	93	89	86	25	73	71	69
Williston	48 1	1877	-28	-21	-17	M	94	90	87	25	71	69	67

Table 12-1. Climatic Conditions for the United States (Cont'd)*,a

Col. 1 State and Station[b]	Col. 2 Latitude[c] ° '	Col. 3 Elev.[d] ft	Winter Col. 4 Median of Annual Extremes	Winter Col. 4 99%	Winter Col. 4 97½%	Col. 5 Coincident Wind Velocity[e]	Summer Col. 6 Design Dry-Bulb 1%	Col. 6 2½%	Col. 6 5%	Col. 7 Outdoor Daily Range[f]	Col. 8 Design Wet-Bulb 1%	Col. 8 2½%	Col. 8 5%
OHIO													
Cincinnati CO	39 1	761	2	8	12	L	94	92	90	21	78	77	76
Norwalk	41 1	720	−7	−1	3	M	92	90	87	22	76	75	74
Toledo AP	41 4	646r	−5	1	5	M	92	90	87	25	77	75	74
OKLAHOMA													
Ardmore	34 2	880	9	15	19	H	103	101	99	23	79	78	77
Oklahoma City AP	35 2	1280	4	11	15	H	100	97	95	23	78	77	76
Woodward	36 3	1900	−3	4	8	H	103	101	98	26	76	74	73
OREGON													
Astoria AP	46 1	8	22	27	30	M	79	76	72	16	61	60	59
Baker AP	44 5	3368	−10	−3	1	VL	94	92	89	30	66	65	63
The Dalles	45 4	102	7	13	17	VL	93	91	88	28	70	68	67
PENNSYLVANIA													
Butler	40 4	1100	−8	−2	2	L	91	89	86	22	75	74	73
Sunbury	40 5	480	−2	3	7	L	91	89	86	22	76	75	74
West Chester	40 0	440	4	9	13	M	92	90	87	20	77	76	75
RHODE ISLAND													
Newport	41 3	20	1	5	11	H	86	84	81	16	75	74	73
Providence AP	41 4	55	0	6	10	M	89	86	83	19	76	75	74

361

Table 12-1. Climatic Conditions for the United States (Cont'd)*,a

Col. 1 State and Station[b]	Col. 2 Latitude[c] °	Col. 3 Elev.,[d] ft	Winter Col. 4 Median of Annual Extremes	99%	97½%	Col. 5 Coincident Wind Velocity[e]	Summer Col. 6 Design Dry-Bulb 1%	2½%	5%	Col. 7 Outdoor Daily Range[f]	Col. 8 Design Wet-Bulb 1%	2½%	5%
SOUTH CAROLINA													
Charleston CO	32 5	9	23	26	30	L	95	93	90	13	81	80	79
Columbia AP	34 0	217	16	20	23	L	98	96	94	22	79	79	78
Rock Hill	35 0	470	13	17	21	L	97	95	92	20	78	77	76
SOUTH DAKOTA													
Aberdeen AP	45 3	1296	−29	−22	−18	L	95	92	89	27	77	75	74
Huron AP	44 3	1282	−24	−16	−12	L	97	93	90	28	77	75	74
Pierre AP	44 2	1718r	−21	−13	−9	M	98	96	93	29	76	74	73
TENNESSEE													
Bristol-Tri City AP	36 3	1519	−1	11	16	L	92	90	88	22	76	75	74
Chattanooga AP	35 0	670	11	15	19	L	97	94	92	22	78	78	77
Tullahoma	35 2	1075	7	13	17	L	96	94	92	22	79	78	77
TEXAS													
Brownsville AP	25 5	16	32	36	40	M	94	92	91	18	80	80	79
Killeen-Gray AFB	31 0	1021	17	22	26	M	100	99	97	22	78	77	76
Pampa	35 3	3230	0	7	11	M	100	98	95	26	73	72	71
UTAH													
Provo	40 1	4470	−6	2	6	L	96	93	91	32	67	66	65
St. George CO	37 1	2899	13	22	26	VL	104	102	99	33	71	70	69
Vernal AP	40 3	5280	−20	−10	−6	VL	90	88	84	32	64	63	62

362

Table 12-1. Climatic Conditions for the United States (Cont'd)*,[a]

Col. 1 State and Station[b]	Col. 2 Latitude[c] °	Col. 3 Elev.[d] ft	Winter Col. 4 Median of Annual Extremes	Winter Col. 4 99%	Winter Col. 4 97½%	Col. 5 Coincident Wind Velocity[e]	Summer Col. 6 Design Dry-Bulb 1%	Col. 6 2½%	Col. 6 5%	Col. 7 Outdoor Daily Range[f]	Col. 8 Design Wet-Bulb 1%	Col. 8 2½%	Col. 8 5%
VERMONT													
Barre	44 1	1120	-23	-17	-13	L	86	84	81	23	73	72	70
Burlington AP	44 3	331	-18	-12	-7	M	88	85	83	23	74	73	71
Rutland	43 3	620	-18	-12	-8	L	87	85	82	23	74	73	71
VIRGINIA													
Charlottesville	38 1	870	7	11	15	L	93	90	88	23	79	77	76
Harrisonburg	38 3	1340	0	5	9	L	92	90	87	23	78	77	76
Norfolk AP	36 5	26	18	20	23	M	94	91	89	18	79	78	78
WASHINGTON													
Bellingham AP	48 5	150	8	14	18	L	76	74	71	19	67	65	63
Moses Lake, Larson AFB	47 1	1183	-14	-7	-1	VL	96	93	90	32	68	66	65
Seattle CO	47 4	14	22	28	32	L	81	79	76	19	67	65	64
WEST VIRGINIA													
Elkins AP	38 5	1970	-4	1	5	L	87	84	82	22	74	73	72
Huntington CO	38 2	565r	4	10	14	L	95	93	91	22	77	76	75
Wheeling	40 1	659	0	5	9	L	91	89	86	21	76	75	74
WISCONSIN													
Ashland	46 3	650	-27	-21	-17	L	85	83	80	23	73	71	69
Beloit	42 3	780	-13	-7	-3	M	92	90	87	24	77	76	75
Sheboygan	43 4	648	-10	-4	0	M	89	87	84	20	76	74	72

Air Conditioning

Table 12-1. Climatic Conditions for the United States (Cont'd)*,a

Col. 1 State and Station[b]	Col. 2 Latitude[c] ° '	Col. 3 Elev.[d] ft	Winter Col. 4 Median of Annual Extremes	99%	97½%	Col. 5 Coincident Wind Velocity[e]	Summer Col. 6 Design Dry-Bulb 1%	2½%	5%	Col. 7 Outdoor Daily Range[f]	Col. 8 Design Wet-Bulb 1%	2½%	5%
WYOMING													
Casper AP	42 5	5319	−20	−11	−5	L	92	90	87	31	63	62	60
Cheyenne AP	41 1	6126	−15	−6	−2	M	89	86	83	30	63	62	61
Lander AP	42 5	5563	−26	−16	−12	VL	92	90	87	32	63	62	60

* Data for U.S. stations extracted from *Evaluated Weather Data for Cooling Equipment Design, Addendum No. 1, Winter and Summer Data*, with the permission of the publisher, Fluor Products Company, Inc. Box 1267, Santa Rosa, CA.

a Data compiled from official weather stations, where hourly weather observations are made by trained observers and from other sources. Table 12-1 prepared by ASHRAE Technical Committee 2.2, Weather Data and Design Conditions. Percentage of winter design data show the percent of a three-month period, December thru February. Canadian data are based on January only. Percentage of summer design data show the percent of a four-month period, June thru September. Canadian data are based on July only.

b When airport temperature observations were used to develop design data, "AP" follows station name, and "AFB" follows Air Force Bases. Data for stations followed by "CO" came from office locations within an urban area and generally reflect an influence of the surrounding area. Stations without designation can be considered semirural and may be directly compared with most airport data.

c Latitude is given to the nearest 10 minutes, for use in calculating solar loads. For example, the latitude for Anniston, Alabama is given as 33.4 or 33°40″.

d Elevations are ground elevations for each station as of 1964. Temperature readings are generally made at an elevation of 5 feet above ground, except for locations marked r, indicating roof exposure of thermometer.

e Coincident wind velocities derived from approximately coldest 600 hours out of 20,000 hours of December through February data per station. VL=Very Light, 70 percent or more of cold extreme hours ≤ 7 mph   M=Moderate, 50 to 74 percent cold extreme hours > 7 mph   L=Light, 50 to 69 percent cold extreme hours ≤ 7 mph   H=High, 75 percent or more extreme hours > 7 mph, and 50 percent are > 12 mph.

f The difference between the average maximum and average minimum temperatures during the warmest month.

† More detailed data on Arizona, California, and Nevada may be found in *Recommended Design Temperatures, Northern California*, published by the Golden Gate Chapter, and *Recommended Design Temperatures, Southern California, Arizona, Nevada*, published by the Southern California Chapter.

tions where hourly observations have been made by trained observers for a sufficient period of years to permit reasonably accurate definition of the climatic range for each location.

The outdoor daily range of dry-bulb temperatures shown in Column 7 gives the difference between the average maximum and average minimum temperatures during the warmest month at each station. Large daily ranges are listed for inland stations at long distances from large bodies of water and for stations at higher elevations.

The design data on dry- and wet-bulb temperatures presented in Columns 6 and 8 represent the highest 1, 2, and 5 percent of all the hours (2928) during the summer months of June through September. In a normal summer there would be approximately 30 hours at or above the 1 percent design value and approximately 150 hours at or above the 5 percent design value.

In terms of enthalpy (heat content), an interval of 3°F dry-bulb temperature is approximately equal to a 1°F interval in wet-bulb temperature in the temperature ranges that prevail during the summer.

All data presented in Columns 6 and 8 are based on the average number of hours at or above a given design value over a period of several seasons. It follows that about half the summers would have more hours at or above a particular design level, and half would have fewer such hours.

INTERPOLATION BETWEEN STATIONS

Data from many weather stations at specific locations and at published elevations furnish a network from which, by interpolation, good estimates can be made of the expected conditions at a precise location:

1. *Adjustment for Elevation* — For a lower elevation, the design values from Table 12-1 should be increased; while for a higher elevation, the values should be decreased. The increments used in these adjustments are:

    Dry-bulb temperature: 1°F/200 feet
    Wet-bulb temperature: 1°F/500 feet

AIR CONDITIONING

2. *Adjustment for Air-mass Modification* — Short distance variations are most extreme near large bodies of water where air moves from the water over the land in the summer. Along the West Coast, both dry- and wet-bulb temperatures increase with distance from the ocean. In the region north of the Gulf of Mexico, dry-bulb temperatures increase for the first 200 or 300 miles, with a very slight decrease in wet-bulb temperature due to mixing and drier inland air. Beyond this 200- to 300-mile belt, both dry- and wet-bulb values tend to decrease at a somewhat regular rate.
3. *Adjustment for Vegetation* — The difference between large areas of dry surfaces and large areas of dense foliage upwind from the site can account for variations of up to 2°F wet-bulb and 5°F dry-bulb temperature. The warmer temperatures are associated with the dry surfaces. If the nearest official weather station is well surrounded by buildings and streets (large dry surfaces) and the precise location where design data are needed is similarly surrounded, no adjustment is needed. But if the local exposure at the location site is quite different from the nearest weather station listed in Table 12-1, an adjustment of 1°F wet-bulb and 2°F dry-bulb temperature may be easily justified.

WEATHER-ORIENTED DESIGN FACTORS

The rational approach to air-conditioning system design involves computation of a peak design load at a condition established using one of the *frequency of occurrence* levels (Table 12-1, Columns 6 and 8) of dry- and wet-bulb weather data published herein.

Although the values enumerated are statistically quite accurate, certain precautions are recommended concerning their use. These figures are frequency-distribution statistics for the ends of the distribution. During the four warmest summer months, these values occur at an hour of the day usually between 2:00 P.M. and 4:00 P.M., suntime. The winter statistics occur during the three coldest months at an hour of the day usually between 6:00 A.M. and 8

A.M., suntime. The daily dry-bulb variation will be of the order of the daily range stated on typical design days; however, statistically the daily range (Table 12-1, Column 7) is the long-term average daily range for only the warmest month. The daily range is generally greater during clear weather and much less during cloudy weather or precipitation.

A very crude approximation of the distribution of the hourly dry-bulb temperatures during a design day can be made by locating the daily maximum at 3:00 P.M. and the minimum (maximum less daily range) at 7:00 A.M. This day basically is made up of 12 or more daylight hours of solar heating and 12 or less hours of nighttime radiant cooling.

Throughout the continental or inland mass of the United States, the maximum dry-bulb and maximum wet-bulb temperatures are not coincident. As a matter of fact, the 1 percent level values rarely occur at the same time in such areas. In maritime areas they tend to be coincident. Typically, the maximum dry-bulb temperature is coincident with a wet-bulb substantially below the maximum wet-bulb temperature and vice versa. The assumption of dry- and wet-bulb coincidence can result in weather-oriented loads up to one-half greater than might otherwise be expected. A rational solution would be to determine whether the structure was the most sensitive to dry-bulb, i.e., extensive exterior, or wet-bulb, i.e., outside-air ventilation. Then the appropriate maximum temperature could be used for design with its corresponding coincident value. This could be ascertained for any station by an analysis involving usually only a day or two of hand calculations. An outline of this approach follows.

From at least 5 years of local weather records, select days containing the maximum wet- and dry-bulb entries in any time frame (hourly, 3-hourly, etc.) for that station. On the published 3-hourly increments, tabulate the dry-bulb peak with the corresponding wet-bulb and the wet-bulb with its corresponding dry bulb. This tabulation will illustrate not only the local coincident data, but also the 3-hourly temperature array for a typical design day which is accurate within a degree or two.

This procedure is certainly within the capability of any engineering office that is involved in several local projects. Where projects of greater significance are involved, there are several alternatives. A longer period of record such as 10 years can be studied. Hourly data instead of the usual 3-hour interval can be used. Machine tabulations can be made on any desired criteria; consulting meteorologists are available to analyze both the raw data and the statistical compilations.

As more complicated and sophisticated applications are utilized to meet critical demands, these design maximums based on peak values will not suffice. Engineers will have to consider offpeak values. Some types of days that must be accommodated are more frequent in occurrence than maximum or minimum design days. Examples of these are: Cloudy—Small temperature change; Windy; Warm A.M.; Cool P.M.; and, of course, Fair and Warm, and Fair and Cool. Quite often due to temperature-control implication, these days must be studied before a final design can be implemented.

Since the advent of machine computations, many system designers are making calculations of dry-bulb temperature with its corresponding wet-bulb maximum, and vice versa, for several hours daily on both a room-by-room basis and a zone or building basis. This can result in more accurate design data as well as more compatible systems in operation. Furthermore, the trend of the industry to attempt energy-consumption estimates in order to establish system design is making some progress.

SUMMARY

The indoor conditions to be maintained within a building are the dry-bulb temperature and relative humidity of the air at a breathing line 3 to 5 feet above the floor in an area that would indicate average conditions at that level.

For summer cooling, a relative humidity of 50 percent may be chosen for the average job in the United States. In arid climates a design relative humidity of 40 percent is more realistic. The relative humidity should not exceed 60 percent at any point in the occupied space and normally should not fall below 20 percent to prevent human nostrils from becoming dry and furniture from drying out.

Adjustments must be made for a number of factors when you are designing an air-conditioning system for human comfort.

The trend of the air-conditioning industry is to attempt energy-consumption estimates in order to establish system design. This trend seems to be making some progress.

REVIEW QUESTIONS

1. What is the distance above the floor that is used as a standard for making comfort measurements in air conditioning?
2. What is the difference between dry- and wet-bulb temperature readings?
3. How long were temperature readings taken in order to establish a standard for the region?
4. Were the temperature readings taken for each locality?
5. During the summer, what is the heat content shown by an interval of 3°F dry-bulb in terms of wet-bulb temperatures?
6. How do you make adjustments for air-mass modification?
7. What is the adjustment made for vegetation?
8. Do the dry- and wet-bulb maximum temperatures coincide throughout the continental or inland mass of the United States?
9. What is the trend of the air-conditioning industry in regard to using energy-consumption figures instead of temperatures to determine design factors and load characteristics for air-conditioning equipment?

CHAPTER 13

Year-Round Central Air Conditioning

Year-round air conditioning is required to warm the air to a comfortable temperature during the winter and, in addition, wash, filter, and humidify it. During the summer, the air-conditioning system is required to cool the air plus perform the necessary humidification or dehumidification as required. A year-round air-conditioning system should accomplish the following:

1. Eliminate the dust, soot, and germs from the air entering the intake ducts
2. Warm the air in winter
3. Cool the air in summer
4. Increase the humidity in winter operation
5. Decrease the relative humidity or moisture content of the air in summer operation
6. Circulate the conditioned air evenly and pleasantly throughout the building at all times

AIR CONDITIONING

Modern year-round central air conditioning systems may be classified with respect to nature of service as: *gas-compression system*, *absorption system*, and *steam-jet vacuum system*.

The gas-compression, or mechanical air-conditioning method consists simply of a refrigeration system employing a mechanical compression device (compressor) to remove the low-pressure refrigerant enclosed in the low-pressure side and deliver it to the high-pressure side of the system.

An absorption air-conditioning system is one in which the refrigerant gas evolved in the evaporator is taken up by an absorber and released in a generator upon application of heat. A steam-jet air-conditioning system is a vacuum-refrigeration system in which high-pressure steam supplied through a nozzle and acting to eject water from the evaporator maintains the requisite low pressure in the evaporator, producing the required pressure on the high side by virtue of compression in a following diffusion passage.

CENTRAL SYSTEM FEATURES

A central air-conditioning system is generally a system designed for assembly in the field rather than in the factory as a unit. The principal advantage in a central air-conditioning system lies in the fact that one condensing unit servicing several rooms will result in lower cost than a number of self-contained units serving single rooms.

Central air-conditioning systems are usually provided with air ducts to the various rooms served in addition to a properly placed exhaust fan to effect complete removal and disposal of any desired proportion of the air.

Air-Supply System

A fresh air intake should be located at some strategic point so that dust, dirt, and fumes from the street or odors from kitchens and surrounding buildings will not be drawn into the house. The air taken into the fresh air intake should be as pure as possible, for while the system will remove the undesirable dust and dirt by means of filters, highly impure air will only place an additional burden on the system and result in an increase in the cost of operation. Figs. 13-1 and 13-2 show the typical equipment for a central year-round air-conditioning system.

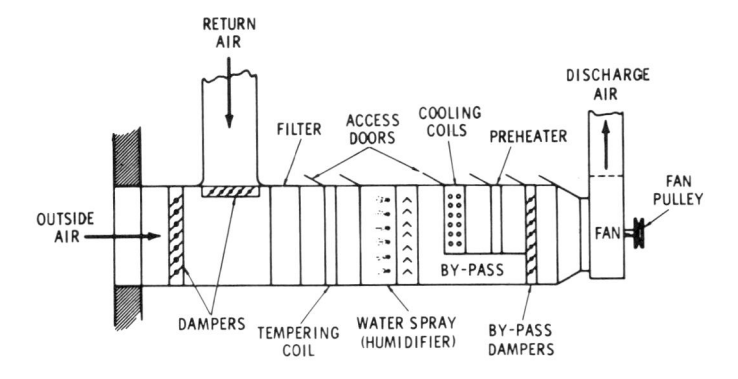

Fig. 13-1. Equipment for a central year-round air-conditioning system.

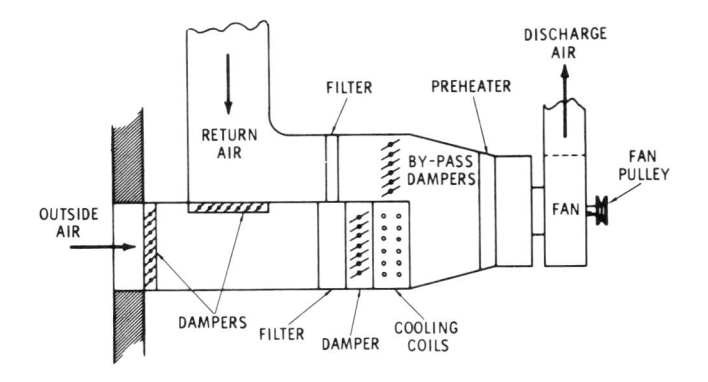

Fig. 13-2. Alternative arrangement of equipment for a central year-round air-conditioning system.

Mixing Chamber

A mixing chamber is generally incorporated in the system for the purpose of recirculating a portion of the air within the building in order to reduce the load and operating costs. In the summer the cooled air and in the winter the heated air is mixed with the

AIR CONDITIONING

fresh air drawn in from outside, so that the load is reduced considerably. A certain amount of fresh air is drawn in and mixed with the recirculated air, which may be accomplished by the use of a simple mixing damper. The size of the mixing chamber depends on the size of the installation and, like the ducts that carry the air, must be made airtight so that little or no air leakage occurs. Concrete, wood, or metal can be used in this construction.

Heating Coils

Where cold weather conditions prevail, it is necessary to make use of a heating or tempering coil to warm the intake air, for in such localities the outside air during the winter months may have such a low temperature that the heat from the recirculated air may not be sufficient to raise the temperature above freezing. As a result, the water used in the sprays may freeze and destroy the efficiency of the washer.

Tempering heaters must be used wherever unusually cold conditions are found. In many cases, these conditions do not occur every year or for any length of time, but if they occur at all, tempering heaters must be used. There are several different types of heaters, depending on the type of heat used in the building. The electric heater requires little space but is expensive to operate unless electricity is available at very low rates. In most instances, a cast-iron heating bank or stack is used and makes use of the steam of the building. In special instances, a boiler is used in direct conjunction with the apparatus.

Preheaters

A preheater is used for adding heat to the airstream when the temperature of the incoming air is so low that the tempering heaters on which the heating load falls are unable to maintain the required 68° or 70°F at the fan.

The preheater is necessary only where extreme temperature conditions are encountered. In many instances, preheaters are eliminated as unnecessary or tempering heaters are constructed with an overabundance of surface or sections, so that, when required, additional heating surface is available to carry the heating load.

Cooling Coils

Cooling coils are used for the purpose of cooling the air during the summer months. Cooling coils may be chilled by means of refrigerants or a pump-forced liquid, such as brine or water. A watertight drainage tank must be installed under the cooling coil, and the accumulated water should be drained by means of a properly air-vented waste trap.

If water of the proper coolness is obtainable in sufficient quantity, as is the case in many northern cities, it may be employed, requiring only a pumping apparatus to lift it out of the ground. Where it is difficult to secure water of proper temperature, or where it exists so deeply underground that pumping costs would be excessive, it then becomes necessary to use some other method of cooling the water used in the washers.

Cold water can be secured by the use of ice or mechanical refrigeration equipment. Mechanical apparatus, while expensive, is available at all times to cope with sudden loads, such as weather-conditions. Ice has no real initial cost and, of course, is dependable, for it can develop no mechanical trouble. However, if not used up, it will melt away without any returns being derived from it.

Air-Supply Fans

Supply fans and blowers are made in various types for specific purposes, and the selection of the proper type is important if the operation is to be economical and efficient. In general, fans used in air-conditioning work may be divided into two types: *propeller* and *centrifugal and turbine.*

Propeller fans are used where there are short ducts or when there are no ducts. The low speed of the blades near the hub, compared with the peripheral speed, makes this form of fan unsuitable when used to overcome the heavy resistance encountered in long ducts. The air driven by the effective blade areas near the rim finds it easier to pass back through the less effective blade areas at the hub than to push against the duct resistance.

Propeller fans are adapted to all installations to either supply or exhaust air without the use of a duct, or at most, a very short duct. Due to noise tendency, this type of fan is more practical for commercial and industrial work. Noise and vibration incident to operation are not objectionable in the commercial field but would be objectionable in other installations.

The majority of air-conditioning installations, other than the portable or self-contained room coolers, require long air ducts to carry the conditioned air to the various parts of the building. Centrifugal and turbine-type fans are employed for the purpose.

Filters

Air filters may be mechanically cleaned, of the replaceable type, or of the electronic type. The air filter is one part of the air-conditioning system that can be operated the year round. There are times when the air washer, cooling, and humidifying apparatus are not needed, but the filter is one part of the system that may be kept in continuous operation.

Air-Washers

The air washer is a unit that brings the air into intimate contact with water or a wet surface. In some cases, the air is passed over the wet surface. A more intimate contact is afforded where the water is divided into a fine spray and the air passed through the spray chamber.

The water is generally circulated by means of a pump, the warm water being passed over refrigerating coils or blocks of ice to cool it before being passed to the spray chamber. The water lost in evaporation is usually automatically replaced by the use of a float arrangement, which admits water from the main as required. In many localities, the water is sufficiently cool to employ just as it is drawn from the ground. In other places, the water is not cool enough and must be cooled by means of ice or a refrigerating machine.

The principal functions of the air washer are to cool the air passed through the spray chamber and control the humidity. In many cases, the cooling coils are located in the bottom of the spray chamber so that, as the warm spray descends, it is cooled and ready to again be sprayed by the pump. As some of the finer water particles tend to be carried along with the air current, a series of

YEAR-ROUND CENTRAL AIR CONDITIONING

curved plates called *eliminators* is generally used, which force the cooled and humidified air to change the direction of flow, throwing out or eliminating the water particles in the process.

Control Method

A control board is frequently a part of a large system so that a record of operations and conditions can be secured. In air-conditioning systems, various automatic controls are used. Wet- and dry-bulb temperatures, which show the temperature and humidity present, are always used. A hygrostat is sometimes used on smaller installations, while most small room coolers are manually controlled since they only operate during certain hours of the day.

Zone Control

Zone control is generally a division of the air supply delivered into zones with varying requirements. A zone is an area among a number of areas, all being conditioned by a single supply system. Each zone can be satisfied simultaneously with the use of either single- or multizone control, depending on the number and size of each zone. The conventional single-zone units must be installed one to each zone, which results in a simpler application in many aspects. Single units are recommended whenever each zone is large enough to justify a separate unit and when the system layout will accommodate this number of units (Fig. 13-3).

The use of multizone control is preferred whenever several relatively small zones can be grouped together and handled by one control. The requirement of treating simultaneously varying loads with one unit leads to the following recommendations for obtaining the most economical and satisfactory results from multizone units:

1. If both heating and cooling are employed and more than one multizone unit is used, the grouping should place on each unit those zones having the least variation with respect to each other.
2. The heating and cooling supply should be capable of capacity reduction to nearly zero.

Air Conditioning

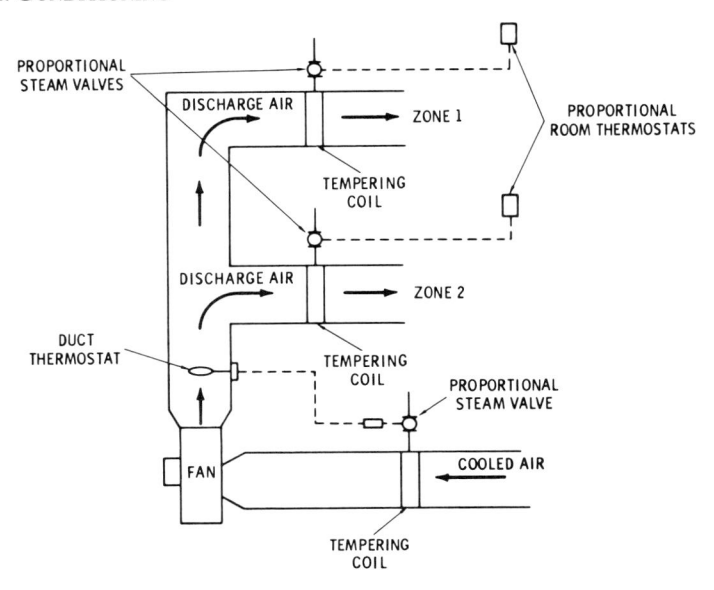

Fig. 13-3. Arrangement for two-zone, constant-volume, variable-temperature control.

The purpose of the multizone arrangement is to divide the total air volume into a number of separate portions and condition each independently in accordance with the requirements of the area it serves. This is accomplished through the operation of the zoning dampers, which provide the necessary individualized control. The air outlet from the unit is partitioned into a number of equal-sized zones, each equipped with a set of dampers. The dampers are composed of two blades set at 90° angles to one another on the same shaft, so that one controls air from the heating pass and the other controls air from the cooling pass. By regulating the position of the shaft, the air through the two passes can be proportioned to maintain the desired final temperature in each zone, providing such temperatures are feasible with existing coil conditions.

In operation, the shaft from each damper may be extended either upward or downward to permit convenient field installation of a damper motor, either electric or pneumatic, as provided. All damper shafts are connected together by one continuous link, which may be kept intact or cut between any or all adjacent

YEAR-ROUND CENTRAL AIR CONDITIONING

zones. Thus, groups of two or more adjacent dampers may be kept interlocked together for identical movement by one common motor. Each damper may also be individually operated by its own motor. This unique, yet simple arrangement provides a maximum of flexibility when supporting the total airflow from the unit according to the various zone requirements.

System Selection

In any year-round air-conditioning application it is necessary first to determine the overall accomplishment required of the unit. This consists of surveying the building details and available services, calculating the needed capacity of conditions of operation, and devising a layout of the system. A typical year-round air-conditioning unit is shown in Fig. 13-4.

A second step is to translate the system-design requirements into

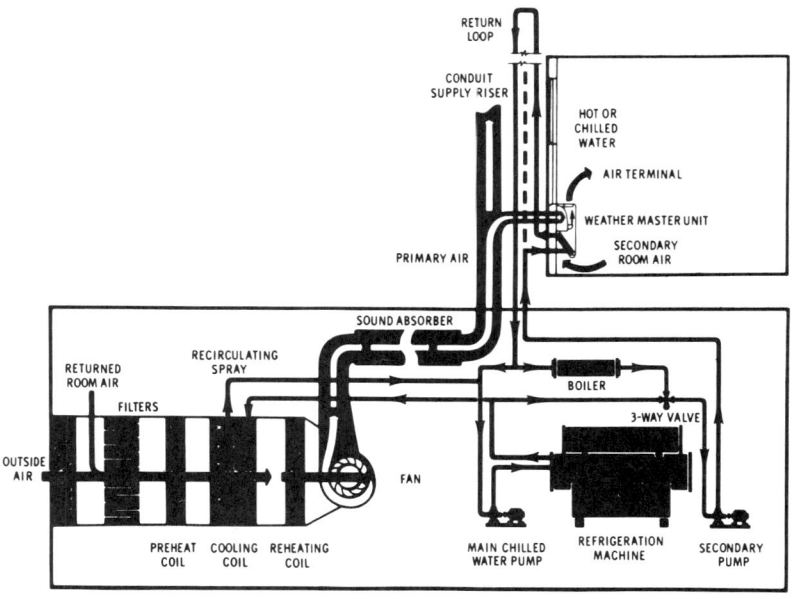

Fig. 13-4. Typical year-round air-conditioning system showing interconnection of conduits and refrigeration piping.

AIR CONDITIONING

specific equipment selections. The general procedure for selecting a unit is first to determine what casing is to be used and then make individual selections of each of the component items (coils, fans, filters, etc.) that are to go with the casing. Because it serves as the enclosure for the unit, the selection of a casing is essentially the determination of the size and basic arrangement of the unit. The factors that lead to this selection are the principal requirements of the job as to airflow, heat transfer, and configuration to match the system layout. Once the casing size is known, the selection of each component becomes a straightforward procedure and can be made from the manufacturer's catalog.

ABSORPTION-TYPE AIR CONDITIONING

Year-round air conditioners operating on the absorption principle differ from the conventional compression type mainly in that they use heat energy instead of mechanical energy to change the condition required in the refrigeration cycle. A comparison between the functioning of an absorption system and a compression system of air conditioning is principally as follows: The absorption system uses heat (usually steam) to circulate the refrigerant, while the compression system employs a compressor to circulate the refrigerant.

Four components of the absorption system may be compared with four components of the compression system, as follows:

1. The boiler or generator compares with the stroke of the compressor.
2. The condenser and evaporator serve the same purpose in an absorption system as in the compression system of refrigeration.

Cycle of Operation

A typical absorption system is shown diagrammatically in Fig. 13-5. The system is charged with lithium bromide and water; lithium bromide is the absorbent, and water is the refrigerant. As noted in Fig. 13-5, the upper vessel contains the absorber and

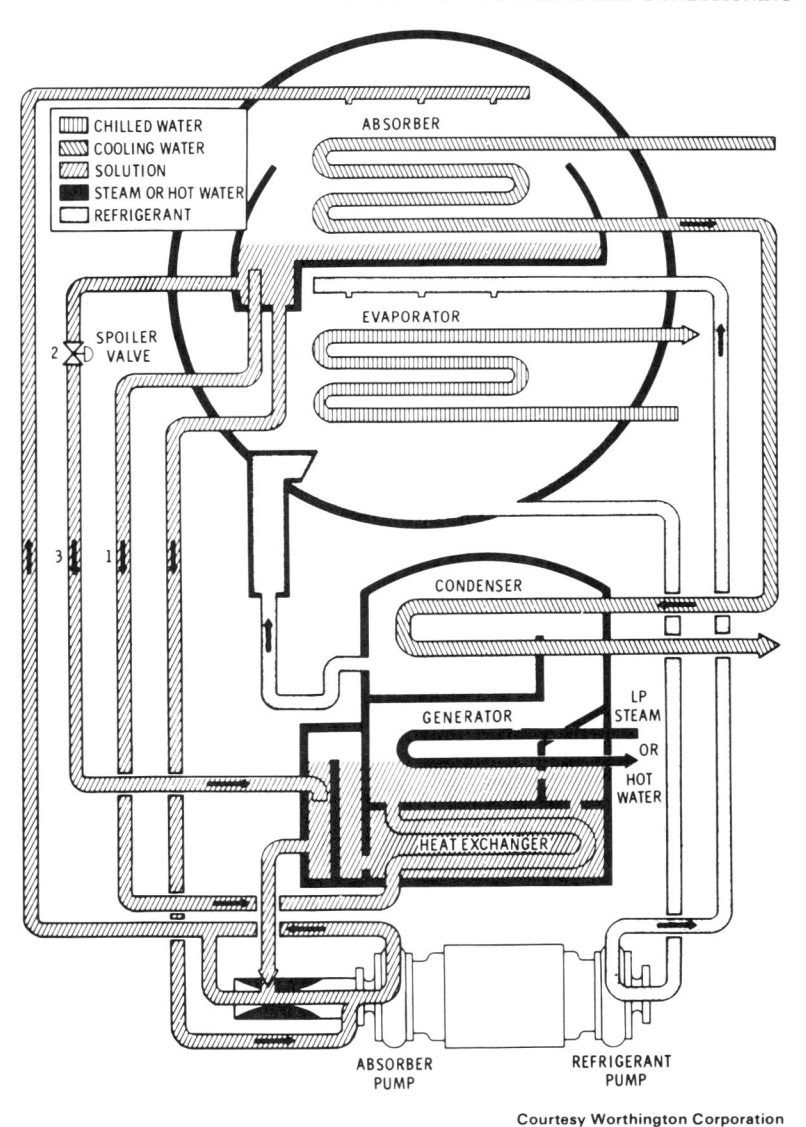

Fig. 13-5. Schematic-flow diagram of absorption-refrigeration cycle.

evaporator, whereas the lower vessel contains the condenser, generator, and heat exchanger. The refrigeration, or chilling effect, takes place in the evaporator. Refrigerant (water) supplied to the evaporator section is sprayed by the refrigerant pump over a tube bundle through which the water passes to be chilled. The refrigerant evaporates on the tube surface. The heat required to evaporate the refrigerant is taken from the water in the tubes, thereby chilling the water. The evaporator is maintained at 0.25 in.Hg, a pressure low enough that the boiling temperature of the refrigerant is sufficiently low to produce the desired chilled-water temperature leaving the unit.

The pressure of the evaporator is maintained at this desired level by drawing off the refrigerant vapor as it is produced. This vapor is produced by locating an absorbent adjacent to the evaporator in the absorber section. The absorbent has a very strong affinity for water vapor and will take this vapor into solution as it is produced. The solution used is lithium bromide brine at about 62% concentration. Lithium bromide is sprayed by the absorber pump into the absorber to provide as large a liquid surface as possible since the absorption process is one of contact with the water vapor. As the solution falls from the sprays, it contacts and passes through a tube bundle, which cools the solution and removes the heat liberated to the solution when the refrigerant vapor is absorbed and returns to its liquid state. It is through this heat exchange that the heat removed from the chilled water is rejected from the machine and passed to a cooling tower or other heat sink.

As the absorption process continues, the solution in the absorber becomes more dilute, and its tendency is to lose its ability to absorb at the same rate as vapor is produced. Therefore, the absorber solution is replenished with a strong solution to maintain the required concentration. An equivalent amount of weak solution from the absorber pan drops down pipe 1 (Fig. 13-5) by gravity and flows through one side of the heat exchanger into the generator, where heat is applied inside a tube bundle from low-pressure steam or hot water. The heat applied raises the weak solution to its boiling point and drives off the water vapor. The solution becomes strong again and flows out of the generator through the other side of the heat exchanger and is drawn into an ejector. It mixes with intermediate-strength solution in the ejector and is pumped back to the absorber to do more work.

The refrigerant vapor driven off in the generator flows into the condenser, where it is condensed. The condensing medium is the same water that was used to cool the solution in the absorber, mentioned previously. The condensed refrigerant accumulates in the condenser pan and is forced by a pressure difference back to the evaporator. The heat exchanger removes heat from the hot, strong solution, which must be cooled to work in the absorber, and transfers it to the weak solution, which must be heated before it can be concentrated. The cycle is now complete, and the process will continue as long as heat is available and liquid flows are maintained.

The system is controlled by matching the absorber capacity to the chiller load. Absorber capacity is varied by modulating the solution concentration, which is controlled by varying the amount of strong solution that may be withdrawn from the generator. The ejector is a constant-flow device and will take whatever solution is available from the ejector weir box. When less than full capacity is required, a weak solution from the absorber is bled into the weir box. This weak solution will displace the strong solution, preventing the strong solution from flowing into the ejector. The end result is a weakening of the absorber-solution strength. This "ejector spoilage" is accomplished by modulating valve 2 (Fig. 13-5) in the bleed line 3 in accordance with the leaving chilled-water temperature.

Actual unit construction includes additional refinements, which increase efficiency, but the cycle just described is typical of the cycle in any absorption system.

REVERSE-CYCLE AIR CONDITIONING (HEAT PUMPS)

Year-round air-conditioning units of this type, usually termed *heat pumps*, operate on the reverse-cycle principle. By means of a specially designed reversing valve, alternate heating and cooling can be obtained. The operation of a typical unit during the *heating* and *cooling cycle*, with reference to Fig. 13-6, is principally as follows.

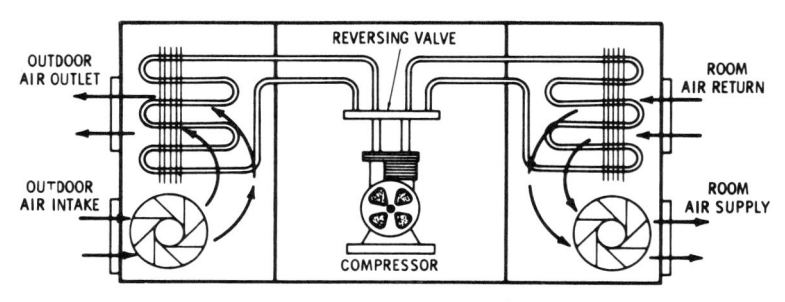

Fig. 13-6. Schematic diagram of a typical reverse-cycle air conditioner. The machinery compartment (usually remotely located) contains the compressor unit, reversing valves and controls.

Heating Cycle

In the outdoor air coil, cooled liquid refrigerant under very little pressure picks up heat from the warmer outdoor air, evaporating the refrigerant into gas. The pump compresses the refrigerant at a high pressure, thus raising its temperature. The hot, compressed refrigerant gives off heat, which is circulated into the space to be heated by an air coil. As the refrigerant loses its heat, it condenses into liquid form, ready to repeat its cycle.

Cooling Cycle

During the cooling cycle, the refrigerant control valves are reversed, and the compressor pumps the refrigerant in the opposite direction; thus the coils that heat the space in cold weather cool it in warm weather. It is in this manner that the outdoor coil functions as a condenser during the cooling cycle and as an evaporator

YEAR-ROUND CENTRAL AIR CONDITIONING

during the heating cycle. The control of the reversing valve is by means of a slide action actuated by a solenoid (Fig. 13-7).

Reversing Valve Operation

The operation of a typical reversing valve during the heating and cooling cycles is principally as follows:

1. With reference to Fig. 13-8, the system is on the heating cycle with the discharge gas flowing through reversing valve ports D to 2, making the indoor coil the condenser. The suction gas is flowing from the outdoor coil (evaporator), through reversing valve ports 1 to S, and back to the compressor.
2. The three-way solenoid pilot is deenergized with the slide positioned so as to close the B port and open the A port. The main slide is in the UP position, sealing off nose valve C and opening nose valve E. Nose valve E is isolated from the suction side by the pilot solenoid valve, and nose valve C is exposed to suction pressure through the pilot solenoid valve.
3. With the valves so positioned, the controlled leakage of the high-pressure discharge gas builds up on both ends around the main slide. The area around the top end of the slide, less the area of nose valve port C, is exposed to discharge pressure, while the area of nose valve C is exposed to the suction pressure. The area on the bottom end of the slide and the area of nose valve E are both exposed to discharge pressure. Thus, the unbalanced force, due to the difference between discharge and suction pressures acting on the area of the nose valves, holds the slide in the UP position.
4. When the coil is energized, the slide in the pilot solenoid valve raises, closing port A and opening port B. With the pilot solenoid valve so positioned, the discharge pressure imposed on the bottom of the main slide and the area of nose valve E will bleed off through the pilot solenoid valve to the suction side of the system. The unbalanced force in a downward direction is then due to the difference between discharge and suction pressures acting on the areas of the ends of the slide and the nose valves.

385

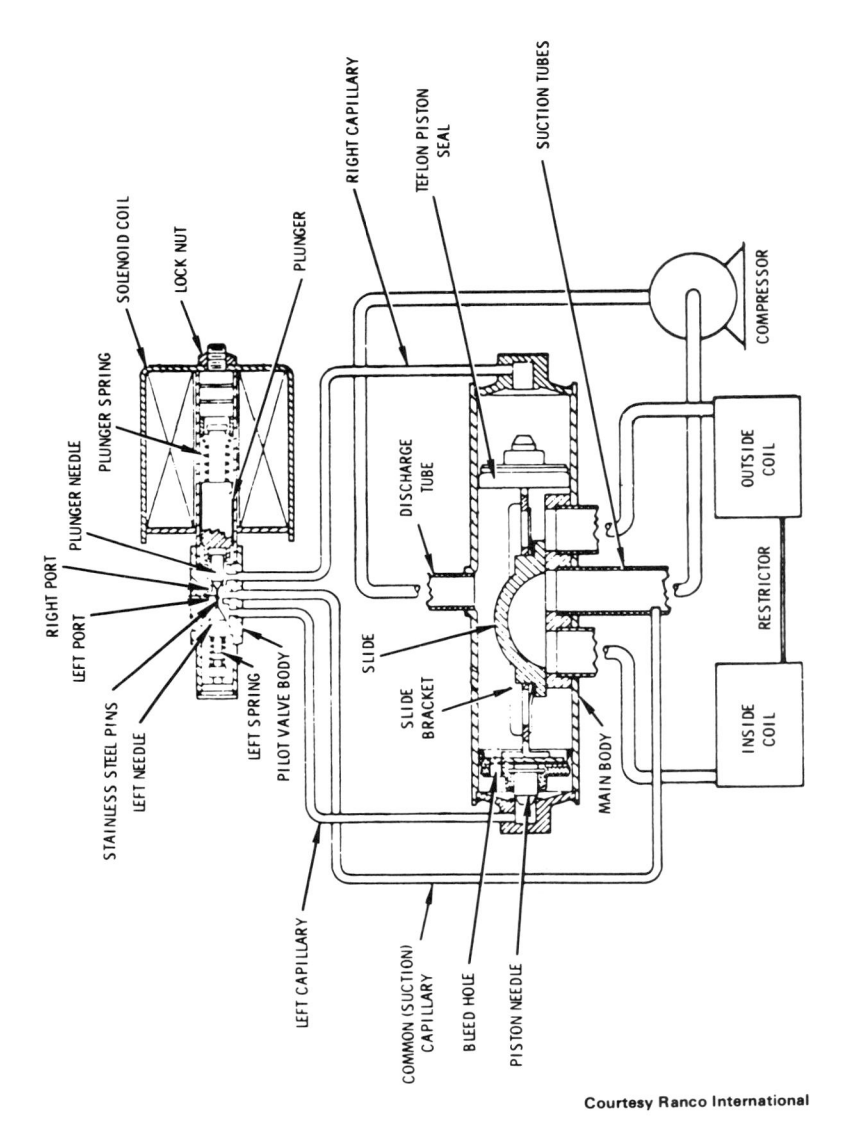

Fig. 13-7. Sectional view of a reversing valve showing the valve connection into a heat-pump system for cooling and heating.

YEAR-ROUND CENTRAL AIR CONDITIONING

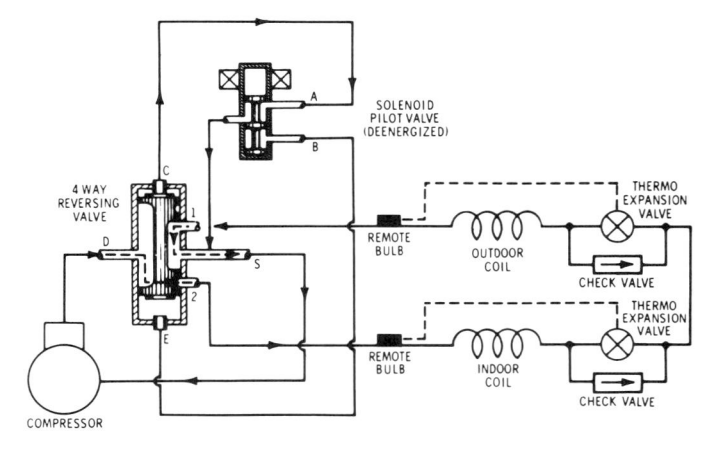

Fig. 13-8. Reversing-valve operation during heating cycle.

5. This unbalanced force moves the main slide to the down position, as shown in Fig. 13-9, and the unbalanced force across the area of the nose valve holds the main slide in the new position.

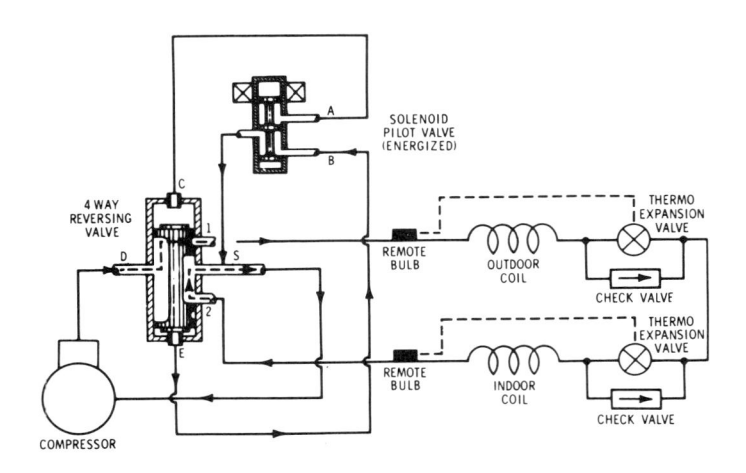

Fig. 13-9. Reversing-valve operation during cooling cycle.

6. The system has now changed over to the *cooling cycle* with the discharge gas flowing through, reversing valve ports D to 1, making the outdoor coil the condenser with the suction gas flowing through reversing valve ports 2 to S, making the indoor coil the evaporator.
7. Depending on the manner in which the reversing valve is piped into the system, power failure to the pilot solenoid valve coil will cause the system to "fail-safe" on either the heating (defrost) or the cooling cycle. In Figs. 13-8 and 13-9 the valve is piped to "fail-safe" on heating. In order to "fail-safe" on cooling, the indoor coil would be connected to reversing valve port 1 and the outdoor coil connected to reversing valve port 2.

Control Equipment

Heat-pump control panels are usually provided with a 24-volt transformer for relay operation in addition to the following:

1. A double-pole, single-throw, changeover relay to actuate the changeover valve and contactor on the heating cycle
2. A single-pole, double-throw, impedance relay to provide electrical lockout and remote reset of the compressor safety controls (also to provide control for a trouble-light circuit to indicate system shutdown)
3. A single-pole, single-throw, fan relay to control the indoor fan
4. A defrost relay to deenergize the outdoor fan and changeover valve and turn on the second-stage heat strip when the defrost control(s) calls for defrost
5. A high or low combination pressure control available for installation on the panel

Defrost-Cycle Operation

The three generally accepted methods for controlling coil-frost accumulation are as follows:

1. *The timer switch starts and the temperature switch stops the defrost cycle.* When the outdoor coil reaches the defrost cut-in temperature, the temperature switch makes contact, energizing the timer motor if the compressor motor is running. After the off-cycle time of the timer has elapsed, the timer contact closes. The defrost relay then is energized through the temperature switch and timer-motor contact. The outdoor fan and changeover valve are deenergized, which reverses the system and allows the outdoor coil to warm up quickly. If auxiliary strip heaters are used, their heating relay(s) can be energized during the defrost cycle to compensate for the indoor cooling effect during that cycle. The timer runs whenever the switch is making contact. The timer contacts operate only once per cycle for a short period of time. The defrost relay remains unenergized through the temperature switch and its own holding contact. When the outdoor coil reaches the temperature-switch cut-out temperature, the timer motor and relay are deenergized, ending the defrost cycle.
2. *The pressure switch starts and the temperature switch stops the defrost cycle.* The operation is the same as previously explained, except that the pressure switch replaces the timer switch. The pressure switch senses the air pressure between the outdoor coil and fan.
3. *The defrost control starts and stops the defrost cycle.* The defrost control senses a frosted coil and energizes the defrost relay. When defrost is completed, the defrost switch opens and ends the cycle.

Installation and Maintenance

Reversing-cycle air-conditioning units are normally located *outside* the building or space to be air conditioned. A unit should be located as near as possible to the indoor section in order to keep the length of the connecting-refrigerant tubing to a minimum. The air inlet and outlet should be at least 30 inches from the wall or any other obstruction. The unit should be oriented so that the prevailing wind is not blowing directly into the outdoor coil. All

AIR CONDITIONING

outdoor units pull air through the coil. Fig. 13-10 shows a typical reverse-cycle air-conditioning unit.

Courtesy Worthington Corporation

Fig. 13-10. Exterior view of a typical outdoor section of a reverse-cycle air conditioner.

The base of the outdoor unit is usually cut out beneath the coil to allow unrestricted disposal of outdoor coil condensate on the heating cycle. In any installation, the condensate should drip in such a manner that ice buildup on the ground (in colder climates) will not damage the unit or interfere with the opening under the outdoor coil. Always mount the unit on a concrete slab in such a position that it may be easily reached for servicing and maintenance, as shown in Fig. 13-11. In colder climates it is recommended to raise (or block up) the unit higher above the ground. Ice buildup on the ground may damage the unit if there is insufficient clearance between the unit and ground.

Indoor Arrangement

Units are normally designed for either free discharge or ductwork application. By rotation of the blower and leg assembly, the

YEAR-ROUND CENTRAL AIR CONDITIONING

Courtesy Carrier Corporation

Fig. 13-11. Typical reverse-cycle central air-conditioning installation. The refrigeration unit (mounted on a sturdy concrete pad) expels unwanted heat and is connected by piping to the cooling coil atop the forced-warm-air furnace (left).

unit can provide vertical or horizontal airflow. In all installations, the unit must be mounted level to ensure proper condensate disposal. The cabinets are well insulated inside the air-conditioned compartment; in most installations, this construction will prevent sweating on the outside of the unit. However, in cases where blower coils are installed in attics above living quarters or in areas where high-humidity conditions are prevalent, it is recommended that an insulated watertight pan with an adequate drain connection be constructed and installed under the cabinet. This drain should extend approximately 2 inches beyond the unit on all sides to ensure collecting any condensate forming on the outside of the cabinet. Fig. 13-12 shows duct arrangement for indoor and outdoor units and condensation drip pan installed under the cabinet.

AIR CONDITIONING

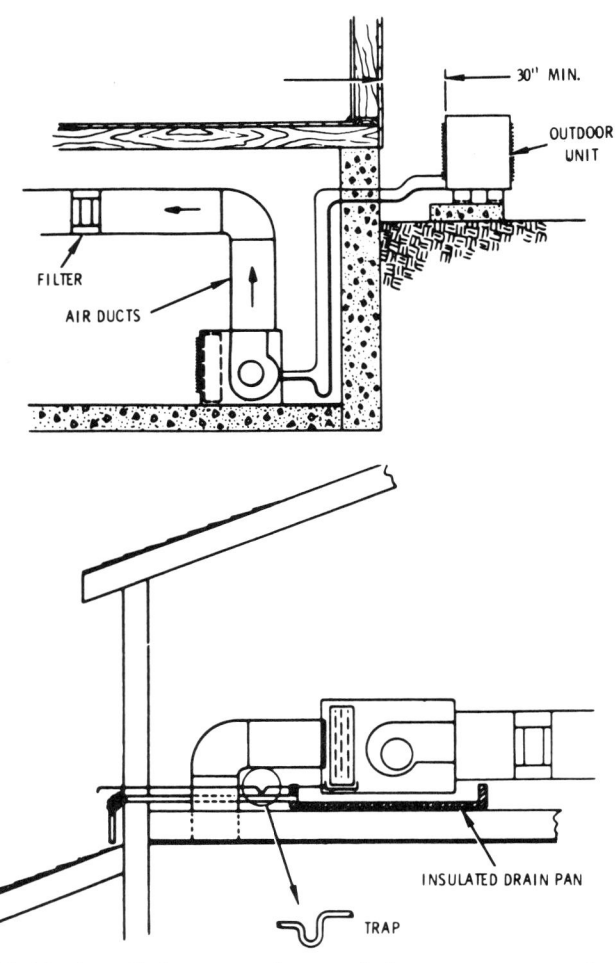

Fig. 13-12. Installation procedure and duct arrangement for outdoor and indoor units with vertical and horizontal discharge, respectively.

Refrigerant Tubing

Locations where copper tubing will be exposed to mechanical damage should be avoided. If it is necessary to use such locations, the tubing should be enclosed in rigid or flexible conduit. The suction line should be insulated to prevent sweating, and horizontal lines should be pitched toward the compressor unit. When the evaporator is higher, or not more than 10 feet lower than the condensing unit, no oil trap is required. When the evaporator is more than 10 feet lower than the condensing unit, an oil trap is required in the suction line to ensure the return of oil to the compressor.

Refrigerant lines are connected as follows, using either a hard or soft-type solder:

1. Unsolder the sealed end of the suction and liquid lines on the *evaporator* unit.
2. Carefully clean the fittings and *solder* the refrigerant lines in place.
3. Remove caps on the suction and liquid lines of the *outdoor* unit. Apply heat to the caps to remove. The caps are soldered to the fittings with soft solder. (*Note:* If the cap on the suction line does not loosen readily, refrigerant oil may be present in the suction line that is conducting the heat away from the cap. To relieve the oil, drill a small hole in the suction-line cap to allow the oil to drain out. After the oil has drained, the suction-line cap should then be easily loosened and removed.)
4. The liquid and suction-line service valves are usually front-seated (closed to the atmosphere). The service port is positioned so that it can be used to pump down the indoor unit after the refrigerant lines are soldered in place, without losing the basic holding charge contained in the unit.
5. Carefully clean the suction and liquid-line fittings on the unit, and solder the refrigerant lines to these fittings. Various stages in making a connection are shown in Fig. 13-13.

System Troubles

One of the most common troubles in a reverse-cycle air-conditioning system is usually caused by incorrect operating pressures, which, in turn, may prevent the reversing valve from shifting properly. For example:

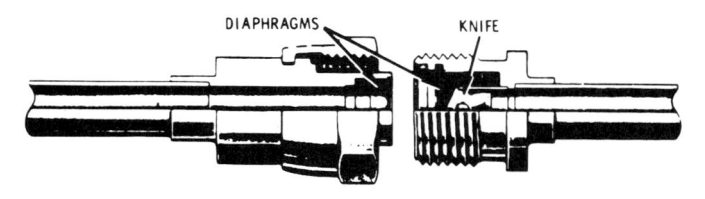

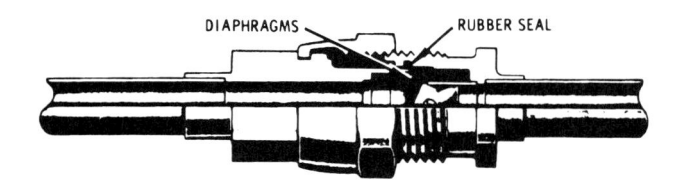

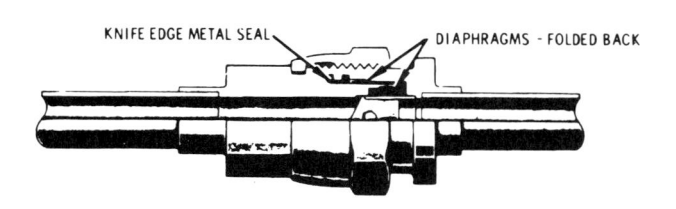

AIR CONDITIONING

Fig. 13-13. Refrigerant-tube couplings and methods of assembly.

1. A leak in the system resulting in loss of charge
2. A compressor that is not pumping properly
3. A check valve that is leaking
4. A defective electrical system
5. Mechanical damage to the reversing valve

Before attempting to diagnose reversing-valve trouble, the following checks on the system should be made:

YEAR-ROUND CENTRAL AIR CONDITIONING

1. Make a physical inspection of the valve and solenoid coil for dents, deep scratches, and cracks.
2. Check the electrical system. This is readily done by having the electrical system in operation so that the solenoid coil is energized. In this condition, remove the locknut to free the solenoid coil. Slide it partly off the stem, and notice a magnetic force attempting to hold the coil in its normal position. By moving the coil farther off the stem, a clicking noise will indicate the return of the plunger to its deenergized position. When returning the coil to its normal position on the stem, another clicking noise indicates that the plunger responds to the energized coil. If these conditions have not been satisfied, other components of the electrical system should be checked for possible trouble.
3. Check the heat-pump refrigeration system for proper operation as recommended by the manufacturer of the equipment.

REVERSING-VALVE TROUBLESHOOTING GUIDE

The following troubleshooting guide will assist in the diagnosis of reversing-valve malfunction and suggest checks and corrections.

System and Possible Cause   *Possible Remedy*

Valve Will Not Shift From Heat to Cool

(a) No voltage to coil.
(b) Defective coil.
(c) Low refrigeration charge.

(d) Pressure differential too high.
(e) Pilot valve operating correctly. Dirt in one bleeder hole.

(a) Repair electrical circuit.
(b) Replace coil.
(c) Repair leak and recharge system.
(d) Recheck system.
(e) Deenergize solenoid, raise head pressure, and reenergize solenoid to break

395

Air Conditioning

System and Possible Cause	Possible Remedy
	dirt loose. If unsuccessful, remove valve and wash out. Check on air before installing. If no movement, replace valve, add strainer to discharge tube and mount valve horizontally.
(f) Piston cup leak.	(f) Stop unit. After pressures equalize, restart with solenoid energized. If valve shifts, reattempt with compressor running. If still no shift, replace valve.
(g) Clogged pilot tubes.	(g) Raise head pressure, and operate solenoid to free. If still no shift, replace valve.
(h) Both ports of pilot open. Back seat port did not close.	(h) Raise head pressure and operate solenoid to free partially clogged port.

Valve Starts to Shift but Does Not Complete Reversal

(a) Not enough pressure differential at start of stroke, or not enough flow to maintain pressure differential.	(a) Check unit for correct operating pressures and charge. Raise head pressure. If no shift, use valve with smaller ports.
(b) Body damage.	(b) Replace valve.
(c) Both ports of pilot open.	(c) Raise head pressure, and operate solenoid. If no shift, replace valve.
(d) Valve hung up at midstroke. Pumping volume of compressor not sufficient to maintain reversal.	(d) Raise head pressure, and operate solenoid. If no shift, use valve with smaller ports.

System and Possible Cause   *Possible Remedy*

Apparent Leak in Heating Position

(a) Piston needle on end of slide leaking.

(a) Operate valve several times; then recheck. If excessive leak, replace valve.

(b) Pilot needle and piston needle leaking.

(b) Operate valve several times; then recheck. If excessive leak, replace valve.

Valve Will Not Shift from Heat to Cool

(a) Pressure differential too high.

(a) Stop unit; will reverse during equalization period. Recheck system.

(b) Clogged pilot tube.

(b) Raise head pressure. Operate solenoid to free dirt. If still no shift, replace valve.

(c) Dirt in bleeder hole.

(c) Raise head pressure; operate solenoid. Remove valve and wash out. Check on air before reinstalling. If no movement, replace valve. Add strainer to discharge tube. Mount valve horizontally.

(d) Piston-cup leak.

(d) Stop unit. After pressures equalize, restart with solenoid deenergized. If valve shifts, reattempt with compressor running. If it still will not reverse while running, replace valve.

(e) Defective pilot

(e) Replace valve.

AIR CONDITIONING

NONPORTABLE AIR CONDITIONERS

Nonportable air-conditioning units are used in stores, restaurants, and offices and are manufactured in a large variety of cooling capacities designed to suit almost any requirement. A typical nonportable self-contained unit is shown in Fig. 13-14. It is essentially a remote vertical-type air conditioner with the condensing unit mounted in a sound-insulated enclosure.

Units of this type may not require condenser cooling-water connections. If the unit is to include winter conditioning, it will also require steam or hot water piping for the heating coil and a water connection for the humidifier. The provision for connecting the

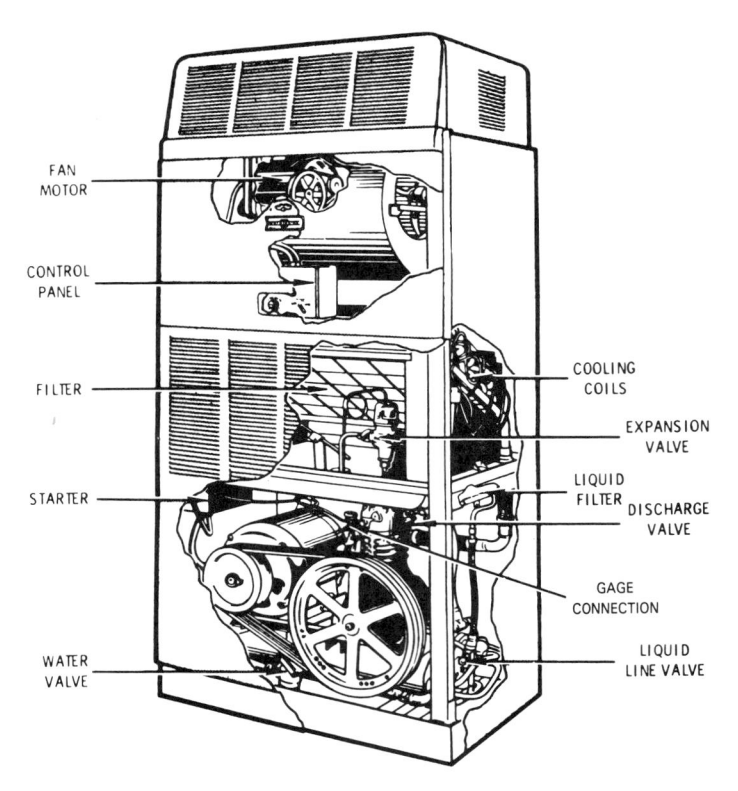

Fig. 13-14. Typical nonportable air-conditioning unit.

YEAR-ROUND CENTRAL AIR CONDITIONING

unit to an outside air duct is usually optional. The heat generated by the compression of refrigerant gases and that given off by the compressor motor is removed from the compressor compartment by means of one of the following:

1. Use of a water coil in the compressor compartment
2. Utilization of the cold suction gases
3. Drawing part of the return air through the compressor compartment
4. Using an air-circulating fan on the motor shaft

In water-cooled units, drain facilities must be available, and although window connections are not required for heat disposal, it is usually desirable to have an outside air connection for ventilation purposes.

Construction Principles

With reference to Fig. 13-14, showing a typical self-contained store-type air conditioner, observe that the unit is divided into three principal compartments, or sections, as follows:

1. The acoustically treated refrigerating section, containing the condensing unit (complete with motor), compressor, receiver, and controls
2. The air-conditioning section, with the cooling centrifugal fan and motor, fan drive, inlet grille with filter, and space for a heating coil
3. The air-distributing section, or hood, with acoustical insulation and air-discharge grilles

The steel hood has air-distribution grilles on all four sides. The top, ends, and rear of the hood are of one-piece welded construction with insulation covering the air-distribution grilles. The front-grille panel of the hood is a separate piece attached to the main hood with screws. A set of vertical adjustable louvers is located behind the front-panel horizontal grilles.

It should be clearly understood that the air-conditioning unit shown in Fig. 13-14 is only typical, and that arrangements and

AIR CONDITIONING

placing of components vary, depending on the particular manufacture. Thus, for example, numerous units of recent make are provided with a hermetically sealed compressor. They all operate on the same principle, having a similar refrigeration cycle, and are controlled by an easily accessible control station. Automatic control is provided by means of the conventional thermostat, which is actuated by the temperature of the room air entering the unit.

Plumbing Connections

The only plumbing connections necessary for summer air conditioning are the condenser water inlet, condenser water outlet, and evaporator condensate drain. If the unit is to include winter conditioning, it will also require steam or hot-water piping for the heating coil and a water connection for the humidifier. Condenser water pipes enter through knockouts provided in the side of the bottom section. For city water applications, the condenser water-inlet and outlet connections are usually $3/4$-inch standard steel or iron pipe for runs up to 50 feet. For greater lengths, 1-inch pipe is used. The evaporator condensate drain should be $1^{1}/_{4}$-inch pipe.

The water leaving the condenser (if it is not to be reused) should be piped to the sewer in accordance with local plumbing regulations. Many codes require an open funnel or sink discharge, and this method is desirable from a service standpoint as it facilitates checking performance. An automatic water-regulating valve is required in the inlet-water line. This valve will operate satisfactorily on inlet-water pressures of from 25 to about 75 psi. A shutoff water-regulating valve is required in the inlet and outlet-water pipes, respectively, as shown in Fig. 13-15.

Water-Cooling Tower

In locations where the use of city water is restricted, circulating water from a cooling tower may be used for condenser cooling. The quantity of water circulated through water-cooling towers usually varies between 3 and 5 gal/min/ton of refrigeration, the larger quantity being used when the lowest possible water temper-

Fig. 13-15. Plumbing connections for a water-cooled condensing unit.

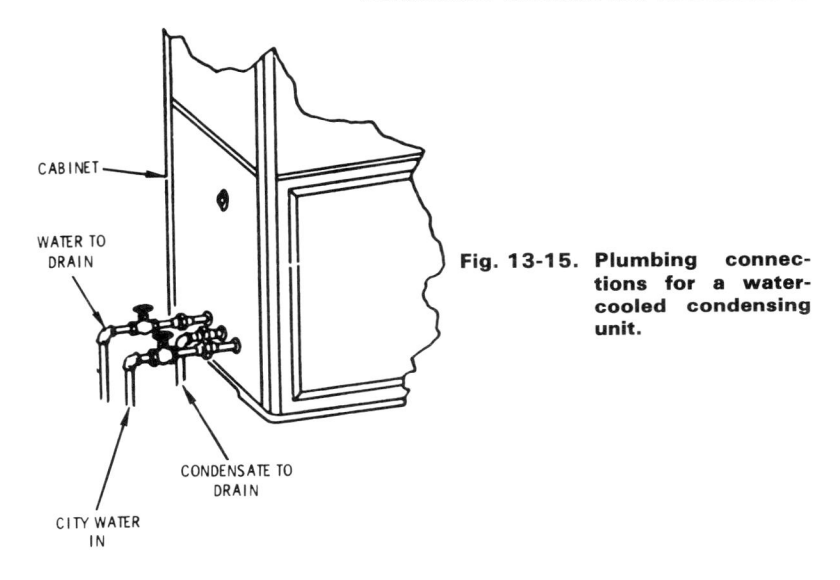

ature is desired. The total head (in feet) against which the pump in a balanced system must work is composed of the following:

1. Condenser head (in feet)
2. Head required at base of tower (in feet)
3. Friction head in piping and fitting (in feet)

Since most store unit air conditioners are shipped from the factory for operation with city water, it is necessary to do the following for application with a water-cooling tower:

1. Remove the water regulating valve.
2. Reconnect the condenser.
3. Connect the pump, and complete all plumbing, as indicated in Fig. 13-16.

Heating Coils

Heating coils are commonly supplied by the manufacturer as extra equipment for field installation directly above the cooling

AIR CONDITIONING

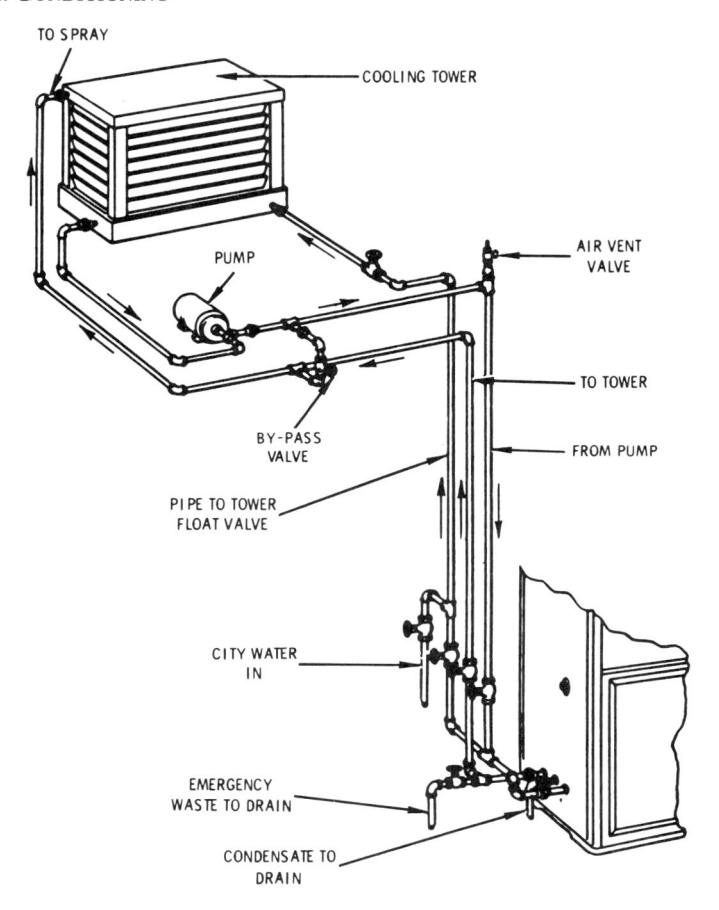

Fig. 13-16. Plumbing connections for a water-cooled condensing unit interconnected to a reciprocating pump and cooling tower.

unit. The controls for heating are generally purchased locally. If space permits, the heater pipes can be brought in from the sides of the cabinet directly opposite the coil connections by drilling holes in the cabinet and notching the side-access plates as necessary. This arrangement, however, is not as desirable as the other from the appearance standpoint.

Heating coils, as supplied by the manufacturer, are usually provided with two $1^{1}/_{2}$-inch female pipe taps in the inlet and outlet headers. The coil should be installed with these openings below the center line of the coil to allow proper drainage. In addition, the coil must be sloped downward approximately $3/8$-inch in the direction of the steam or water flow. The coil outlet is always brought into the opposite header from the inlet and in the opposite location front to back, so that the heating medium passes through the coil in a zig-zag path to obtain the most satisfactory distribution. For hot water, a manual-type vent valve is used; and for steam, an automatic vent valve is used.

In a one-pipe steam system, a bypass connection around the heater is required to return the condensed steam to the supply riser. This connection should pass down from the outlet header at one end and around the bottom of the cooling coil and into the supply pipeline on the opposite end. It is necessary that this line be properly sized to maintain a water seal against the incoming steam pressure, and that it be tapped into the supply riser for returning the condensate.

SUSPENDED-TYPE AIR CONDITIONERS

Suspended-type air conditioners are used in stores, garages, and establishments with limited floor space. They are designed to suit various applications requiring dehumidification, cooling, and/or heating combinations, as conditions require.

Suitable locations for store units are placement above the door of restaurants or above the kitchen floor. They may also be incorporated in wall structures, particularly where false walls are employed or where the wall is a partition into another room of the same establishment. Because of the weight of the unit, careful consideration must be given to the strength of the location where it will be installed. Some smaller suspended-type units use propeller fans with trailing-edge blades in order to obtain the required pressure characteristics.

Casings

Suspended-type air-conditioner casings are usually made of sheet metal supported by suitable angle frames, with removable

AIR CONDITIONING

panels permitting access to the interior components for proper maintenance. The compressor compartment is normally insulated to prevent noise from escaping into the room. Drain pans are usually made of 14-gage or heavier sheet metal and should be hot-dipped galvanized after fabrication in order to resist corrosion. The drain connections should be readily accessible for cleaning; in air-conditioner work, they should be properly sized and trapped according to plumbing regulations in the area of installation.

Coils

The cooling and dehumidifying coils used in suspended-type air conditioners do not differ in any important respect from those used in the self-contained units previously described. The face of the coil area is usually fixed, and the number of coils in depth in the airflow direction determines the capacity. A typical installation is shown in Fig. 13-17. Heating coils of unit air conditioners are normally of the conventional blast coil-type and are usually furnished separately for installation on the premises. They usually match the cooling coils in face area and may be one or two rows in depth, depending on heating requirements.

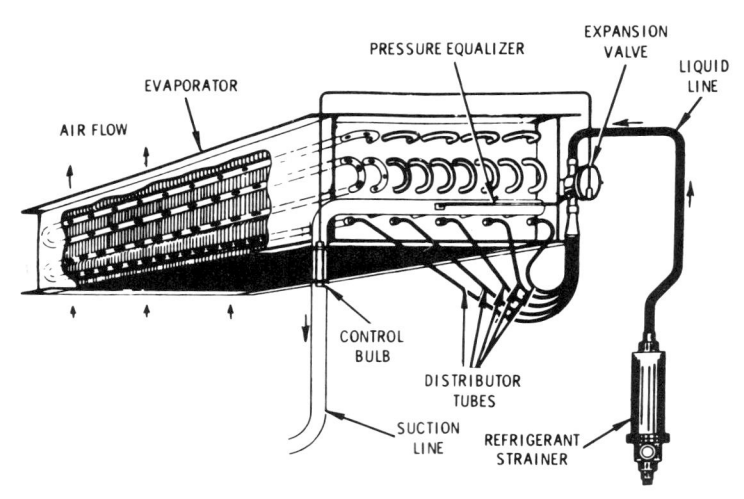

Fig. 13-17. Cooling-coil arrangement of a typical suspended-type air conditioner.

404

Filters

Air filters of the viscous domestic type are usually furnished on suspended-type air conditioners. These may be the throwaway type, although cleanable filters may be used on the larger units. The refrigerant may be controlled by the conventional thermostatic expansion valve or restricter tubing, as conditions dictate. Wall thermostats are commonly used for temperature control in conjunction with solenoid valves. The use of solenoid valves in the liquid line between the strainer and expansion valve depends on whether or not the unit is to be installed in a multiple system.

Where the unit is connected to a single dome-head condensing unit, the thermostat and manual switch may be connected directly to the low-pressure switch circuit. Solenoid-valve-refrigerant control, however, is usually employed where these units are used in conjunction with the piston-valve-type compressor to prevent flooding of the crankcase during the off cycle of the compressor.

Controls

Where the suspended-type air conditioner is equipped with blast-type heating coils, two methods of automatic control are generally used: (1) the control of steam flow and air temperature by automatic modulating valves, and (2) the control of temperature by automatic modulating face and bypass dampers. Both control methods are used to a considerable extent, but the modulating-valve method is preferred.

Modulating valves proportion, or meter, the steam supply to the coils according to the exact heating demand at any time; they provide heating as required by the modulating thermostat to maintain the desired conditions. The quantity of air passing through the coil is constant. Face and bypass dampers are actuated by a modulating motor, which is controlled by a modulating thermostat. The dampers are positioned to permit the quantity of air necessary for maintaining the desired temperature to pass through the coils for heating. The air quantity through the coils changes as the demand changes with the remainder of the air being bypassed. Finally, when no heating is required, the face dampers close and all

AIR CONDITIONING

air is bypassed. It is usual practice to provide an automatically operated steam valve for disconnecting the steam supply from the coil bank when the face dampers are closed.

TROUBLESHOOTING GUIDE

The troubleshooting guides on the following pages will assist in troubleshooting air-conditioning and self-contained air-conditioner systems. It should be noted, however, that at times the trouble may differ from that given in this section. Experience will usually reveal the exact cause, after which the remedy and its correction may be carried out.

SELF-CONTAINED AIR-CONDITIONER TROUBLESHOOTING GUIDE

System and Possible Cause   *Possible Remedy*

Unit Fails to Start

(a) Starting switch off.
(b) Reset button out.
(c) Power supply off.
(d) Loose connection in wiring.
(e) Valves closed.

(a) Place starting switch in start position.
(b) Push reset button.
(c) Check voltage at connection terminals.
(d) Check external and internal wiring connections.
(e) See that all valves are opened.

Motor Hums but Fails to Start

(a) Motor single-phasing on three-phase circuits.
(b) Belts too tight.

(a) Test for blown fuse and overload tripout.
(b) See that the motor is full floating on trunnion base. See that the belts are in the

System and Possible Cause　　*Possible Remedy*

　　　　　　　　　　　　　　　　pulley groove and not binding.

(c) No oil in bearings. Bearings tight from lack of lubrication.　　(c) Use proper oil for motor.

Unit Fails to Cool

(a) Thermostat set wrong.　　(a) Check thermostat setting.
(b) Fan not running.　　(b) Check fan-motor electrical circuit. Also determine if fan blade and motor shaft revolve freely.
(c) Coil frosted.　　(d) Dirty filters; restricted airflow through unit may be caused by some obstruction at air grille. Fan not operating. Attempting to operate unit at too low a coil temperature.

Unit Runs Continuously but No Cooling

(a) Shortage of refrigerant.　　(a) Check liquid-refrigerant level. Test for leaks, repair, and add refrigerant to proper level.

Unit Cycles Too Often

(a) Thermostat differential too close.　　(a) Check differential setting of thermostat and adjust setting.

Unit Vibrates

(a) Not setting level.　　(a) Level all sides.
(b) Shipping bolts not removed.　　(b) Remove all shipping bolts and steel bandings.
(c) Belts jerking.　　(c) Motor not floating freely.

System and Possible Cause   *Possible Remedy*

(d) Unit suspension springs not balanced.
(d) Adjust unit suspension until unit ceases vibrating.

Condensate Leaks

(a) Drain lines not properly installed.
(a) Drainpipe sizes, proper fall in drain line, traps, and possible obstruction from foreign matter should be checked.

(b) Organic slime formation in pan and drain lines and sometimes present on evaporator fins.
(b) This formation is largely biological and usually complex in nature. Different localities produce different types. It is largely a local problem to combat and should be observed as such. Periodic cleaning will tend to reduce the trouble but will not eliminate it totally. Filtering air thoroughly will also help, but at times some capacity must be sacrificed when doing this.

Noisy Compressor

(a) Too much vibration in unit.
(a) Check for point of vibration in set-up.

(b) Slugging oil.
(b) Low suction pressure.

(c) Bearing knock.
(c) Liquid in crankcase.

(d) Oil level low in crankcase.
(d) Pumpdown system and add oil if too low.

SUMMARY

Central air conditioning has the advantage of servicing several rooms with lower cost than several self-contained single-room units. During the winter, central air conditioning is required to warm the air to a comfortable temperature. During the summer it is required to cool the air, plus humidify and dehumidify, as required.

Year-round air-conditioning systems may be classified as gas-compression, absorption, and steam-jet vacuum. A gas-compression air-conditioning system is a simple refrigeration system employing a compressor to remove low-pressure refrigerant enclosed in the low-pressure side and transfer it to the high-pressure side of the system. The absorption system is one in which the refrigerant gas in the evaporator is taken up by an absorber and released in a generator upon application of heat. A steam-jet system is vacuum refrigeration in which high pressure is supplied through a nozzle acting to eject water from the evaporator.

The basic difference between absorption- and compression-type air conditioning is that the absorption system uses heat or steam to circulate the refrigerant while the compression system employs a compressor to circulate the refrigerant.

Window air-conditioning units provide cooling for a small portion of a home where it is not desirable to install a central unit. The advantage in window units is their low cost of operation, portability, and ease of installation. Various conditions should be considered when planning the use of a window unit. These conditions are room size, wall construction, wall exposure, and ceiling height. Other factors should be taken into account, such as number of people in the room, miscellaneous heat loads, such as lamp wattage and radio and television sets used in the room, and proportion of outside-wall area that is glass.

REVIEW QUESTIONS

1. What are the fundamental requirements for year-round air conditioning?
2. What are the principal advantages of year-round air conditioning?
3. What is meant by the *mixing chamber* and what is its purpose?

AIR CONDITIONING

4. What are the means used for cooling in year-round air conditioning?
5. Describe blower types and methods of installation in an air-conditioning system.
6. What is the function of air filters in air conditioning?
7. Describe what is meant by zone control, and state its principal advantages.
8. What are the fundamental requirements in selecting a year-round air conditioner?
9. In what respect does an absorption-type air conditioner differ from that of a conventional compression system?
10. Describe the principles of operation of an absorption-type air conditioner.
11. Describe the operating principles of a reverse-cycle air-conditioning system.
12. What is the function of the reversing valve?
13. What is the vacuum-pump method of purging the reverse-cycle air-conditioning unit?
14. What are the most common troubles in reverse-cycle air-conditioning systems?
15. What is meant by a *self-contained* air-conditioning unit?
16. Describe the function and operation of a window-type air conditioner.
17. Describe the electrical control system of a window-type air conditioner.
18. Describe the plumbing methods in a nonportable air conditioner equipped with a water-cooled condenser.
19. Where are heating coils normally located in nonportable year-round air conditioners?
20. What type of air conditioners are usually found in stores and garages with limited floor space?

CHAPTER 14

Automobile Air Conditioning

The automobile air conditioner is basically no different from any other type of air-conditioning system. The major components, such as the compressor, condenser, and evaporator, are utilized in primarily the same manner as the common room air conditioner. Certain variations and differences are unique to automobile air-conditioning systems, such as the power source, method and type of controls, and component design. A typical automobile air-conditioning installation is shown in Fig. 14-1. Power is supplied to the compressor directly from the crankshaft of the engine by means of a V-belt assembly and a series of pulleys, as shown in Figs. 14-1 and 14-2.

Two *temperature-control* methods are frequently used most by automotive air-conditioning manufacturers: the magnetic (or electric) clutch and the refrigerant-flow control. Normally, the *magnetic clutch* utilizes a stationary electromagnet, which is at-

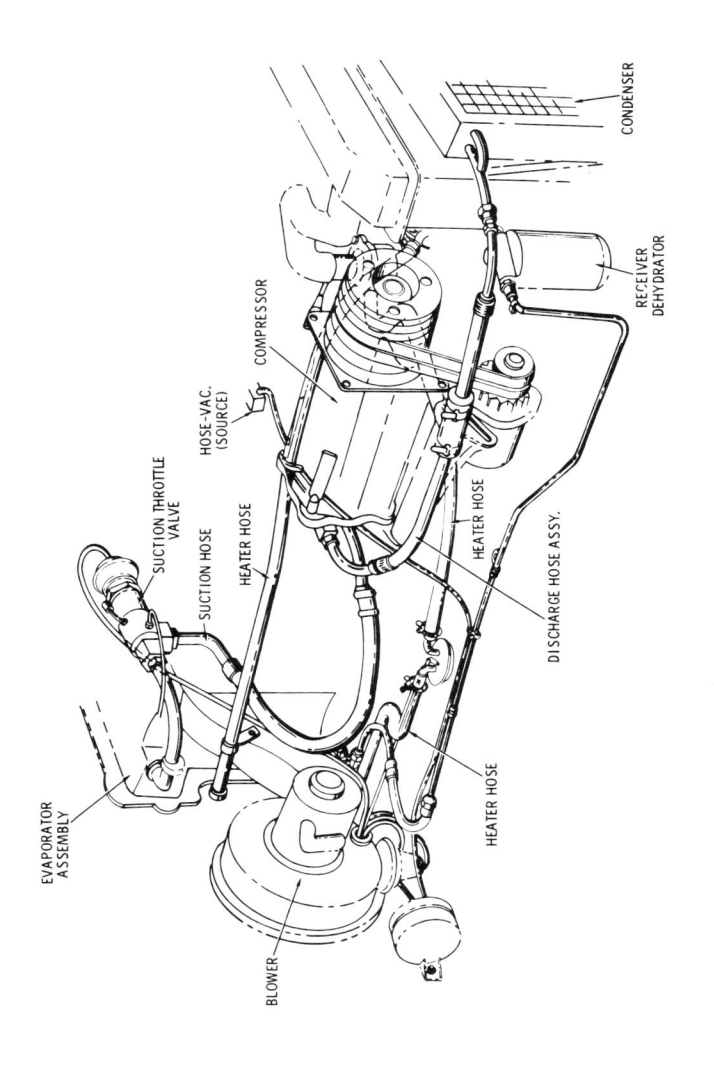

Fig. 14-1. Typical automobile air-conditioning system showing location of main components.

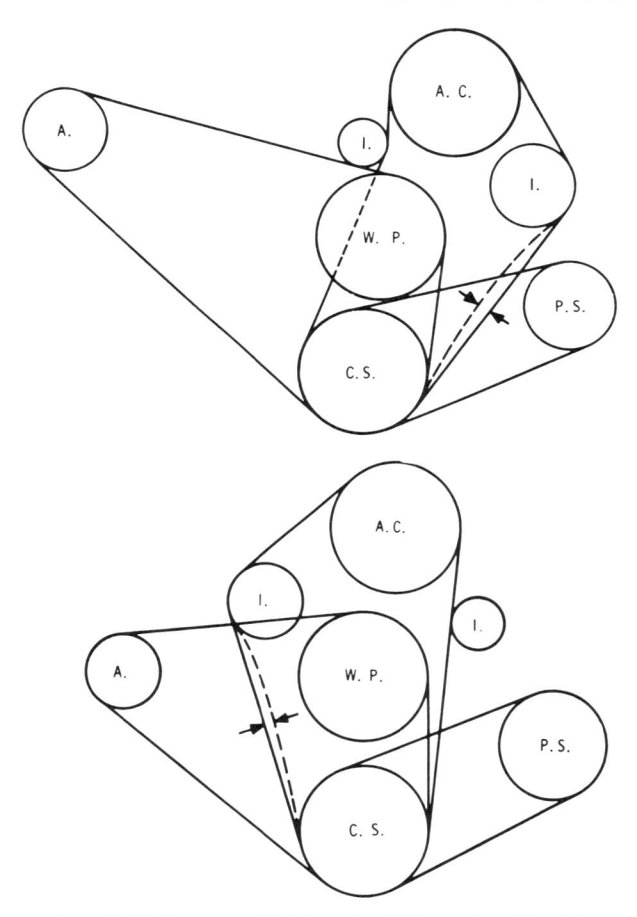

AUTOMOBILE AIR CONDITIONING

Fig. 14-2. Typical belt arrangement in a modern automobile engine.

CS = CRANKSHAFT; A = ALTERNATOR; WP = WATER PUMP; PS = POWER STEERING; AC = AIR-CONDITIONER COMPRESSOR; I = IDLER PULLEYS

Courtesy General Motors Corporation

tached to the compressor. The brushes and collector rings can be eliminated with this arrangement since the electromagnet does not rotate.

When a high temperature is present in the passenger compartment of the car, a thermostat (set by the driver to a specified temperature) closes the electrical circuit that connects the battery power supply to the stationary magnetic field. This field exists around the electromagnet that is mounted on the shaft of the compressor. A continuously rotating clutch plate then engages the compressor drive shaft and causes the compressor to operate at a higher rate of speed. This clutch plate is mechanically attached to the compressor flywheel, which is driven by the V-belt assembly. When the interior of the car reaches the low-temperature setting of the thermostat, the thermostat will open the electrical circuit; the current will cease flowing, and the clutch plate will disengage the flywheel from the compressor. The flywheel will continue to rotate freely until the interior compartment temperature again rises above the thermostat setting. A typical compressor with magnetic-clutch assembly is shown in Fig. 14-3.

The *refrigerant-flow* control maintains a constant rate of refrigerant flow in the refrigeration system. This equalizing effect is extremely necessary since the compressor does not operate at a constant rate of speed. The control is located in the suction, or low, side of the refrigeration system; its spring-loaded bellows are responsive to pressure changes that take place in the system. With a decrease in pressure, which accompanies a proportional decrease in interior temperature, the valve closes and thereby restricts the refrigerant flow. This restriction affects the operation of the system by limiting or completely halting the cooling effect, as the case may be. The refrigerant-flow control is also responsive to the flowrate of the refrigerant. The valve is closed by an increase in flowrate; this increase causes the refrigerant pressure surrounding the valve stem to decrease and thereby affect the control valve, independent of the low-side pressure.

Another type of cooling control is the *hot-gas bypass valve.* This control regulates the temperature by constantly varying the amount of refrigerant allowed to bypass the cooling network, thereby increasing or decreasing the cooling capacity of the system. Utilization of this control valve permits constant compressor operation and also provides an evaporator antifreeze device. By regulating the internal pressure of the evaporator, the hot-gas valve affects the cooling capacity of the evaporator and prevents the pressure from decreasing to a pressure (29.5 psi) that will cause the evaporator core to freeze. If this freeze-up occurs, the

AUTOMOBILE AIR CONDITIONING

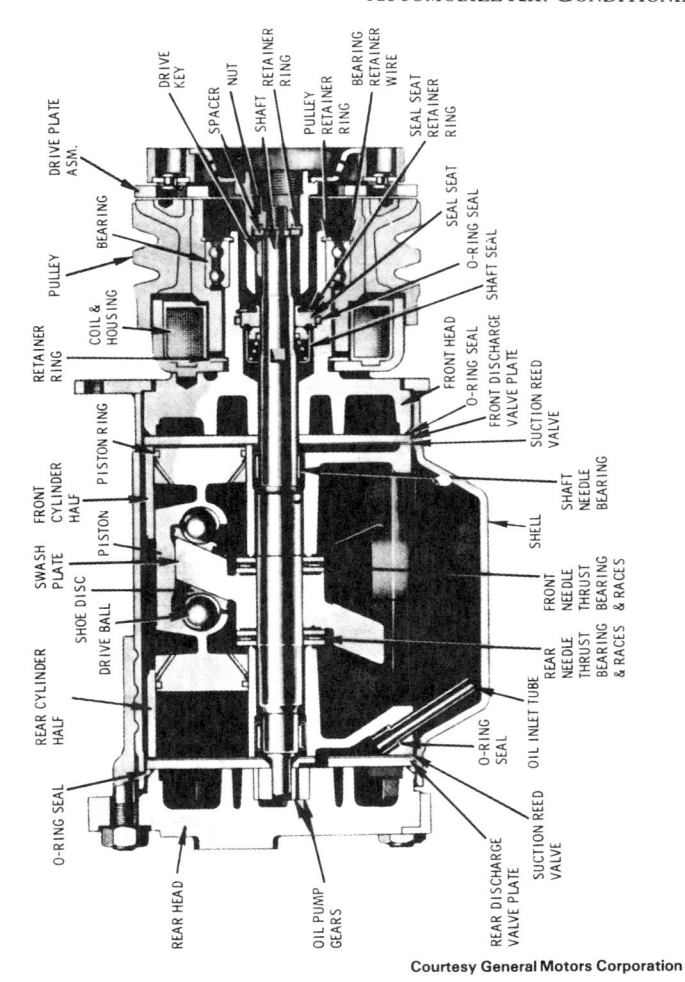

Fig. 14-3. Sectional view of an air-conditioner compressor with double-action piston design and magnetic-clutch coil.

evaporator will not function and the cooling effect of the air conditioner will be greatly lowered or halted completely. The reason the evaporator will not function under these conditions is because the air cannot pass through the evaporator core to provide the necessary cooling in the interior of the automobile.

415

AIR CONDITIONING

Sealed systems are highly improbable for use on automobiles because all systems are, at present, powered by the engine crankshaft, which is an external source. The typical compressor used in the automobile refrigeration system has a large refrigerating capacity as compared with the average room air conditioner (15,000 to 20,000 Btu/hr as opposed to 6000 to 10,000 Btu/hr). This large cooling capacity is due primarily to the large bore and stroke design of the automotive refrigerating compressor, and it is a necessity for automobiles because of the temperature range and poor insulation of car interior. The refrigerant most commonly used in automobile air-conditioning systems is Freon-12. Extreme caution should be exercised when working on or in the vicinity of the refrigeration system. Freon-12 becomes a highly poisonous gas when it comes in contact with an open flame. For this reason, the system should be completely discharged when any repairs or replacements are made on the refrigeration unit.

AIR-CONDITIONING SYSTEM OPERATION

Cool refrigerant gas is drawn into the compressor from the evaporator and pumped from the compressor to the condenser under high pressure (Fig. 14-4). This high-pressure gas will also have a high temperature as a result of being subjected to compression. As this gas passes through the condenser, the high-pressure, high-temperature gas rejects its heat to the outside air as the air passes over the surfaces of the condenser. The cooling of the gas causes it to condense into a liquid refrigerant. The liquid refrigerant, still under high pressure, passes from the bottom of the condenser into the receiver dehydrator. The receiver acts as a reservoir for the liquid. The liquid refrigerant flows from the receiver dehydrator to the thermostatic expansion valve. The thermostatic expansion valve meters the high-pressure refrigerant flow into the evaporator. Since the pressure in the evaporator is relatively low, the refrigerant immediately begins to boil. As the refrigerant passes through the evaporator, it continues to boil, drawing heat from the surface of the evaporator core warmed by air passing over the surfaces of the evaporator core.

In addition to the warm air passing over the evaporator reject-

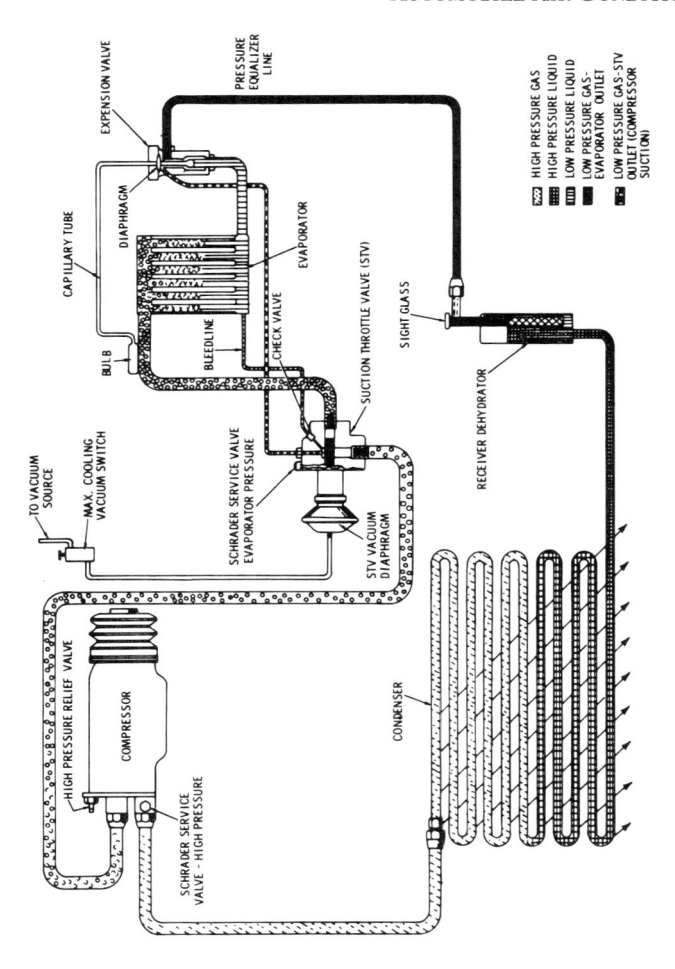

AUTOMOBILE AIR CONDITIONING

Fig. 14-4. Refrigerant-circuit arrangement in a typical automobile air-conditioning system.

ing its heat to the cooler surfaces of the evaporator core, any moisture in the air condenses on the cool surface of the core, resulting in cool dehydrated air passing into the inside of the car. By the time the gas leaves the evaporator, it has completely vaporized and is slightly superheated. Superheat is an increase in temperature of the gaseous refrigerant above the temperature at which the refrigerant vaporized. The pressure in the evaporator is controlled by the suction-throttle valve. Refrigerant vapor passing through the evaporator flows through the suction-throttle valve and is returned to the compressor, where the refrigeration cycle is repeated.

COMPONENT DESCRIPTION

The problems encountered in automobile air conditioning will perhaps be understood by a thorough study of the basic components comprising the system. Figure 14-4 shows the typical components of an automobile air-conditioning unit.

Compressors

In the early days of automobile air conditioning, modified commercial-type compressors were used. Because of the increase in engine speed, the design and development of precision high-speed compressors became necessary. All air-conditioning system compressors are presently driven by a belt from the engine crankshaft. The compressor speed varies with the speed of the automobile engine, which is considerable. For example, at idling speeds, the compressor turns at a rate of about 400 rpm and develops a refrigerating effect in the area of $3/4$ ton. In an automobile traveling 55 mph, this same compressor may be turning as high as 3500 rpm and developing capacities in the neighborhood of 3 to $3^{1}/_{2}$ tons. This varying capacity has dictated the design of present automotive control systems.

To obtain a more favorable speed characteristic, a different solution to the problem could be utilized, such as a drive-shaft take-off or an electric-motor drive through an alternator. Although some of these methods of operating the compressor might prove more desirable in forms of operating efficiency, the additional cost of application in addition to space requirement is, without doubt, an important consideration.

The compressor has two functions:

1. To pump refrigerant through the system
2. To raise the pressure of the refrigerant gas received from the cooling coil so that it will condense more readily and give up heat as it passes through the condensing coil

A typical automotive air-conditioning compressor (six-cylinder rotary double-action piston-type) is shown in Fig. 14-3. The reed-type suction and discharge valves are mounted in a valve plate between the cylinder assembly and the head at the end of the compressor. The ends are connected with each other by gas-tight passageways that direct refrigerant gas to a common output. Each cylinder head contains suction and discharge cavities. In addition, the reed head contains an oil-pump cavity in the center of the suction cavity to house the oil-pump gears, which are driven by the compressor main shaft. The suction cavity is in the center and indexes with the suction reeds. The discharge cavity is around the outside and indexes with the discharge needs.

These cavities are separated from each other with a Teflon seal molded to the cylinder head. The discharge cavity is sealed from the outside of the compressor by an O-ring seal, which rests in a chambered relief in the cylinder head and compresses against the compressor body. An oil pump mounted at the rear of the compressor picks up oil from the bottom of the compressor and pumps it to the internal parts. The inner gear fits over a matching D flat on the main shaft. The outer driven gear has internal teeth that mesh with the external teeth on the inner drive gear.

Condensers

Air-conditioning condensers are similar to ordinary automobile radiators but are designed to withstand much higher pressure. A condenser is normally mounted in front of the car radiator so that it receives a high volume of air. Air passing over the condenser cools the hot high-pressure refrigerant gas causing it to condense into low-pressure liquid refrigerant.

Receiver-Dehydrator

The purpose of the receiver-dehydrator assembly is to ensure a solid column of liquid refrigerant to the thermostatic expansion valve at all times (provided the system is properly charged). The dehydrator (drier) part of the assembly is to absorb any moisture that might be present in the system after assembly. Also it traps foreign material that may have entered the system during assembly. A liquid indicator or sight glass is a part of most systems and is an integral part of the outlet pipe of the receiver dehydrator. The appearance of bubbles or foam in the sight glass when the ambient temperature is higher than 70°F indicates air or a shortage of refrigerant in the system.

Thermostatic Expansion Valves

The function of the thermostatic expansion valve is to automatically regulate the flow of liquid refrigerant into the evaporator in accordance with the requirements of the evaporator. The valve is located at the inlet to the evaporator core. It consists essentially of a capillary bulb and tube connected to an operating diaphragm (sealed within the valve itself) and an equalizer line connecting the valve and low-pressure suction throttling-valve outlet. The thermal bulb is attached to the evaporator-outlet pipe.

The thermostatic expansion valve is the dividing line in the system between high-pressure liquid refrigerant supplied from the receiver and relatively low-pressure liquid and gaseous refrigerant in the evaporator. It is designed so that the temperature of the refrigerant at the evaporator outlet must have 4°F of superheat before more refrigerant is allowed to enter the evaporator. Superheat is an increase in temperature of the gaseous refrigerant above the temperature at which the refrigerant vaporized.

A capillary tube filled with carbon dioxide and the equalizer line provide temperature regulation of the expansion valve. The capillary tube is fastened to the low-pressure refrigerant pipe coming out of the evaporator so that it samples the temperature of the refrigerant at this point to the expansion valve. If the temperature differential between the inlet and outlet decreases below 4°F, the expansion valve will automatically reduce the amount of refrigerant entering the evaporator. If the temperature differential increases, the expansion valve will automatically allow more refrigerant to enter the evaporator. A typical thermostatic expansion valve is shown in Fig. 14-5.

AUTOMOBILE AIR CONDITIONING

The temperature of the air passing over the evaporator core determines the amount of refrigerant that will enter and pass through the evaporator. When the air is very warm, the heat transfer from the air to the refrigerant is great and a greater quantity of refrigerant is required to cool the air and to achieve the proper superheat on the refrigerant gas leaving the evaporator. When the air passing over the evaporator is cool, the heat transfer is small and a smaller amount of refrigerant is required to cool the air and achieve the proper superheat on the refrigerant gas leaving the evaporator.

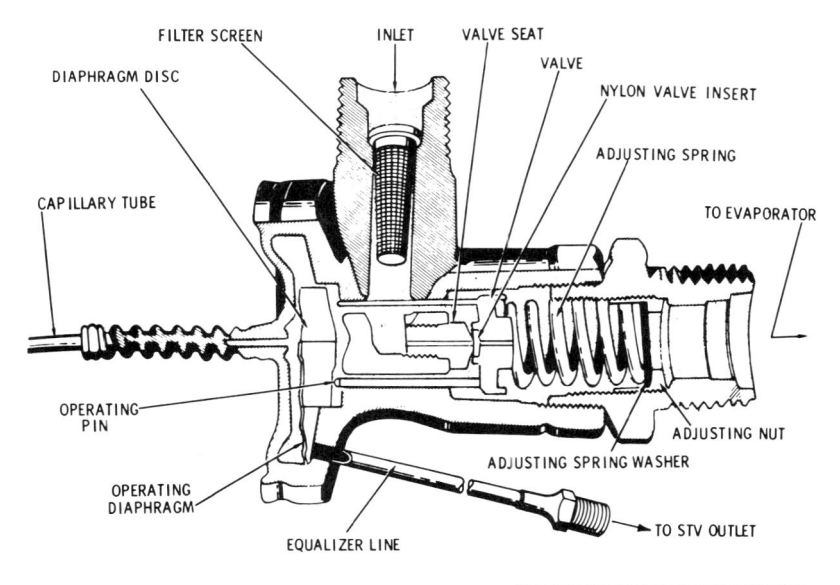

Courtesy General Motors Corporation

Fig. 14-5. Typical internally adjusted automobile air-conditoning thermostatic expansion valve.

A mechanical adjusting nut located within the valve is provided to regulate the amount of refrigerant flow through the valve. When adjusted, the spring seat moves to increase or decrease the tension on the needle-valve-carriage spring. By varying the tension on this spring, it is possible to regulate the point at which the needle valve begins to open or close, thereby regulating refrigerant flow into the evaporator. Since this adjustment feature is inside the valve, *no external adjustment is possible.* All valves are preset at the time of manufacture.

When the air-conditioning system has not been operating, all pressures within the thermostatic expansion-valve assembly will have equalized at the ambient (surrounding air) temperature; thus pressure above and below the operating diaphragm and at the inlet and outlet side of the valve will be equal. Pressure under the diaphragm is evaporator pressure. It reaches this area by means of clearance around the operating pins, which connect the area under the diaphragm with the evaporator-pressure area. While pressures in the expansion valve are almost equal, the addition of the valve-adjusting-spring pressure behind the needle will hold the needle valve over to close the needle-valve orifice.

When the air-conditioning system first begins to operate, the compressor will immediately begin to draw refrigerant from the evaporator, lowering the pressure in the evaporator and in the area under the operating diaphragm. As the pressure in this area decreases, the pressure above the diaphragm exerted by the carbon dioxide in the capillary tube will overcome spring pressure and push the diaphragm against the operating pins, which, in turn, will force the needle off its seat. Refrigerant will then pass through the expansion valve into the evaporator, where it will boil at a temperature corresponding to the pressure in the evaporator. This will begin to cool the air passing over the evaporator, as well as the evaporator-outlet pipe.

The valve-adjusting spring is calibrated so that the pressure of the refrigerant in the evaporator-outlet pipe and equalizer line to the valve, plus the spring force, will equal the force above the operating diaphragm when the temperature of the refrigerant in the evaporator outlet is 4°F above the temperature of the refrigerant entering the evaporator. In other words, the refrigerant should remain in the evaporator long enough to completely vaporize and then warm (superheat) to 4°F.

If the temperature differential begins to go below 4°F (outlet pipe too cold), carbon dioxide pressure in the capillary tube and the area above the diaphragm will decrease, allowing the valve-adjusting spring to move the needle valve toward its seat, closing off the flow of refrigerants past the needle valve. If the temperature differential begins to go above 4°F (outlet pipe too warm), the pressure in the capillary tube and the area above the operating diaphragm will increase, pushing this diaphragm against the operating pins to open the needle valve further, admitting more refrigerant to the evaporator.

The equalizer line permits the suction-throttle-valve-outlet pressure to be imposed on the expansion valve diaphragm, thus overriding its normal control of liquid refrigerant. As the compressor capacity becomes greater than the evaporator load, the drop in compressor suction-line pressure forces the expansion valve to flood liquid through the evaporator and the suction-throttle valve, thus preventing the suction pressure from dropping below a predetermined pressure. The equalizer line is used primarily to prevent prolonged and constant operation of the compressor in vacuum conditions. This operation is considered undesirable because of the noise angle and the possibility of subjecting the compressor to reduced oil return. Second considerations for having the external equalized expansion valve are to maintain a full evaporator during throttling and also guard against noncondensibles entering the system, especially through loosened fittings.

Evaporators

The purpose of the evaporator core is to cool and dehumidify the air flowing through it when the air conditioner is in operation. High-pressure liquid refrigerant flows through the orifice in the thermostatic expansion valve into the low-pressure area of the evaporator. This regulated flow of refrigerant boils immediately. Heat from the core surface is lost to the boiling and vaporizing refrigerant, which is cooler than the core, thereby cooling the core. The air passing over the evaporator loses its heat to the cooler surface of the core. As the process of heat loss from the air to the evaporator-core surface is taking place, moisture in the air condenses on the outside surface of the evaporator core and is drained off.

Since Freon-12 will boil at −21.7°F at atmospheric pressure, and since water freezes at 32°F, it becomes obvious that the temperature in the evaporator must be controlled so that the water

Air Conditioning

collecting on the core surface will not freeze in the fins of the core and block off the air passages. In order to control the temperature, it is necessary to control the amount of refrigerant entering the core and the pressure inside the evaporator. To obtain maximum cooling, the refrigerant must remain in the core long enough to completely vaporize and then superheat to a minimum of 4°F. If too much or too little refrigerant is present in the core, maximum-cooling efficiency is lost. A thermostatic expansion valve in conjunction with the suction-throttling valve is used to provide this necessary refrigerant-volume control.

Solenoid Control

In early automobile air-conditioning systems, a solenoid bypass arrangement (Fig. 14-6) was used for maintaining comfortable interior temperatures. As noted in Fig. 14-6, this system consisted of connecting a solenoid valve into the system between the condenser and the suction line. When the valve is open, hot gas leaves the condenser and travels through the valve into the suction line, thus nullifying the compression effect of the compressor. The solenoid units were controlled initially by a manually operated toggle

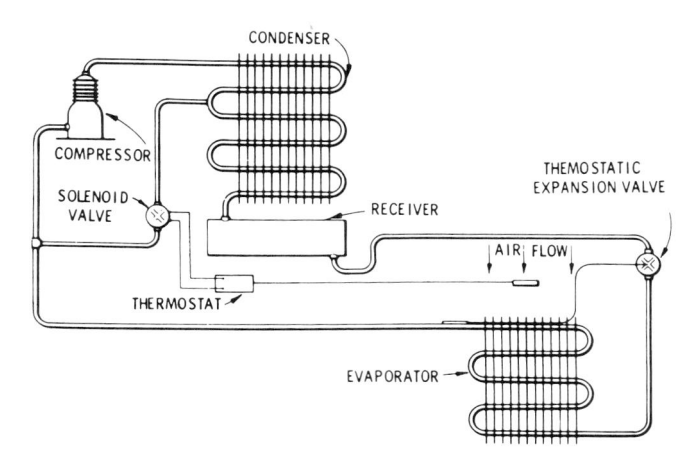

Fig. 14-6. Refrigerant circuit showing a solenoid hot-gas bypass-control system used on early automobile air-conditioning systems.

switch, but subsequently a thermostat was placed in the return airstream to provide automatic on-and-off control.

A later development brought about the installation of a bypass valve in the system. This application is the same as the solenoid valve, with one exception: it is an automatic-type valve that provides a modulating control. As the suction pressure is lowered due to increased compressor speed, the bypass valve opens at its setting and bypasses a certain portion of the hot gas back into the suction line. The higher the speed of the compressor, the wider the valve opens in an attempt to maintain constant suction pressures.

This bypass system is still used in many cars today. In a current refinement of this system, an operating cam is installed on the valve and connected with a wire so that the operator may adjust this valve to control the automobile interior temperature. The bypass system is based on constant compressor operation as long as the automobile engine is running. This situation has proven objectionable to car owners.

Magnetic-Clutch Control

The pulley assembly contains an electrically controlled magnetic clutch, permitting the compressor to operate only when air conditioning is actually desired. When the compressor clutch is not engaged, the compressor shaft does not rotate, although the pulley is being rotated by the belt from the engine. The clutch armature plate, which is a movable member of the drive-plate assembly, is attached to the drive hub through driver springs and riveted to both driver and armature plate. The hub of this assembly is pressed over the compressor shaft and aligned with a square drive key located in the keyway on the compressor shaft. The pulley assembly consists of three units: pulley rim, power-element ring, and pulley hub.

A frictional material is molded between the hub and rim of the pulley, which make up the magnetic-clutch assembly. A power-element rim is embedded in the forward face of the molded material that houses the electrical coils and components that make up the electromagnet circuit. When air-conditioner controls are set for cooling, current flows through the coils, creating a magnetic force that draws the armature plate forward to make contact with

AIR CONDITIONING

the pulley. This action magnetically locks the armature plate and the pulley together as one unit to start compressor operation.

An illustration of a typical magnetic clutch assembly is shown in Fig. 14-7. Application of a thermostat to the return air in the evaporator provides an off-on cycle. Under extreme conditions, the rapid heat gain of the average automobile interior causes the clutch to cycle quite often, occasionally several times a minute.

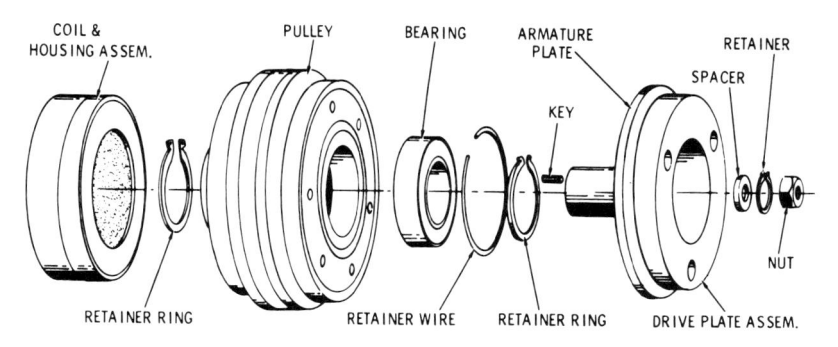

Courtesy General Motors Corporation

Fig. 14-7. Typical magnetic-clutch assembly used on an automobile air-conditioner compressor.

Suction-Throttle Control

Some of the latest automobile air-conditioning systems incorporate an evaporating regulating valve with the inlet connected to the evaporator outlet, and the outlet connected to the compressor suction port. A valve of this type, usually termed a *suction-throttle valve,* is shown in Fig. 14-8 and consists essentially of a valve body, piston, piston diaphragm, control spring, diaphragm cover, diaphragm cap, and vacuum diaphragm. The suction-throttle valve controls the evaporator pressure and evaporator outlet temperature. The inside of the piston is hollow and is open to the piston diaphragm through small holes in the end of the piston. Located in the lower extremity of the piston is a fine mesh screen held in place by a retainer. The purpose of this screen is to prevent any foreign particles from entering the piston and lodging in the holes drilled in the piston walls.

426

AUTOMOBILE AIR CONDITIONING

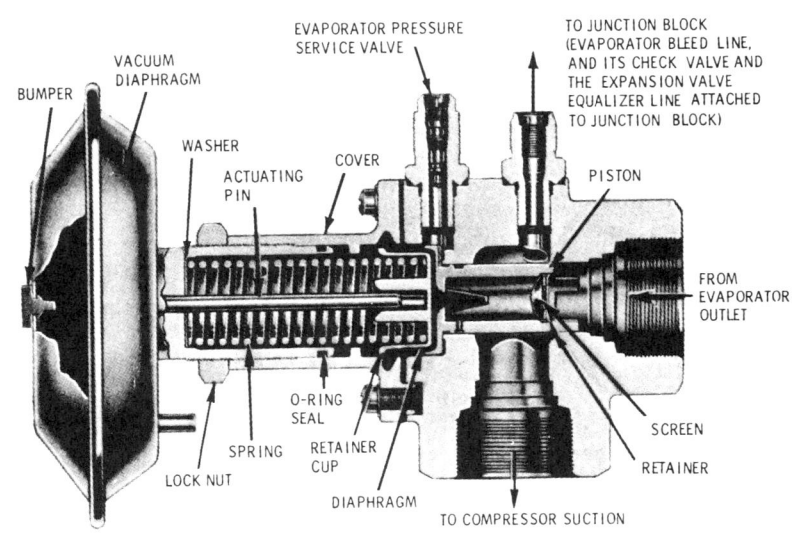

Fig. 14-8. Typical suction-throttle valve.

Courtesy General Motors Corporation

The piston diaphragm is held in position by the piston on the front side and by a retainer cup and spring on the rear side. The vacuum-diaphragm actuating pin fits in the end of the cup. The body of the vacuum diaphragm threads into the valve cover and determines the amount of spring tension on the cup. The vacuum diaphragm is locked to the cover after it has been set by a locknut. A vacuum connection on the vacuum diaphragm housing is connected to the vacuum modulator on the instrument panel by a small hose. When a vacuum is present on diaphragm, it is pulled toward the piston and adds spring pressure to the piston diaphragm.

The flow of the low-pressure vapor from the evaporator to the compressor is determined and controlled by the position of the piston in the valve body of the suction throttle valve. The position of the piston in the body is determined by the balance of the forces that are applied to the piston diaphragm. These forces consist of the refrigerant-vapor pressure returning from the evaporator, on one side, and the spring tension plus the force of the actuating pin if vacuum is present at the vacuum diaphragm, on the other side.

Air Conditioning

Movement of the piston permits vapor to pass by scallops in the piston skirt and then on to the compressor inlet.

During the time that maximum cooling is being produced, the suction throttle-valve vacuum diaphragm does not have engine vacuum applied to it. The full flow of low-pressure refrigerant vapor is being returned to the compressor to permit it to exert its full capacity on the evaporator and produce maximum cooling. Under most all operating conditions, the suction-throttle-valve inlet and outlet pressures will not be the same because there will be some throttling to prevent evaporator icing.

When the operator desires to raise the temperature within the car, the controls are changed to apply engine vacuum to the vacuum diaphragm. This checks, or throttles, the flow of low-pressure vapor returning to the compressor, resulting in a higher pressure to be maintained in the evaporator assembly. The suction-throttle-valve-outlet pressure will also increase, but the differential between inlet and outlet will be much greater than when the suction-throttle valve is at maximum cooling.

Air Distribution

The air is introduced in the car by various outlets, depending upon the various types and models. The air-conditioner distribution system also varies but is entirely separate from the car-heater distribution system. The air-conditioner distribution ducts are usually located forward of the car-heater ducts, with one end positioned against the evaporator housing.

Air-Conditioner Control System

The operator controls the unit through switches located on the instrument panel or on the evaporator case, depending upon the installation procedure. (Follow the manufacturer's instructions for proper operation.)

SERVICE AND MAINTENANCE

Since automobile air conditioners do not differ in any important respect from home air conditioners insofar as the refrigeration cycle is concerned, such common procedures as charging, leak detection, compressor testing, and other standard refrigeration service operations will not be discussed in this chapter. Only servicing and troubleshooting problems peculiar to automobile air conditioning will be treated.

Insufficient or No Cooling

In a system incorporating a magnetic clutch and thermostatic switch, the first check is to make certain that the compressor is running. The electrical system should also be checked. The condition and tension of the belt should be checked; in fact, correct tension on the belt is one of the most important service checks to be made. The tremendous load and high speed of these belts mean they must be kept very tight. Deflection of the belt between pulleys should not be over $1/4$ inch. If a belt is replaced, this deflection should be checked after about $1/2$ hour running period, and tightened again if necessary.

Air Output Not Normal

When it has been determined that the compressor is operating properly, blowers should be checked to make certain they are delivering air. The blower control should be set at high. The temperature-control level (if there is one) should be set at maximum cooling. Insufficient discharge of air indicates an obstruction in the evaporator or electrical problems with the blower motor.

Air Output Normal

Assuming that actual air delivery at the discharge grill is about normal (around 300 cu ft/min), the next check is the position of the bypass dampers. It is a well-known fact that the condensation of the refrigerant depends upon air passing over the condenser as the automobile travels on the road. To simulate this condition in the shop, a large fan should be placed in front of the car. Head-pressure readings will vary somewhat with ambient air temperatures. A fast idle (about 1500 rpm) and an ambient temperature of

AIR CONDITIONING

75°F should indicate head pressures from 100 to 130 psig. For every 10°F increase in the ambient-air temperature, an increase of about 20 psig will occur. Extreme ambient temperatures, such as 110°F or higher, may send the pressure as high as 300 psig. Extreme head pressure could indicate the presence of dirt, air, or debris in the system.

Complete removal of air from the refrigerating system is essential for two reasons:

1. To prevent excessive head pressures when air volume over the condenser is low
2. To prevent decomposition of the oil, which is accelerated by the normally higher operating temperatures encountered in these systems

It is not too uncommon to find that an inexperienced person has serviced the air-conditioning system and overcharged it, causing a high head pressure.

Checking Suction Pressures

Suction pressures normally run about 16 to 25 psig at 75°F. Any wide variation from these pressures indicates the usual service problems resulting from this condition. Abnormally high suction pressures indicate wide-open expansion valves, a loose feeler bulb, or leaky compressor valves. Valve plates usually can be replaced on these compressors. The normal refrigeration procedures for checking compressor operation and maintenance are recommended.

Checking Control Systems

Systems equipped with either the evaporator regulator, bypass-valve type of controls, or the expansion valve should present no difficult problems to qualified servicemen. Special features of these valves may dictate variations from normal refrigeration service practice.

Air Distribution

If an air-conditioning distribution system is not functioning properly, first check the vacuum-hose connections, control cable, and vacuum-switch adjustments. If the controls and cables operate properly, check for vacuum at valve diaphragms, proper functioning of vacuum switches, and correct position of air valves. If the airflow changes or shuts off when the car is accelerating, check for a faulty valve at the intake manifold. If there is no vacuum to the suction-throttle valve when the cool switch is on, check for a disconnected vacuum hose at the vacuum modulator.

Defective Compressor

Compressor malfunctioning will appear in one of four ways: noise, seizure, leakage, or low-discharge pressures. Resonant compressor noises are no cause for excessive alarm, but irregular noises or rattles are likely to indicate broken parts. Seizure will be indicated by the failure of the compressor to operate if the clutch is in good operating condition and there is no break in the electrical system. A wiring diagram of a typical automobile air conditioner is shown in Fig. 14-9. Continued operation of a partially seized compressor will result in damage to the clutch. To check for sei-

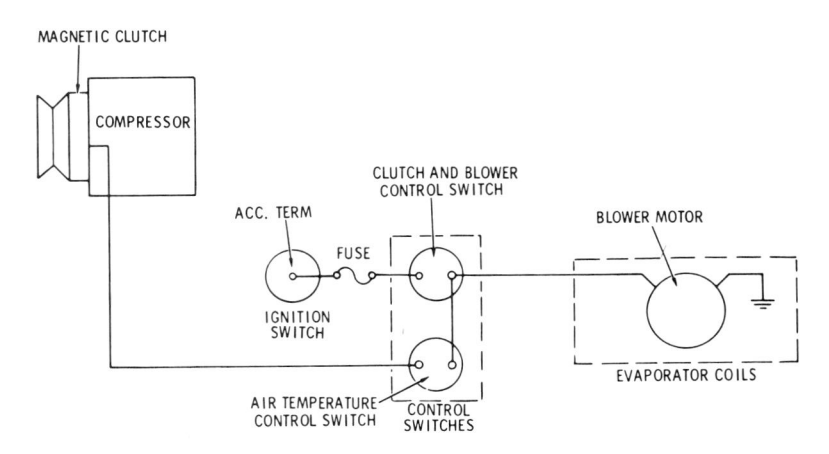

Fig. 14-9. Electrical circuit in a typical automobile air conditioner.

zure, deenergize the clutch and attempt to rotate the compressor by using a wrench on the compressor shaft. If the shaft will not turn, the compressor is seized.

Leakage of compressor refrigerant may be detected through routine leak detection. Low-discharge pressures may also be caused by insufficient refrigerant or a restriction elsewhere in the system. These should be checked out prior to compressor servicing.

Compressor Clutch

If the compressor is inoperative, the electrical lead to the clutch should be checked first. If there is current to the clutch and the compressor is not seized, the clutch is defective and should be repaired.

Condenser

There are two types of possible condenser malfunctions. The condenser may leak, resulting in loss of refrigeration and low system pressure. The condenser may also have a restriction, resulting in excessive compressor-discharge pressures and inadequate cooling. If a restriction occurs and some refrigerant passes the restriction, icing or frost may occur on the external surface of the condenser in the area of the restriction. If the airflow through the condenser is restricted or blocked, high-discharge pressures will result. It is important that the external fins of the condenser and radiator core are not plugged with bugs, dirt, etc.

Thermostatic Expansion Valve

If malfunction of the valve is suspected, make sure the power element bulb is in proper position, securely attached, and well insulated from the outside air temperatures. If the thermostatic expansion valve fails, it usually fails in the power element and the valve remains closed. This will be indicated by low high-side or discharge pressures. Also the inlet screen could be plugged. The screen can be cleaned with liquid refrigerant.

Evaporator

Dirt or other foreign matter on the core surface or in the evaporator housing will restrict the airflow. A cracked or broken housing can result in leakage of cold air and in insufficient air or warm air being delivered to the passenger compartment.

Refrigerant Lines

Restrictions in the refrigerant lines may be indicated as follows:

1. High-pressure liquid line (low head pressure, no cooling)
2. Suction line (low-suction pressure, low-head pressure, little or no cooling)
3. Discharge line (compressor blowoff)
4. Receiver-drier (leakage of refrigerant indicates a defective unit. This cannot easily be checked, but if the system has been exposed to outside air for a considerable length of time, the unit should be replaced.)

Restrictions in the receiver-drier can also cause system malfunction. If the inlet tube is blocked, it is likely to result in high head pressure. If the outlet tube is blocked, head pressure is likely to be low and there will be little or no cooling.

Suction-Throttle Valve

If the suction-throttle valve is defective, it may cause evaporator pressure to be too high (air-outlet temperature too warm), or it could cause the evaporator pressure to be too low (air-outlet temperature too cold, which may cause icing of the evaporator core). If the vacuum diaphragm of the suction-throttle valve is defective, there would be no means to change (increase) the air-outlet temperature. Refrigerant leakage of the suction-throttle valve may be detected through routine leak detection.

Before servicing the suction-throttle valve, it should be determined that it is actually the cause of the complaint. If evaporator pressure remains too high when checking and adjusting the suction-throttle valve, the low-pressure-gage line should be attached to the valve located on the compressor-suction line. If the compressor-suction pressure is also too high, the compressor or possibly the thermostatic expansion valve may be the cause of the trouble.

Use of Receiver Sight Glass for Diagnosis

A clear sight glass will indicate a properly charged refrigeration system. The occurrence of slow moving gas bubbles or a broken column of refrigerant for momentary periods during normal operation should not be considered an indication of refrigerant shortage if the sight glass is generally clear and performance is satisfactory. The tendency of the sight glass to indicate refrigerant shortage when the system is under light load should be considered.

If the sight glass consistently shows foaming or a broken liquid column, it should be observed after partially blocking the air to the condenser. If under this condition the sight glass clears and the performance is otherwise satisfactory, the charge shall be considered adequate. In all instances where the indications of refrigerant shortage continues, additional refrigerant should be added in $1/4$-lb increments until the sight glass is clear. An additional charge of $1/2$ lb should be added as a reserve. In no case should the system be overcharged.

INSTALLATION PROCEDURE

Two installation methods for automobile air conditioning are: factory installation and universal-type installation. Factory-installed air conditioners are furnished with the automobiles, usually with instructions for proper maintenance and operation. In such installations, several methods are used on evaporator assemblies. Many such installations combine the heating system with the cooling system, as shown in Fig. 14-10, with the evaporator coil and heater core located in the same enclosure. One significant service problem that should be pointed out here is the function of the bypass dampers.

A common complaint of no cooling is corrected by proper ad-

AUTOMOBILE AIR CONDITIONING

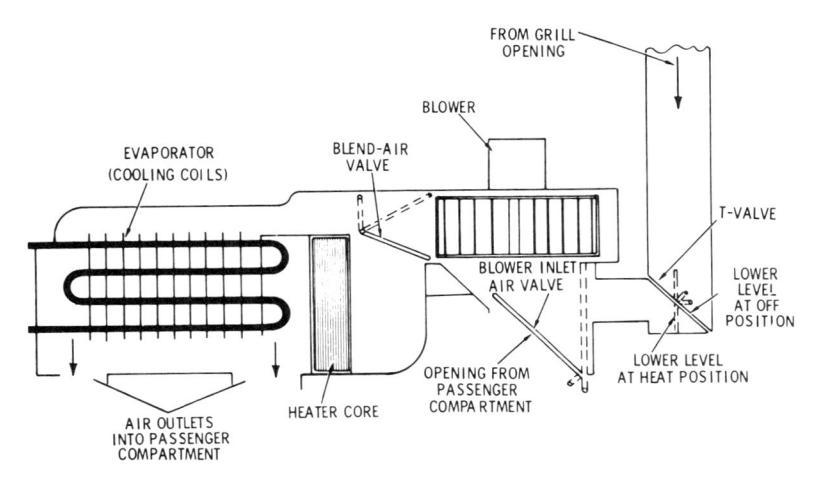

Fig. 14-10. Schematic diagram showing heating- and cooling-coil arrangements.

justment of the bypass dampers, which, in many installations, are controlled by cables that operate dampers to furnish outside or recirculated air. These control cables are connected to a control panel in such a way that the combined functions are accomplished or controlled through a single lever. A more recent refinement in damper controls is shown in Fig. 14-11. Here, vacuum actuators are used with pushbutton control. Installations of this sort mean that servicemen must thoroughly acquaint themselves with the basic operating principle governing this type of control. These controls are similar in many respects to the controls used in pneumatic-controlled air-conditioning systems.

The evaporator housings resemble the old-fashioned automobile heaters that were suspended under the dash. This construction, popular with independent automotive air-conditioner manufacturers, simplifies installation. Its operating characteristics are the same as factory-installed units. Fig. 14-12 shows a typical automotive air-conditioning-system condenser, which is usually located in front of the radiator. Circulation of air over the condenser depends upon either ram air, resulting from the forward movement of the car, or from air drawn across the unit by the engine fan.

AIR CONDITIONING

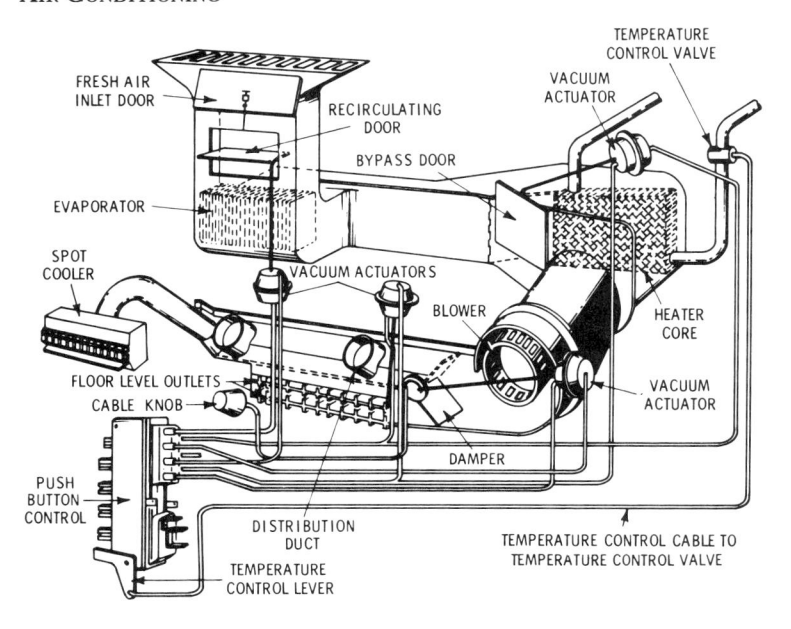

Fig. 14-11. Typical damper-control assembly showing vacuum actuators with pushbutton controls.

It is interesting to note that head pressures vary on the highway when the car travels with or against the wind. When traveling "downwind," heat pressure usually runs 10 to 15 lbs higher than when traveling against the wind. Most road tests, however, have proven that the condenser does not materially affect temperature of the car engine. One major problem encountered is the head pressure increasing substantially when the engine is left idling, such as in slow-moving or stalled traffic. Normally, air circulation over the condenser is supplied by the forward movement of the car; at high speeds, the ram effect of this air does an adequate job of cooling the condenser. At more moderate speeds, the fan draws sufficient air over the condenser to keep head pressures normal, but when the car is stalled or is traveling very slowly, air volume is not sufficient to keep head pressures within operating limits. Some systems provide a *fast-idling* control, which increases the engine speed sufficiently to supply air over the condenser.

Location of the condenser in front of the car contributes to ac-

AUTOMOBILE AIR CONDITIONING

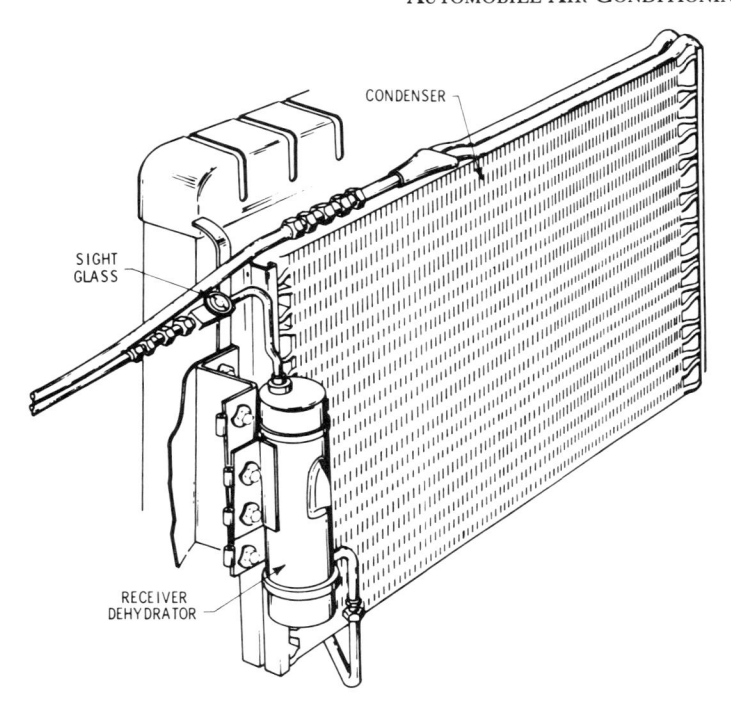

Fig. 14-12. Condenser and receiver installation in a typical automobile air-conditioner system.

cumulation of dirt and debris on the finned surface. Recently developed service tools for effectively cleaning finned surfaces are available. Also shown in Fig. 14-12 is the system receiver. An innovation is encountered here through the frequent practice of placing a drier core within the receiver, thus resulting in a combination receiver-drier unit. This construction may pose a question to some servicemen. With the drier located within the receiver, how can this component be replaced?

Since present automobile air-conditioning systems make no provision to "pump down" the unit, there is no need for a liquid-line valve. The only valves available are the suction and discharge service valves at the compressor. This simply means that if any service operation is required away from the compressor itself, the refrigerant must be removed from the system. The drier cartridge inside the receiver also usually provides a filter. Many times this is the only filter or screen in the system since the expansion valve is not always equipped with a screen. This fact makes such a filter vital to system operation.

TROUBLESHOOTING GUIDE

The following troubleshooting guide lists the most common operating faults along with possible causes and suggested checks and correction for each. This guide is intended to provide a quick reference to the cause and correction of a specific fault.

AUTOMOBILE AIR CONDITIONING TROUBLESHOOTING GUIDE

Symptom and Possible Cause   *Possible Remedy*

Insufficient Cooling

(a) Low airflow.   (a) Check blower operation. Check for obstruction in air-distribution system. Check for clogged evaporator. If iced, deice core and check adjustment and operation of suction-throttle valve.

(b) Defective heater-temperature control valve.   (b) Check operation of valve. Adjust or replace as necessary.

(c) Heater controls or ventilator control not in the "off" position.   (c) Advise operator of correct operation of controls.

Compressor Discharge Pressure Too High

(a) Engine overheated.   (a) Check for possible cause.
(b) Overcharge of refrigerant or air in system.   (b) Systems with excess discharge pressures should be

AUTOMOBILE AIR CONDITIONING

Symptom and Possible Cause   *Possible Remedy*

slowly depressurized.
(i) If discharge pressure drops rapidly, it indicates air (with possibility of moisture) in the system. When pressure levels drop but still indicate in excess of specifications, slowly bleed system until bubbles appear in the sight glass and stop. Add refrigerant until bubbles clear: then add $1/2$ lb of refrigerant. Recheck operational pressures. If system pressures still remain above specifications and the evaporator pressure is slightly above normal, then a restriction exists in the high-pressure side of the system.
(ii) If discharge pressure drops slowly, it indicates excessive refrigerant. If pressures drop to specifications and sight glass remains clear, stop depressurizing and recheck operational pressures. If pressures are satisfactory, depressurize until bubbles appear in the sight glass, stop depressurizing, then add $1/2$ lb refrigerant. Recheck operational pressures.
(iii) If discharge pressure remains high after depres-

AIR CONDITIONING

Symptom and Possible Cause *Possible Remedy*

	surizing the system, continue depressurizing until bubbles appear in the sight glass. If evaporator pressures also remain high, there is a possibility of a restriction in the high-pressure side of the refrigeration system, or the suction throttle valve may require adjustment.
(c) Restriction in condenser or receiver liquid indicator.	(c) Remove parts, inspect, and clean or replace.
(d) Condenser airflow blocked.	(d) Clean condenser and radiator core surfaces as well as the space between the condenser and radiator.
(e) Evaporator pressure too high.	(e) See Evaporator Pressure Too High (below).

Compressor-Discharge Pressure Too Low

(a) Insufficient refrigerant.	(a) Check for presence of bubbles or foam in liquid indicator. If bubbles or foam are noted (after 5 minutes of operation), check system for leaks. If no leaks are found, refrigerant should be added until sight glass clears; then add an additional $1/2$ lb.
(b) Low-suction pressure.	(b) See Evaporator Pressure Too Low (below).
(c) Defective compressor and/or broken compressor reed valves.	(c) Repair compressor.

AUTOMOBILE AIR CONDITIONING

Symptom and Possible Cause   *Possible Remedy*

Evaporator Pressure Too High

(a) Thermostatic expansion valve capillary-tube bulb not tight to evaporator outlet tube.

(a) Check for tightness.

(b) Thermostatic expansion valve improperly adjusted or inoperative.

(b) Replace valve.

(c) Suction-throttle valve adjusted improperly or defective.

(c) Check operation of suction-throttle valve. Repair valve, if necessary.

(d) Vacuum modulator defective.

(d) There should he no vacuum to the suction throttle valve when "cool" level on instrument panel is at maximum "on" position. Replace vacuum modulator if defective. *Note:* If compressor-suction line from suction-throttle valve is extremely colder than suction throttle-valve-inlet line from evaporator, this indicates that suction-throttle-valve-outlet pressure is much lower than inlet pressure, and the suction-throttle valve may be defective.

Evaporator Pressure Too Low

(a) Thermostatic expansion valve capillary tube broken, inlet screen plugged, or valve otherwise failed.

(a) Replace valve or clean inlet screen of valve.

441

AIR CONDITIONING

Symptom and Possible Cause

(b) Restriction in system tubes or hoses.
(c) Suction-throttle valve adjusted improperly or defective.

Possible Remedy

(b) Replace kinked tube or restricted hose.
(c) Check operation of suction throttle valve. Repair if necessary.

SUMMARY

The automobile air-conditioning system is no different from any other type of air-conditioning system. The major components are the same.

Two temperature controls are most frequently used on automobile air-conditioning systems: the magnetic clutch, operated by electricity, and the refrigerant-flow control.

Today all automobile air-conditioner compressors are driven by a belt from the engine crankshaft. The compressor speed varies with the speed of the automobile engine.

Air-conditioning condensers are similar to the ordinary automobile radiator but are designed to withstand much higher pressures. They are normally mounted in front of the car radiator.

The function of the thermostatic expansion valve is to automatically regulate the flow of liquid refrigerant into the evaporator in accordance with the requirements of the evaporator. The temperature of the air passing over the evaporator core determines the amount of refrigerant that will enter and pass through the evaporator.

When the compressor clutch is not engaged, the compressor shaft will not rotate. However, the pulley is being rotated by the belt from the crankshaft pulley. The clutch is engaged when the electric solenoid of the magnetic clutch is energized. This causes the magnetic properties of the clutch assembly to attract the pulley section and cause it to engage the section that connects to the compressor shaft.

Some recent automobile air-conditioning systems incorporate an evaporating regulating valve with the inlet connected to the evaporator outlet, and the outlet connected to the compressor suction port. This type of valve is called a *suction-throttle valve*.

REVIEW QUESTIONS

1. How does the automobile air-conditioning system differ from the air conditioner sticking out of the living room window?
2. How does the automobile air-conditioning system operate?
3. Why are automobile compressors different from those used in the home air-conditioning unit?
4. Where is the condenser located on an automobile to take advantage of airflow?
5. What is the difference between the solenoid hot-gas bypass-control system and the older solenoid-control system?
6. Explain the difference between the magnetic clutch and the suction-throttle control.
7. What does abnormally high suction pressure indicate on an automobile air-conditioning system?
8. What are four ways that the compressor system indicates malfunctioning?
9. What are the two possible condenser malfunctions?
10. What trouble does a defective suction valve cause to an air conditioning system?
11. What are the two methods of acquiring automobile air-conditioning?
12. Why is the receiver filter so important in an automobile air-conditioning system?

CHAPTER 15

Motors and Motor Controls

The electric motor is the prime mover in most air-conditioning applications. The type of motor used for a particular condensing unit depends the following essential factors:

1. Power source (whether direct or alternating current)
2. Voltage and frequency of current
3. Voltage regulation
4. Phases (whether single or three-phase)

In selecting a motor for a particular application, the first consideration should be the available power source. Since 60-Hertz alternating current is available in most locations throughout the country, a satisfactory performance usually can be obtained by employing ac motors with accompanying controls. In locations where only direct current is available, shunt and compound-wound motors with suitable controllers may be employed. Series-wound dc motors, however, are not suitable for compressor drive because of their special speed characteristics.

MOTOR VOLTAGE

As noted previously, the power source in practically all locations is alternating current. Isolated plants usually generate direct current for small loads in confined areas. The conventional three-phase ac system is normally used for operation of large motors, whereas single-phase ac is used for motors having fractional-horsepower ratings. Standard conditions of voltage and frequency are the values listed on the motor nameplate.

AC Motors

Alternating-current motors may be divided into several classifications, depending on power supply and winding types: *polyphase* and *single phase*. Where polyphase power is available, it is usually more economical to use polyphase motors instead of single-phase motors because of their higher efficiency and higher power factor. Single-phase motors are universally employed in fractional-horsepower sizes and in portable and central air-conditioning systems for homes and small industrial plants.

POLYPHASE MOTORS

Polyphase motors are classified according to winding methods as: *squirrel-cage induction, wound-rotor induction,* and *synchronous.*

Squirrel-Cage Motors

The squirrel-cage motor derives its name from the fact that the rotor or secondary resembles the wheel of a squirrel cage. Its universal use lies in its mechanical simplicity, ruggedness, and the fact that it can be manufactured with characteristics to suit most industrial requirements. In a squirrel-cage motor, the stator, or primary, consists of a laminated sheet-steel core with slots in which the insulated coils are placed. The coils are so grouped and connected as to form a definite polar area, which produces a rotating magnetic field when connected to a polyphase ac circuit.

MOTORS AND MOTOR CONTROLS

The rotor, or secondary, is also constructed of steel laminations, but the windings consist of conductor bars placed approximately parallel to the shaft and close to the rotor surface. These windings are short-circuited or connected together at each end of the rotor by a solid ring. The rotors of large motors have bars and rings of copper connected together at each end by a conducting end ring made of copper or brass. The joints between the bars and end rings are usually electrically welded into one unit with blowers mounted on each end of the rotor. In small squirrel-cage rotors, the bars, end rings, and blowers are aluminum casted in one piece instead of being welded together.

Squirrel-cage motors are classified by NEMA according to their electrical characteristics as follows:

Class A: Normal-torque, normal starting current
Class B: Normal-torque, low starting current
Class C: High-torque, low starting current
Class D: High-slip

Class-A Motors—Class-A motors are the most popular of all types, employing a squirrel-cage winding having relatively low resistance and reactance. These motors have a normal starting torque and low slip at the rated load and may have sufficiently high starting current to require a compensator or resistance starter for motors above $7^1/2$ horsepower in most cases.

Class-B Motors—These motors are built to develop normal starting torque with relatively low starting current, and they can be started at full voltage. The low starting current is obtained by the design of the motor to include inherently high reactance. The combined effect of induction and frequency on the current is termed *inductive reactance*. The slip at rated loads is relatively low.

Class-C Motors—These motors are usually equipped with a "double squirrel-cage" winding and combine high starting torque with a low starting current. These motors can be started at full voltage. The low starting current is obtained by design of motors to include inherently high reactance. The slip at rated load is relatively low.

Class-D Motors—Class-D motors are provided with a high-

447

resistance squirrel-cage winding, giving the motor a high starting torque, low starting current, high slip (15 to 20 percent) and low efficiency.

As noted in the preceding classifications, Class-A motors provide normal starting torque at starting currents in excess of Class B motors and are suitable for constant-speed application to equipment such as fans and blowers. Class-C motors provide high starting torque with low starting current (same as Class B) and are used on compressor applications started without unloaders and on reciprocating pumps. Class-D motors have a high slip and are usually provided with flywheels. They are used mainly on reciprocating compressors and pumps. Construction of Class-A, -B, -C, and -D motors is shown in Fig. 15-1. Figure 15-2 depicts a squirrel-cage induction motor.

Wound-Rotor Motors

In the squirrel-cage motors described previously, the rotor winding is practically self-contained and is not connected either mechanically or electrically with the outside power source or control circuit. It consists of a number of straight bars uniformly distributed around the periphery of the rotor and short-circuited at both ends by end rings to which they are integrally joined. Since the rotor bars and end rings have fixed resistances, such characteristics as starting and pullout torques, rate of acceleration, and full-load operating speed cannot be changed for a given motor installation.

In wound-rotor motors, the rotor winding consists of insulated coils of wire, not permanently short-circuited, as in the case of the squirrel-cage rotors, but connected in regular succession to form a definite polar area having the same number of poles as the stator. The ends of these rotor windings are brought out to collector rings, or slip rings, as they are commonly termed (Fig. 15-3).

By varying the amount of resistance in the rotor circuit, a corresponding variation in the motor characteristics can be obtained. Thus, by inserting a high external resistance in the rotor circuit at starting, a high starting torque can be developed with a low starting current. As the motor accelerates, the resistance is gradually

Motors and Motor Controls

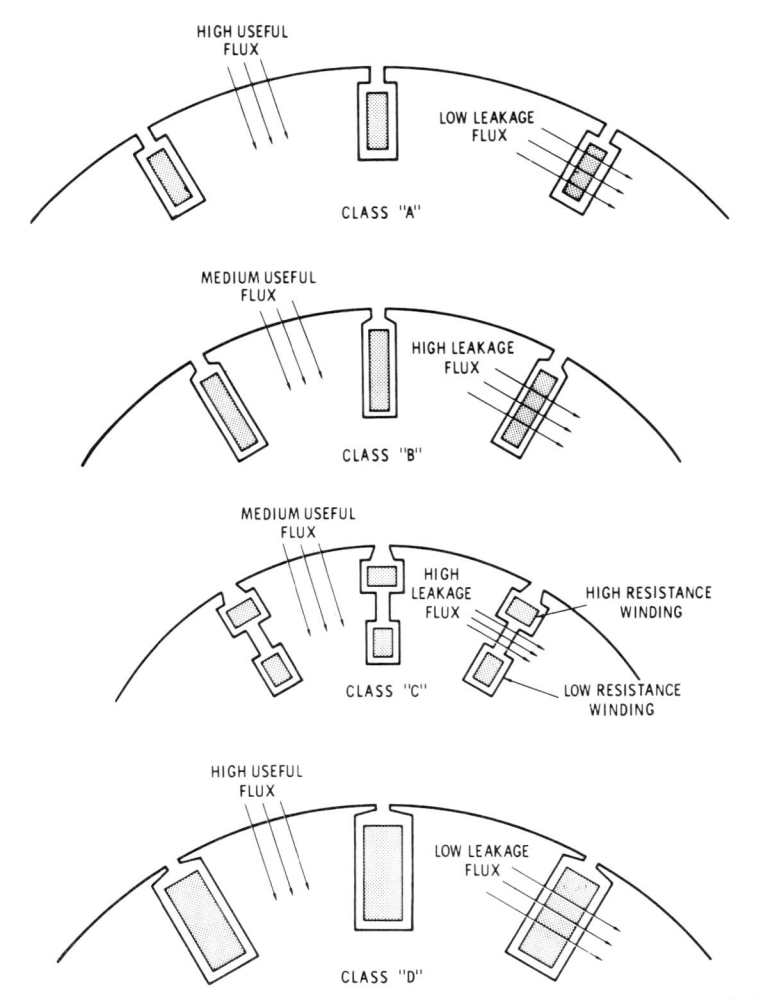

Fig. 15-1. Squirrel-cage rotor construction for Class-A, -B, -C, and -D motors, respectively.

cut out until, at full speed, the resistance is entirely cut out and the rotor windings are short-circuited.

The wound-rotor motor is used in applications requiring a high starting torque at a comparatively low starting current. Thus, the wound-rotor motor has found its use in cranes, hoists, and eleva-

449

AIR CONDITIONING

Courtesy General Electric Company

Fig. 15-2. Sectional view of totally enclosed fan-cooled squirrel-cage induction motor.

Courtesy Mueller Brass Company

Fig. 15-3. Typical wound-rotor induction motor (rotor removed showing construction).

450

tors that are operated intermittently and not over long periods. Exact speed regulation and loss in efficiency are of little consequence. If, however, lower speed is required over longer periods, poor speed regulation and loss in efficiency may become prohibitive.

Synchronous Motors

Synchronous motors are of the constant-speed type and are seldom used in the field of residence air conditioning. Their use is normally limited to large compressors, pumps, and blowers where efficiency, economy, and power-factor control are important.

A synchronous motor may have either a revolving armature or a revolving field, although most are of the revolving-field type. The stationary armature is attached to the stator frame, while the field magnets are attached to a frame that revolves with the shaft. The field coils are excited by direct current, either from a small dc generator (usually mounted on the same shaft as the motor) called an *exciter*, or from some other dc source. The speed of the synchronous motor is determined by the frequency of the supply current and number of poles in the motor. Thus, the operating speed is constant for a given frequency and number of poles. Some motors are required to operate at more than one speed but are constant-speed machines at a particular operating speed. When a speed ratio of 2:1, for example, is required, a single-frame, two-speed, synchronous motor may be suitable. Four-speed motors are used when two speeds not in the ratio of 2:1 are desired. Fig. 15-4 shows stators and rotors of various low-speed synchronous motors.

The single-frame, two-speed motor is usually of salient-pole construction, with the number of poles corresponding to the low speed. High speed is obtained by regrouping the poles so as to obtain two adjacent poles of the same polarity followed by two poles of opposite polarity, which gives the effect of reducing the number of poles on the rotor by one-half for high-speed operation. Corresponding changes in the stator connections are also made. This switching is usually accomplished automatically by means of magnetic starting, or manually operated pole-changing equipment may be used.

AIR CONDITIONING

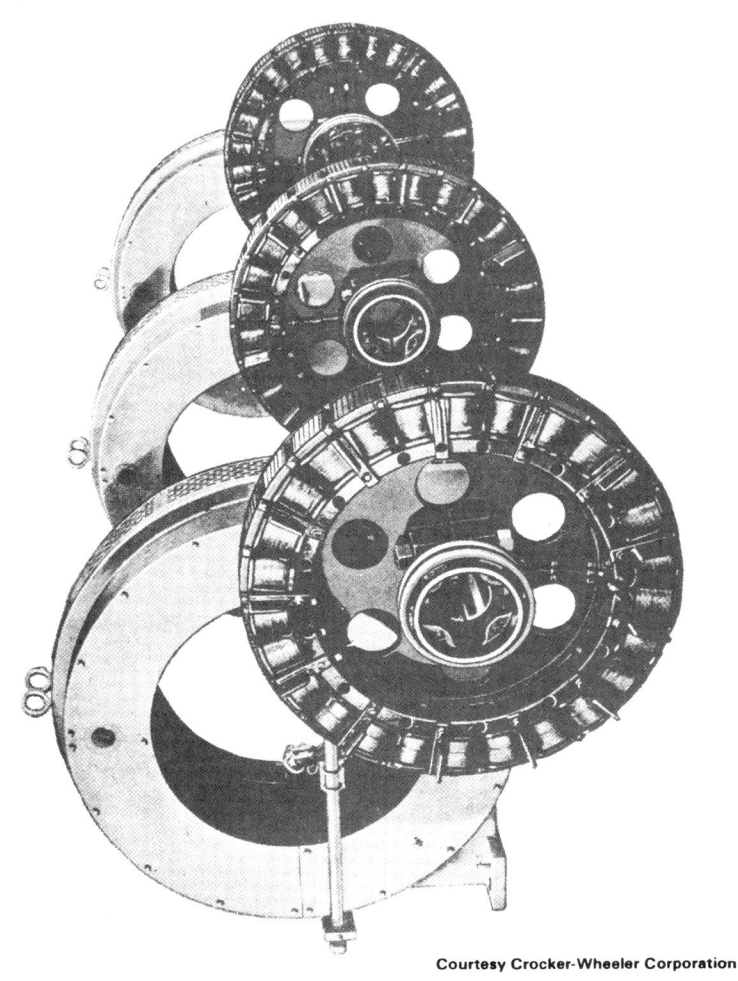

Courtesy Crocker-Wheeler Corporation

Fig. 15-4. Stators and rotors of three low-speed synchronous motors.

SINGLE-PHASE MOTORS

Single-phase induction motors are provided with auxiliary windings or other devices for starting and are used on all types of home appliances. They are also made for a great variety of domes-

452

MOTORS AND MOTOR CONTROLS

tic and industrial applications in sizes up to about 5 horsepower. Single-phase motors are classified according to their construction and method of starting as:

1. Split-phase
2. Capacitor-start
3. Two-valve capacitor
4. Permanent split-capacitor
5. Repulsion-induction
6. Repulsion-start induction

Split-Phase Motors

The split-phase motor consists essentially of a squirrel-cage rotor, two stator windings, a main winding, and an auxiliary or starting winding (Fig. 15-5). The main winding is connected across the supply lines in the usual manner and has a low resistance and a high inductance. The starting or auxiliary winding, which is physically displaced in the stator from the main winding, has a high resistance and a low inductance. This physical displacement, in addition to the electrical phase displacement produced by the relative electrical resistance values in the two windings, produces a weak rotating field that is sufficient to provide a low starting torque.

After the motor has accelerated to 75 or 80 percent of its synchronous speed, a starting relay opens its contacts to disconnect

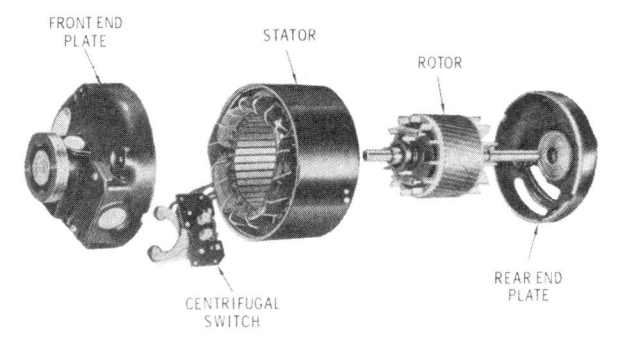

Fig. 15-5. Disassembled view of a typical single-phase induction motor.

AIR CONDITIONING

the starting winding (Figs. 15-6 and 15-7). The function of the starting relay is to prevent the motor from drawing excessive current from the line and also protect the starting winding from damage due to heating. The motor may be started in either direction by reversing either the main or the auxiliary winding. The split-phase motor is most commonly used in sizes ranging from $1/30$ to $1/2$ horsepower for fans, business machines, automatic musical instruments, buffing machines, grinders, and numerous other applications.

Fig. 15-6. Schematic diagram that shows connections of a typical split-phase induction motor with associated potential relay whose function is to disconnect starting windings when operating speed has been reached.

Fig. 15-7. Schematic diagram of a typical split-phase induction motor with a current-type relay to disconnect the starting winding at approximately 85 percent of rated speed.

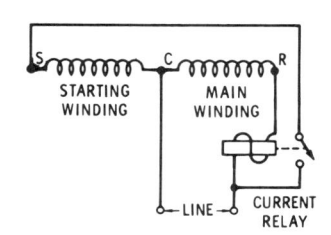

Capacitor-Start Motors

The capacitor-start motor is another form of a split-phase motor but having a capacitor connected in series with the auxiliary winding. The auxiliary circuit is opened when the motor has attained a predetermined speed. The rotor is of the squirrel-cage type, as in other split-phase motors. The main winding is connected directly across the line, while the auxiliary or starting winding is connected through a capacitor, which may be connected into the circuit through a transformer with suitably designed windings; the two windings will be approximately 90°

Motors and Motor Controls

apart electrically. Capacitor-start motors have a comparatively high starting torque accompanied by a high power factor and are used in fractional-horsepower sizes for constant-speed drive on fans, blowers, and centrifugal pumps. Figures 15-8 and 15-9 show capacitor-start split-phase induction motors.

Fig. 15-8. Capacitor-start, split-phase induction motor with potential relay.

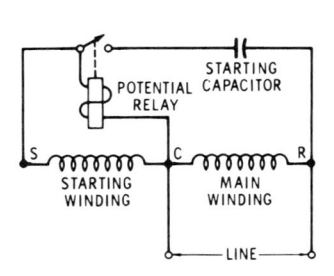

Fig. 15-9. Capacitor-start, split-phase induction motor with current-type relay.

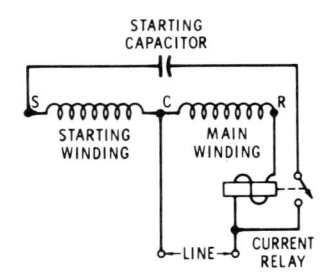

Two-Value Capacitor Motors

In the two-value capacitor motor, the main winding is connected directly to the power supply and auxiliary winding is permanently connected in series with a running capacitor (Fig. 15-10). The running capacitor provides high efficiency at normal speed. The starting capacitor gives high starting ability and is cut out of the circuit by a centrifugal switch after the motor has attained a predetermined speed, usually about 80 percent of its running speed. These motors develop a high starting torque in the fractional-horsepower sizes and are used on compressors, reciprocating pumps, and similar equipment that may start under heavy load.

455

AIR CONDITIONING

Permanent Split Capacitor Motors

In the permanent split-capacitor motor, the main winding is connected directly to the power source and the auxiliary winding is connected in series with a capacitor (Fig. 15-11). The capacitor and auxiliary winding remain in the circuit while the motor is in operation. There are several types of capacitor motors, differing from one another mainly in the number and arrangement of capacitors employed.

The running characteristics of the permanent split-capacitor motor are extremely favorable, and the torque is fixed by the amount of additional capacitance added to the auxiliary winding during the starting period. Its operation is similar to the capacitor-start motor, except that the capacitor remains in the circuit when running. The motor is ideally suitable for small fan drives and other appliances where only a limited amount of starting torque is required.

Fig. 15-10. Two-valve, capacitor-start, split-phase induction motor with potential relay.

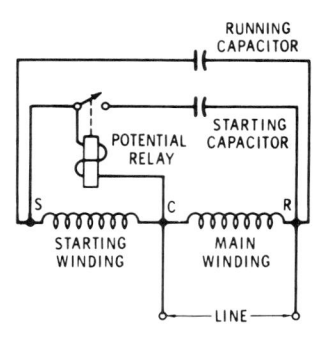

Fig. 15-11. Typical permanent split-capacitor induction motor.

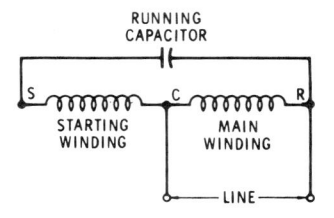

Repulsion Induction Motors

Repulsion induction motors are equipped with two rotor windings: a squirrel cage for running, and a wound-rotor winding connected to a commutator for starting. Both of those rotor windings function during the entire operation period of the motor. There are no automatic devices, such as the starting switch of the split-phase motor or the short-circuiting device of the repulsion-start induction motor.

The cage winding is located in slots below those containing the commutator winding. The slots containing the two windings may or may not be connected by a narrow slot. Usually there are the same number of slots in the two windings. However, it is not absolutely essential that they be the same. At full-load speed, which is slightly below synchronism, the reactance of the cage winding is low and most of the mutual flux passes beneath the cage winding.

Both windings produce torque, and the output of the motor is the combined output of the cage winding and the commutated winding. The repulsion induction motor is especially suitable for household refrigerators, water systems, garage air pumps, gasoline pumps, compressors, and similar applications.

Repulsion-Start Induction Motors

The repulsion-start induction motor is similar to the repulsion induction motor, except that it has only a commutator winding. Motors of this type are provided with a short-circuit mechanism working on the centrifugal principle. When nearly synchronous speed is attained, the commutator bars are short-circuited, resulting in a winding similar to the squirrel cage in its function. Since the motor starts on the repulsion principle, it has the same starting characteristics as the repulsion motor described previously, namely, high starting torque and low starting current.

The repulsion-start-induction motor (Fig. 15-12) is suitable for industrial compressors and similar applications. The efficiency and maximum running torque of the repulsion-start induction-run motor are usually less than those of a cage-wound induction motor of comparative size. In other words, the repulsion-start induction motor must be larger than cage-wound motors of the same rating to give the same performance. Repulsion and repulsion-start induction motors are being used less and less; they are being replaced by capacitor-start motors, which require less maintenance.

AIR CONDITIONING

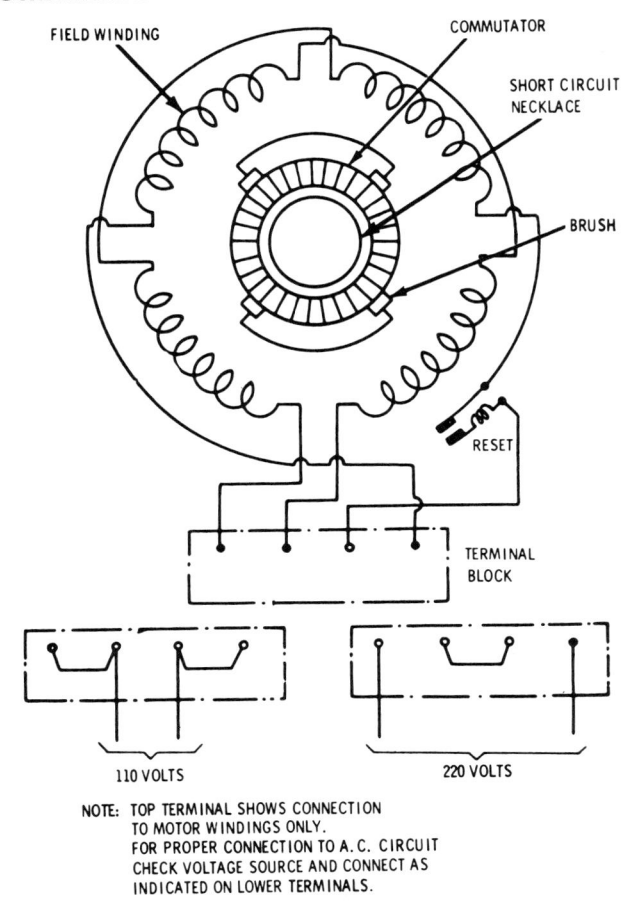

Fig. 15-12. **Winding relationship in a single-phase, repulsion-start induction motor.**

Capacitors

Capacitors are the devices used on single-phase induction motors (usually termed *capacitor motors*) to increase the starting torque and improve the power factor and efficiency. Capacitors are employed primarily to displace the phase of the current pass-

MOTORS AND MOTOR CONTROLS

ing through the starting winding. The capacity of a capacitor is expressed in microfarads (μF) and is dependent on the size and construction of the capacitor.

The voltage rating of a capacitor indicates the nominal voltage at which it is designed to operate. There are two types of capacitors used on single-phase induction motors—*starting* and *running* capacitors—depending on whether the capacitor is connected in the starting or running winding. The voltage that a capacitor is subjected to is not the line voltage, but the back electromotive force generated in the starting winding. It may be as high as twice the line voltage, depending on the compressor speed and winding characteristics. A typical running capacitor is shown in Fig. 15-13.

MANUFACTURER	MARKING
GEN. ELEC	DOT
CORNELL DUBILIER	DASH
SPRAGUE	DOT OR ARROW

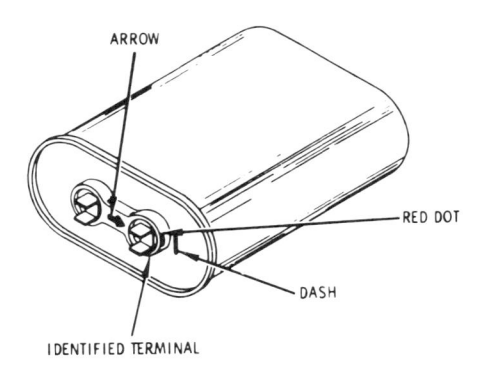

Fig. 15-13. Typical running capacitor.

Capacitors, either start or run, can be connected either in series or parallel to provide the desired characteristics if the voltage and capacitance are properly selected. When two capacitors having the same μF rating are connected in series, the resulting total capacitance will be one-half the rated capacitance of a single capac-

459

itor. The formula for determining capacitance (μF) when capacitors are connected in series is as follows:

$$\frac{1}{\mu F_{total}} = \frac{1}{\mu F_1} + \frac{1}{\mu F_2}$$

When capacitors are connected in parallel, their rating is equal to the sum of their individual capacities, while their voltage rating is equal to the smallest rating of the individual capacitors.

Example: What is the combined capacity of three capacitors of 7-, 8-, and 9-μF capacity, respectively, when connected in series?
Solution: The formula for series connected capacitors is:
(using a calculator)

$$\frac{1}{C_T} = \frac{1}{7} + \frac{1}{8} + \frac{1}{9}$$

or

$$\frac{1}{C_T} = 0.014285 + 0.125 + 0.011111$$

$$\frac{1}{C_T} = 0.037896$$

or

$$C_T = 2.6387435 \ \mu F$$

Example: What is the combined capacity of four capacitors of 10-, 15-, 25-, and 30-μF capacity, respectively, connected in parallel?
Solution: Since

$$C = C_1 + C_2 + C_3 + C_4$$

substituting numerical values gives

$$C = 10 + 15 + 25 + 30 = 80 \ \mu F$$

The capacitor for capacitor-start motors is an electrolytic type, and the number of starts and stops per minute is limited. Running capacitors, however, are generally oil-filled and may stay on the line continuously.

AC MOTOR CONTROLS

Squirrel-Cage Motors

The functions of control apparatus for squirrel-cage motors are:

1. Starting the motor on full voltage or reduced voltage
2. Stopping the motor
3. Disconnecting the motor on voltage failure
4. Limiting the motor load
5. Changing the direction of rotor rotation
6. Starting and stopping the motor at fixed points in a given cycle of operation, at the limit of travel of the load, or when selected temperature or pressures are reached

Where multispeed motors are involved, the following functions may be added:

7. Changing the speed of rotation (rpm)
8. Starting the motor with a definite speed sequence

Control devices that permit motors to be stopped are as follows:

1. Under the direction of an operation
2. Under the control of a pilot-circuit device, such as a thermostat, pressure regulator, float switch, limit switch, or cam switch
3. Under the control of protective devices that will disconnect the motor under detrimental overload conditions or upon failure of voltage or lost phase (Fig. 15-14)

Starting squirrel-cage motors in industrial plants is usually accomplished by means of one of the following methods:

AIR CONDITIONING

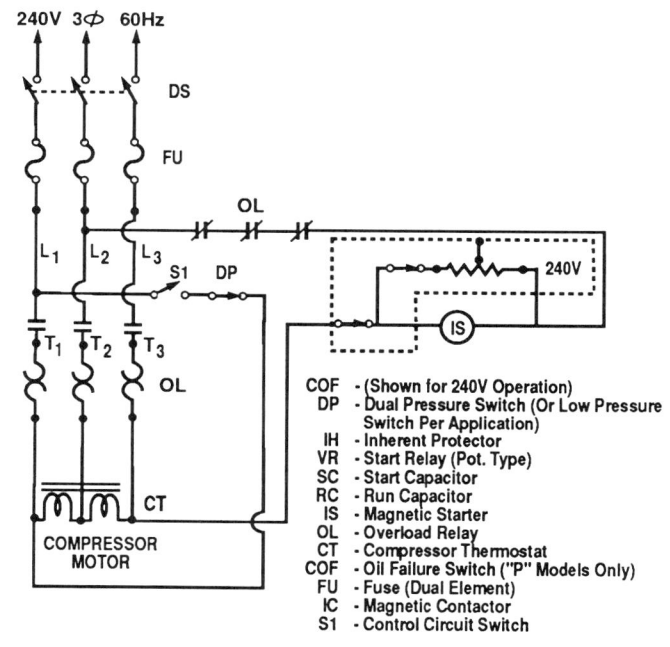

Fig. 15-14. Schematic wiring diagram showing the protective devices for a three-phase, squirrel-cage compressor motor.

1. Directly across the line
2. Use of autotransformers (compensators)
3. By resistance in series with the stator winding
4. Use of a stepdown transformer

Squirrel-cage motors of 5 horsepower or less are generally started with line switches connecting the motors directly across the line. The switches are usually equipped with thermal devices that open the circuit when overloads are carried beyond predetermined values. Large motors usually require various voltage-reducing methods of the type mentioned previously.

All types of squirrel-cage motors may be connected directly across the line provided the starting currents do not exceed those permitted by local power regulations. When started in this manner, a magnetic contactor is usually employed and operated from a start-stop push button or automatically from a float switch, thermostat, pressure regulator, or other pilot-circuit control device (Fig. 15-15).

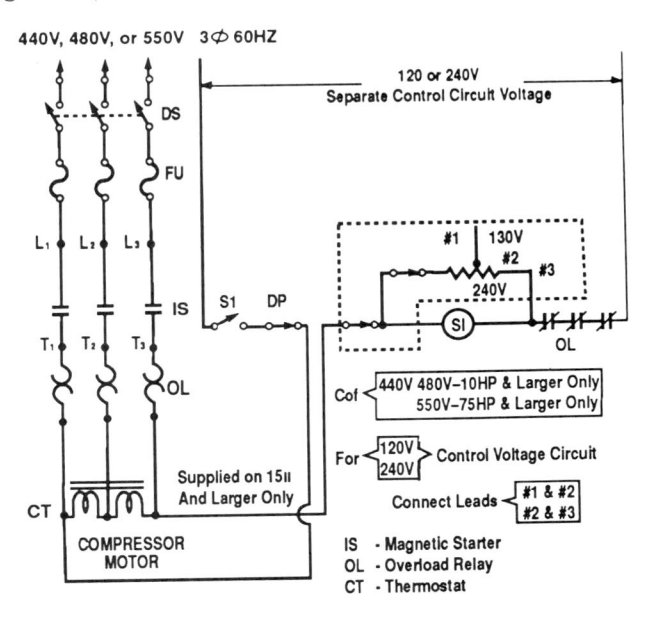

Fig. 15-15. Schematic wiring diagram showing the wiring arrangement for a typical compressor motor with separate control-circuit voltage.

Wound-Rotor Motors

The wound-rotor motor requires control of both primary and secondary circuits. Although the control of the primary circuit may be similar to that of squirrel-cage motors, the secondary control provides means of starting and speed control by varying the rotor resistance. The functions of controllers used with wound-rotor motors are:

1. To start the motor without damage or undue disturbance to the motor, driven machine, or power supply
2. To stop the motor in a satisfactory manner
3. To reverse the motor

AIR CONDITIONING

4. To run the motor at one or more predetermined speeds below synchronous
5. To handle an overhauling load satisfactorily
6. To protect the motor

The various types of controllers utilized may be divided into the following groups, depending on the size and function of the motor:

1. Face-plate starters
2. Face-plate speed regulators
3. Multiswitch starters
4. Drum controllers
5. Motor-driven drum controllers
6. Magnetic starters

By the use of external resistors in the slip-ring rotor windings, a wide variation in rotor resistance can be obtained with a resultant variation in acceleration characteristics. Thus, a heavy load can be started as slowly as desired, without a jerk, and can be accelerated smoothly and uniformly to full speed. It is merely a matter of supplying the necessary auxiliary control equipment to insert sufficiently high resistance at the start and gradually reduce this resistance as the motor picks up speed. An automatic controlled starter for wound-rotor motor operation is shown in Fig. 15-16.

Many power companies have established limitations on the amount of current motors may draw at starting. The purpose is to reduce voltage fluctuations and prevent flickering of lights. Because of such limitations, the question of starting current is often the deciding factor in choosing wound-rotor motors instead of squirrel-cage motors. Wound-rotor motors with proper starting equipment develop a starting torque equal to 150 percent of full-load current. This compares very favorably with squirrel-cage motors, one type of which requires a starting current of as much as 600 percent of full-load current to develop the same starting torque of 150 percent.

This type of motor is suitable where the speed range required is small, where the speeds desired do not coincide with a synchronous speed of the line frequency, or where the speed must be grad-

MOTORS AND MOTOR CONTROLS

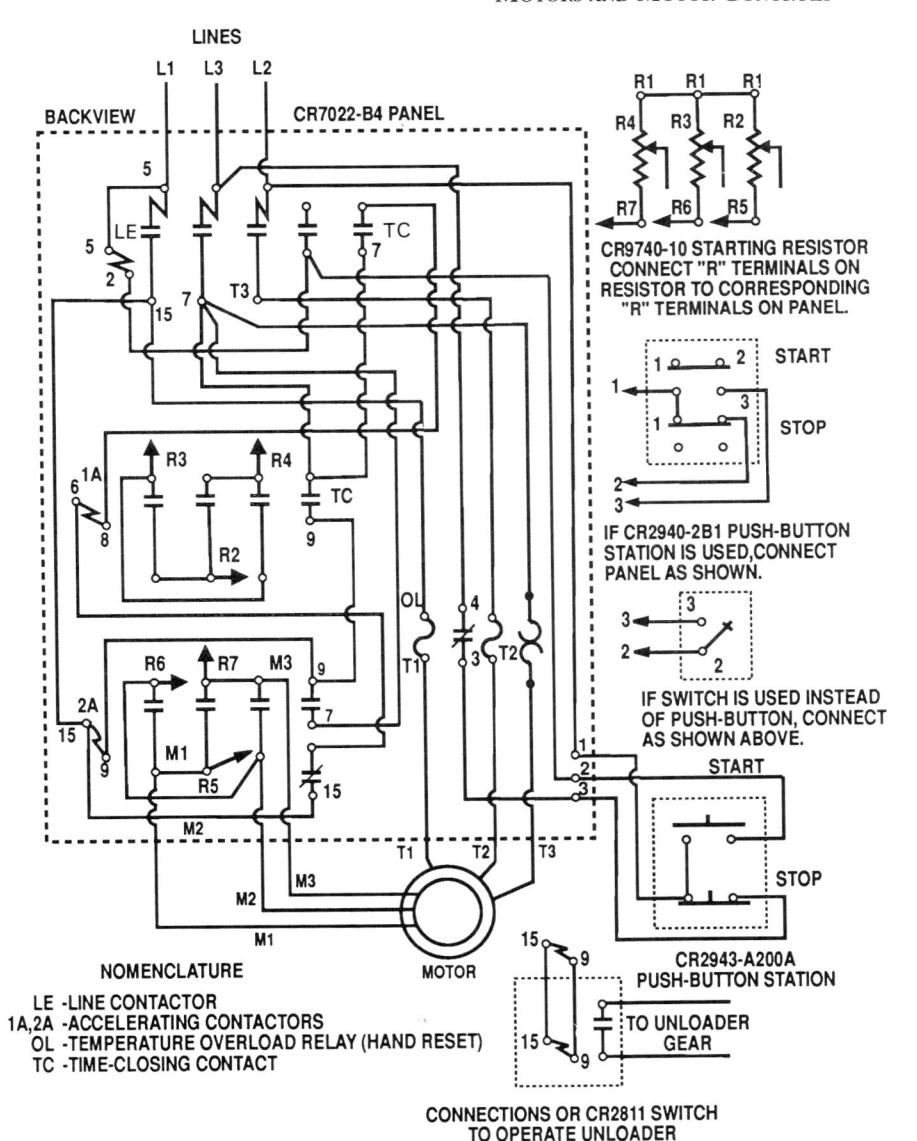

Fig. 15-16. Wiring diagram showing automatic magnetic remotely controlled starter for wound-rotor motor operation.

ually or frequently changed from one value to another. The wound-rotor motor also gives high starting torque with a low current demand from the line, but it is not efficient when used a large proportion of the time at reduced speed. Power corresponding to the percent drop in speed is consumed in the external resistance at slow speed without doing any special work.

Synchronous Motors

To make a synchronous motor self-starting, a squirrel-cage winding is usually placed on the rotor of the machine. After the motor comes up to the speed that is slightly less than synchronous, the rotor is energized. When synchronous motors are started, their dc fields are not excited until the rotor has reached practically full synchronous speed. The starting torque required to bring the rotor up to this speed is produced by induction.

In addition to the dc winding on the field of synchronous motors, they are generally provided with a damper or amortisseur winding. This winding consists of short-circuited bars of brass or copper joined together at either end by means of end rings and embedded in slots in the pole faces. This winding is usually termed a *squirrel-cage winding* and enables the motor to obtain sufficient starting torque necessary to start under load. The starting torque that is sufficient to bring the motor up to synchronous speed is termed the *pull-in torque*. The maximum torque that the motor will develop without pulling out of step is termed the *pull-out torque*.

Reduced-voltage starting differs from the across-the-line method in that reduced voltage is first applied to the motor by means of a transformer (usually an autotransformer). In this method of starting, full voltage is applied only after the motor has reached nearly synchronous speed.

Reactance-motor starting is similar to reduced-voltage methods, except that the first step is obtained by reactance in series with the motor armature instead of by autotransformers. In the reactance method, more current is required from the line for the same torque on the first step than when compensators are used. It has an advantage, however, in that no circuit opening is required when the motor is transferred to running voltage, the transfer being accomplished by short-circuiting the reactance.

SINGLE-PHASE MOTOR CONTROLS

Single-phase motors are usually controlled by means of line starters or contactors, which may be either manual or magnetic. The compressor contactor is a switch that is used on both single- and three-phase units to open or close the power circuit to the compressor. Unlike a starter, a contactor does not include overload protective devices. Contactors are used on single-phase systems and on some three-phase systems. The trend is to eliminate contactor overload protectors and eventually use only inherent compressor-motor protective devices on all systems, including both single- and three-phase units. Under normal operating conditions, the contactor is controlled by the thermostat, although the contactor coil can be deenergized by the pressure switch or compressor inherent overload protectors (see Fig. 15-17).

The motor starter is a switch that performs the same function as a contactor, except that it includes (as an integral part) overload protective devices. These overload protective devices are normally referred to as *heaters* because, with abnormal motor amperage, the device warps from the heat developed by the additional current and breaks a contact, thereby opening the compressor control circuit. This, in turn, allows the starter to break its contacts and stop the compressor.

Overload Protectors

Two types of inherent overload protection are provided on motor compressors: *thermal* and *combined thermal and current*. The thermal type is sensitive to temperature only, whereas the thermal and current type is sensitive to both motor current and temperature.

The current and thermal type of overload protector is affected by both compressor-shell temperature and motor current. This type of protector is designed to carry full motor current and is therefore wired in the power circuit of the compressor.

AIR CONDITIONING

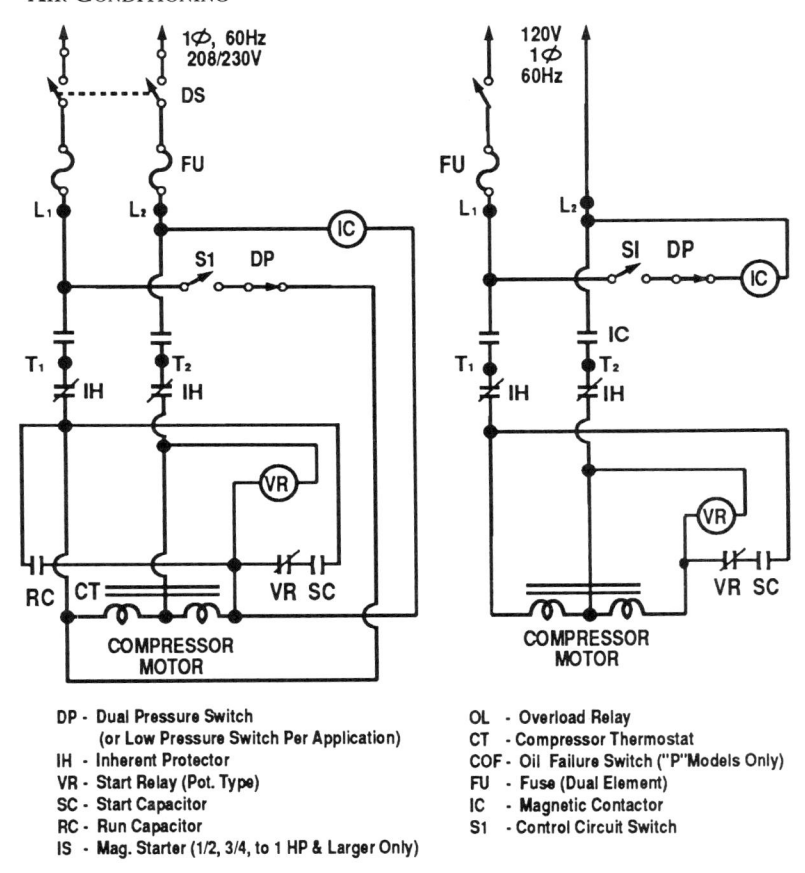

Fig. 15-17. Wiring diagram showing single-phase-capacitor motor with starting and running protective devices.

DP - Dual Pressure Switch
  (or Low Pressure Switch Per Application)
IH - Inherent Protector
VR - Start Relay (Pot. Type)
SC - Start Capacitor
RC - Run Capacitor
IS - Mag. Starter (1/2, 3/4, to 1 HP & Larger Only)

OL - Overload Relay
CT - Compressor Thermostat
COF - Oil Failure Switch ("P"Models Only)
FU - Fuse (Dual Element)
IC - Magnetic Contactor
S1 - Control Circuit Switch

Starting Relays

Single-phase motor relays are employed to obtain automatic starting. Since it is necessary to provide current through the starting winding under predetermined conditions, there must be an automatic device for accomplishing this. This device is the *starting relay*. With the current-type relay, a spring is used to hold the contact points open. When the motor is inoperative, the contact

468

points are open. As current starts to flow through the relay and running winding, the holding coil is energized, thus overcoming the spring tension and closing the contact points. At this time, current is then also directed through the starting capacitor and starting winding. As the speed of the motor increases, the holding-coil current reduces until it reaches the point where the spring tension overcomes the voltage and opens the contact points. The running winding alone remains in the circuit to operate the motor. Normally, one side of the current coil is connected inside the relay, and only three external terminals are furnished. However, when a three-terminal protector is used, both current coil connections must be made externally, and a relay having four terminals or wires is used.

With the potential-type relay, the contact points are held closed by means of spring tension. Therefore, when the motor is not in operation, the contact points are always closed. As the current starts to flow through the closed contacts of the relay, it is directed immediately through the starting capacitor and starting winding as well as through the running winding. The voltage in the holding coil, which is connected across the motor starting winding, increases with increase in speed. As the motor approaches full speed, the holding-coil voltage overcomes the spring tension and opens the contact points, thus removing the starting capacitor and starting winding from the circuit.

COMPRESSOR MOTOR CONTROLS

Compressor motor controls may be divided into two classes: *temperature* and *pressure*.

Temperature-Motor Control

The temperature-motor control serves as a connecting link between the motor and the evaporator. This control mechanism consists essentially of a thermostatic bulb clamped to the evaporator, together with a capillary tube and attached bellows. These three interconnected elements are assembled and charged with a few drops of refrigerant, such as sulfur dioxide or similar highly volatile fluid. An electrical switch is operated through linkage by the movement of the bellows. In this manner the electrical motor circuit is closed on rising temperature and opened on falling temperature.

In operation, as the temperature of the bulb increases, gas pressure in the bulb bellows assembly increases and the bellows pushes the operating shaft upward, operating the toggle or snap mechanism. Consequently, the upward travel of the shaft finally pushes the toggle mechanism off center and the switch snaps closed, starting the compressor. As the compressor runs, the control bulb is cooled gradually, reducing the pressure in the bulb-bellows system. This reduction of bellows pressure allows the spring to push the shaft slowly downward until it has finally traveled far enough to push the toggle mechanism off center in the opposite direction, snapping the switch open and stopping the compressor. The control bulb then slowly warms up until the motor again starts and the cycle repeats itself. The thermostatic control is set to close the compressor circuit at a certain evaporator temperature and open the circuit at another predetermined temperature. In this manner, any desired cabinet temperature may be maintained.

Pressure-Motor Control

The pressure-motor control device is similar in construction to the temperature control, with the exception that the temperature bulb and capillary tube are replaced by a pressure tubing connected directly to the low-pressure control source and an electrical switch operated through linkage by the movement of the bellows. In the pressure-motor control method, the compressor operation is only indirectly dependent on the temperature of the refrigeration load but is controlled by refrigerant pressure at the point of control location.

Solenoid Valves

Solenoid valves are frequently used in refrigeration systems for control of gas and liquid flow. In general, solenoid valves are used to control the refrigerant flow into the expansion valve or the gas flow from the evaporator when the evaporator or fixture has reached the desired temperature. When used as liquid stop valves, they should be installed in the liquid line and as close to the expansion valve as possible to prevent bubbling and subsequent reduction in expansion-valve capacity.

In some installations, a solenoid valve is operated directly by a thermostat located at the load point, and the compressor motor is controlled independently by a low-pressure switch. It is important to furnish the solenoid-valve manufacturer with the electrical characteristics of the circuit on which the valve is to operate. The maximum pressure at which the valve is to open should also be specified because excessive pressure may prevent the valve from opening.

Condensing-Water Controls

Automatic control of condenser-water flow is often necessary in order to prevent water wastage. Such control may be provided by using solenoid valves or a pressure control valve. In the case of solenoid-valve installation, the operation is dependent upon the starting and stopping of the compressor motor; whereas with a pressure-control valve, the operation is dependent entirely upon condenser pressure.

BELT DRIVES

Open-type motors usually require a belt connection between the motor and compressor since the motor speed is normally too high to allow a direct compressor coupling. For belt-drive compressors, speed is determined by the size of the motor pulley since the compressor pulley (flywheel) is normally fixed. The relative speeds of the motor and compressor are in direct relation to the size of the motor pulley and the compressor flywheel. The desired pulley size, or the resulting compressor speed, may be calculated from the following relation:

$$\text{Compressor rpm} = \frac{\text{motor-pulley diam.} \times \text{motor speed (rpm)}}{\text{compressor-pulley diam.}}$$

$$= \frac{\text{compressor speed (rpm)} \times \text{compressor-pulley diam.}}{\text{motor speed (rpm)}}$$

Example — Find the motor-pulley diameter required when a 1750-rpm motor is to be used to drive a compressor having an 8-inch diameter pulley when the compressor speed is 500 rpm.
Solution — The diameter of the motor pulley is obtained by substituting values as follows:

$$\text{Motor-pulley diameter} = \frac{500 \times 8}{1750} \text{ or } 2\ ^{1}/_{4} \text{ inches (approx.)}$$

The desired driven speed in the preceding formula should be increased by 2 percent to allow for belt slip. The diameter of any suitable motor pulley at any motor speed may easily be calculated by inserting values of compressor-pulley diameter in a manner similar to the previous example.

MOTOR MAINTENANCE

Modern methods of design and construction have made the electric motor one of the least complicated and most dependable forms of machinery in existence. Proper maintenance is very important and must be given careful consideration if the best performance and longest life are to be expected from the motor. From the standpoint of performance, the two major features are proper lubrication and the care given to the insulation.

Lubrication

The design of bearings and bearing housings of motors has improved in the last few years. The point has now been reached where the bearings of modern motors, whether sleeve, ball, or roller, require only infrequent attention. This advance in lubrication methods has not yet been fully appreciated.

Although there may have been some necessity for more frequent attention in the case of older designs with housings less tight than

MOTORS AND MOTOR CONTROLS

those on modern machines, oiling and greasing of new motors are quite often entrusted to uninformed and careless attendants, with the result that oil or grease is copiously and frequently applied to the outside as well as the inside of bearing housings. Some of the excess lubricant is carried into the machine and lodges on the windings, where it catches dirt and thereby hastens the ultimate failure of the insulation.

Modern design provides for a plentiful supply of oil or grease held in dust- and oil-tight housings. If the proper amount of a suitable lubricant is applied before starting, there should be no need to refill the housings for several months even in dusty places. Infrequent though periodic and reasonable attention to modern bearings of any type will lead toward longer life of both bearings and insulation.

Horsepower Calculations

The formulas dealing with calculations of electric-motor capacity requirements under various conditions of service are given in the following paragraphs.

Motor Horsepower Calculated from Meter Readings

DC Motors:

$$Hp = \frac{\text{volts} \times \text{amperes} \times \text{efficiency}}{746}$$

Single-Phase AC Motors:

$$Hp = \frac{\text{volts} \times \text{amperes} \times \text{efficiency} \times \text{power factor}}{746}$$

Two-Phase AC Motors:

$$Hp = \frac{\text{volts} \times \text{amperes} \times \text{efficiency} \times \text{power factor} \times 2}{746}$$

Three-Phase AC Motors:

$$Hp = \frac{\text{volts} \times \text{amperes} \times \text{efficiency} \times \text{power factor} \times 1.73}{746}$$

AIR CONDITIONING

Horsepower Calculated from Load (Mechanics' Data)

Constant Speed:

1. Rotational horsepower:

$$Hp = \frac{torque\ (lb/ft) \times speed\ (rpm)}{5250}$$

2. Prony-brake horsepower:

$$Hp = \frac{2 \times 3.1416 \times lb\ applied\ at\ 1\text{-ft}\ radius \times rpm}{33,000}$$

3. Linear horsepower:

$$Hp = \frac{force\ (lb) \times velocity\ (ft/min)}{33,000}$$

Acceleration from Zero to Full Speed:

1. Rotational horsepower:

$$Hp = \frac{inertia\ (WR^2) \times rpm}{1.62 \times 10^6 t\ (in\ seconds\ to\ come\ up\ to\ speed)}$$

2. Linear horsepower:

$$Hp = \frac{inertia\ (W) \times rpm^2}{6.38 \times 10^7 t\ (in\ seconds\ to\ come\ up\ to\ speed)}$$

Horsepower Calculations for Specific Applications (Approximate)

Pumping Water:

$$Hp = \frac{gallons/min \times total\ dynamic\ head\ (ft)}{3960 \times pump\ efficiency}$$

Moving Air:

$$Hp = \frac{volume \times pressure\ (lb/sq\ ft)}{33,000 \times efficiency}$$

$$Hp = \frac{5.2 \times volume\ (cu\ ft/min) \times head\ (in.\ H_2 0)}{33,000 \times efficiency}$$

MOTORS AND MOTOR CONTROLS

Compressing Gases:

For one working stroke per revolution (single-acting):

$$Hp = \frac{PLAN}{33,000 \times 0.90}$$

where P =   effective pressure in cylinders, psi
L =   length of stroke, ft
A =   area of piston, in sq in.
N =   no. of rpm

Power-Conversion Factors

1 kW = 3412.75 Btu
= 860 cal
= 1.34048 hp
1 hp = 0.746 kW
= 2546 Btu
= 642 cal

TROUBLESHOOTING GUIDE

The troubleshooting guides on the following pages will assist in troubleshooting air-conditioning motors and their controls. It should be noted, however, that although the trouble may differ at times from that given in this section, experience will usually reveal the exact cause, after which the possible remedy and its correction may be carried out.

ELECTRICAL SYSTEM TROUBLESHOOTING GUIDE

System and Possible Cause   *Possible Remedy*

Compressor Does Not Operate

(a) Electrical controls.   (a) Check fuses, relays, and capacitors.

AIR CONDITIONING

Symptom and Possible Cause   *Possible Remedy*

Unit Runs but Fails to Cool

(a) Insufficient warmup period of unit.
(b) Compressor motor reset out. Unit has operated for some time, but little or no cooling is experienced.

(a) Allow 30 min running time before a decision is made.
(b) Push in reset button for compressor. Make unit replacement only after a complete check is made of system.

Condensate Fan Will Not Run

(a) Reset button may be stuck.
(b) Fan striking shroud.
(c) Bearings tight. Motor burned out.

(a) Check reset button located on fan motor.
(b) Turn fan by hand to see if it rotates freely.
(c) Replace fan motor.

Evaporator Fan Will Not Run

(a) Reset button out.
(b) Faulty capacitor.

(c) Motor burned out.

(a) Check motor-reset button.
(b) Use test-cord procedure for testing evaporator fan-motor circuit.
(c) Replace motor.

Unit Vibrates

(a) Unit seems noisy and vibrates too much.

(a) Check to see if all shipping bolts have been removed. Check for loose fan on motor shaft. Unit must be level on all four sides and bearing on four corners.

MOTORS AND MOTOR CONTROLS

Symptom and Possible Cause *Possible Remedy*

Unit Does Not Cool Properly

(a) Compressor seems to be operating but little cooling is experienced.

(a) Check intake-air filter for excess dust. Check thermostatic control setting with a good thermometer.

(b) Low-line voltage.

(b) Low-line voltage results in high-current drain and high-head pressures, which ultimately reduces capacity of the compressor.

(c) Dirty air filter.

(c) Causes frosting of evaporator coil; replace with a new filter.

(d) Low evaporator-fan speed.

(d) Check fan for tightness on shaft and free rotation.

(e) Low condensing-fan speed.

(e) Make same check as above.

(f) Restricted airflow in condensing air circuit.

(f) Check condensing-air circuit complete from intake at window duct, through the unit, to outlet at window duct. Window screens hamper free airflow in window duct.

(g) Poorly fitted duct.

(g) Allows discharge air to short-circuit into supply duct causing high condensing medium. Refit duct.

(h) Cooling capacity check.

(h) Take thermometer wet-bulb reading of room air intake at grille. Take wet-bulb reading of discharge room air at discharge grille. Take dry-bulb reading of condensing medium at air-intake window duct.

477

Air Conditioning

Symptom and Possible Cause *Possible Remedy*

Normal wet-bulb depression at the valves found should be checked with capacity. If unit is low in capacity after all other possibilities have been checked, unit must be replaced.

Shortage of Refrigerant

(a) Low capacity.
(b) Indications of oil at leak.

(a) Make capacity check.
(b) Check for oil spots and test with leak test torch. If leak is found, unit must be replaced.

Noisy Compressor

(a) Loose mounting bolts
(b) Unit striking frame.

(a) Tighten mounting bolts.
(b) See that compressor-unit suspension is absolutely free to swing without striking frame.

(c) Compressor unit itself noisy.

(c) Operate compressor a short time with fans off to determine if noise is emanating from compressor.

(d) Unit not level.

(d) Check complete unit for leveling and stability. If compressor still proves noisy, complete unit must be replaced.

Unit Fails to Start

(a) Low voltage.

(a) Check voltage at plug receptacle. Check wire size and length of run. Check

(b) Blown fuses.

power-supply voltage at service switch or meter.
(b) Check fuses.

SUMMARY

Direct-current motors are generally classified as series-, shunt-, and compound-wound. Series motors are those in which the field coil and armature are connected in series and current flows through both the field and armature. In shunt-wound motors, the field coil and armature are connected in parallel. The compound-wound motor has separate field windings. One is usually a predominant field and is connected in parallel with the armature circuit; and the other is connected in series.

Alternating-current motors are divided into two general classes: polyphase and single-phase. Polyphase motors are classified according to windings as squirrel-cage induction, wound-rotor induction, and synchronous motors. Single-phase motors are classified according to their construction as split-phase, capacitor-start, permanent-split capacitor, repulsion induction, and repulsion-start induction.

There are two basic types of overload protection devices: the thermal type, and the thermal and current type. The thermal type is sensitive to temperature only, and the thermal and current type is sensitive to both temperature and current. Temperature-motor controls are also used in connection with protection devices. This control mechanism consists essentially of a thermostatic bulb clamped to the evaporator together with a capillary tube. This assembly is charged with a few drops of refrigerant fluid, which operates switches through proper linkage and attached bellows.

Solenoid valves are used frequently in refrigeration systems to control gas and liquid flow. The solenoid valves are used to control refrigerant flow into expansion valves. In some installations, the valves are operated directly by thermostats.

REVIEW QUESTIONS

1. What precautions should be observed when selecting a motor for compressor devices?
2. Why is alternating current preferred for compressor devices instead of direct current?
3. How are ac motors generally classified?
4. Why are single-phase, fractional-horsepower motors used in preference to polyphase motors in home air conditioning?
5. Describe the construction principles of a squirrel-cage motor.
6. In what respect does the rotor winding in a cage motor differ from the rotor winding in a wound-rotor motor?
7. Why are synchronous motors seldom used in the field of residence air conditioning?
8. What is the normal size limit of single-phase motors?
9. How are single-phase motors designed?
10. What is the function of the auxillary winding in a split-phase motor?
11. Explain the operation of a capacitor motor.
12. What are the operating characteristics of a repulsion-start induction motor?
13. What is the combined capacity of three capacitors, each having a capacity of $20\mu F$, when connected in series and when connected in parallel?
14. Name the various control methods used on squirrel-cage and on wound-rotor motors.
15. How are single-phase, fractional-horsepower motors usually controlled?
16. Name two types of overload protectors as employed in compressor-motor drives.
17. How do the temperature controls function to regulate motor operation?

CHAPTER 16

Maintenance

One of the essential elements in analyzing air-conditioning problems is a thorough understanding of variations in *pressure, temperature, volume,* and *heat-absorbing characteristics* of the refrigerant in the various parts of the system. A malfunction may be caused by one portion of the system or a combination of several portions. For this reason, it is necessary and advisable to check the more obvious causes first.

Each part of the system has a certain function to perform; and if any individual part does not function correctly and efficiently, the performance of the entire air-conditioning system will be affected. The following procedures will be of assistance to the serviceman when analyzing air-conditioning-system troubles.

GENERAL PROCEDURE

When checking the refrigerant or electrical circuit of a unit, the following information should be obtained first. This information

AIR CONDITIONING

will indicate whether the unit is performing properly and if not, in which part of the unit the trouble lies. Obtain:

1. Operating-head pressure
2. Operating-suction pressure
3. Amperage
4. Voltage
5. Evaporator-air cu ft/min and dry- and wet-bulb temperature differences
6. Condenser-air cu ft/min and temperature differences

The suction and discharge pressures are checked by attaching the test manifold to the service valve and the discharge gage to the discharge service valve. Before reading the gage pressure, the service valves should be front-seated approximately one turn and the gage hoses purged through the center tap of the test manifold. Voltage and amperage can be checked with a combination volt-ammeter at the compressor terminals. Evaporator and condenser air can be checked using a draft gage. The air-temperature drop through the evaporator and the temperature rise through the condenser can be obtained by the use of thermometers. A sling psychrometer can be used to obtain the wet-bulb temperature of the evaporator air. Before any action is taken, based on suction or head pressure, the unit should operate or stabilize for at least 20 minutes.

HEAD PRESSURES

Checking the head pressure of a condensing unit is fully as important as checking the suction pressure. Too often this item is overlooked when checking a job, and it may be the direct cause of the trouble. There is no shortcut or easy method of estimating what the head pressure should be because it is affected by speed, temperature of the condensing medium, and suction pressure.

To check for air or other noncondensable gases in a condensing unit, it is necessary to test the head pressure while the condensing unit is idle and after the liquid refrigerant in the receiver has cooled to the temperature of the condensing medium. On water-cooled condensing units, it is possible to open the water valve and allow an excess amount of water to pass through the condenser. On air-cooled condensing units, the belts may be removed to allow the fan on the motor to cool the condenser to room temperature.

A high pressure will indicate the presence of noncondensable gases which must be purged. This purging should be done from a point near the top of the condenser, and in many instances must be repeated several times in order to remove all the air or other gases. Purging heavily as soon as the condenser stops is especially effective because the noncondensable gases separate from the refrigerant only while the compressor is in operation. After properly purging a unit, it is still possible to operate with head pressures that are higher than normal, which may be caused by dirty air-cooled condensers or by limed-up water-cooled condensers. In either instance, the condenser should be cleaned in order to get the proper transfer of heat between the refrigerant and the condensing medium.

Another cause of high head pressure is sometimes found with an excess charge of refrigerant in the system. On air-cooled condensing units, if the liquid refrigerant completely fills the receiver and also partially fills the condenser, it will cut down on the effective condenser surface and high head pressure will result. On water-cooled units, this condition is also true where the liquid level is high enough to submerge part of the condenser tubes. In a great many instances, condensing-unit capacities have been so greatly reduced through excessive head pressures that it is impossible to obtain automatic defrosting between cycles. The reason for this is that the running time increases to the point where no off-time is available.

On air-cooled condensing units, the head or discharge pressure will increase with an increase in ambient air temperature. Under normal operating conditions, the condensing temperature will be from 20 to 30°F higher than the ambient temperature. If the difference between the ambient and the condensing temperatures is more than 30°F, the unit is operating with higher than normal head pressure.

In view of the preceding information, causes of high head pressure are usually as follows:

1. Air or other noncondensables in the system
2. Inoperative condenser blower
3. Dirty air-cooled condensers or limed-up water-cooled condensers
4. Excess of refrigerant in the system

SUCTION PRESSURE

In checking commercial installations it is found that many units are operating at suction pressures far below manufacturers' specifications owing to incorrect setting of expansion valves or switches. Evaporators that operate in a starved condition will invariably oil log and cause a shortage of oil in the compressor crankcase, which burns out connecting rods, main bearings, and seals. Many times a serviceman has assumed that the seal was the first failure and, as a result, the refrigerant and oil charge were lost with subsequent bearing failures. What actually happened was that the oil supply was low in the crankcase due to oil-logged evaporators, thus causing the compressor body failures. To correct this condition, the proper procedure is to check the superheat settings on the evaporator and either add refrigerant or properly adjust the expansion valve to prevent this oil logging. Usually, when this is done, the oil will return to the compressor crankcase and the difficulty will be remedied. Changing the compressor body will not help the situation.

Operating at low-suction pressures also causes the compressor to operate at much higher head temperatures, owing to the amount of vapor being condensed and also the amount of heat contained in this vapor. Where this heat becomes excessive, it will tend to increase a breakdown of oil in the refrigerant as well as cause some gasket materials to flow, thus causing head leaks. When operating a condensing unit at certain specified low temperatures, the motor and compressor pulleys must be the proper size to obtain rated capacities.

BACK PRESSURE

Normally, the suction temperatures will be about 45°F with a suction pressure of about 42 psi for Freon-12 and about 76 psi for

Freon-22. Light evaporator loading will cause a suction-pressure decrease, while heavy evaporator loading will cause a suction pressure increase. It should be noted that if high-suction pressure is accompanied by high-discharge pressure and the temperature of the air over the evaporator and condenser is high, high inside and outside air temperatures may be at fault and no malfunction may be indicated.

Causes of low back pressures are usually as follows:

1. Inoperative evaporator blower
2. Obstructed or dirty evaporator
3. Dirty air filter
4. Low refrigerant charge
5. Faulty expansion valve
6. Incorrect superheat setting

REFRIGERANT CHARGE

The receiver in the air-cooled system stores additional refrigerant and may fill the receiver to a point where the liquid refrigerant will back up in the lower condenser tubes. This will reduce the heat-transfer efficiency of the condenser causing high head pressure and, in turn, reduce the cooling capacity of the unit. There is no accurate way to tell if the unit is overcharged; however, if overcharging is suspected, the most effective method of checking and correcting it is to discharge some of the refrigerant from the system until a low charge is indicated and then add refrigerant until a full charge is indicated. A low-refrigerant charge is indicated by low-suction pressure, a bubbling sight glass, or a hissing expansion valve. A low-refrigerant charge will reduce the capacity or output of the refrigerating unit.

To check the refrigerant charge in units equipped with *sight glasses*, the procedure is usually as follows; The appearance of large masses of bubbles indicates the presence of both vapor and liquid refrigerant, which means that the system is not fully charged and requires additional refrigerant. When the system is fully charged, the sight glass should be completely clear or just show intermittent groups of fine bubbles.

AIR CONDITIONING

In units equipped with liquid-level test valves, proceed as follows: If the unit is not furnished with a sight glass, the receiver may be equipped with a liquid-level test valve. After operating the compressor for 20 minutes, open this valve slightly. The appearance of liquid and frosting of the valve indicates an adequate charge. A discharge of gas only or gas and liquid intermittently indicates a low charge. A discharge of gas will not feel cold while a discharge of liquid will. The high side is shipped with the proper charge of refrigerant, and normally will not need additional refrigerant. The charge is given on the nameplate of the unit. An overcharged system will result in higher than normal head pressure. (This condition can be determined from the procedure outlined in the section "Head Pressures.")

COMPRESSOR SERVICE VALVES

Compressor suction and discharge valves (Fig. 16-1) are shutoff valves with a manually operated stem. The suction and discharge service valves are used to isolate the compressor from the system for servicing or check the gage pressure on both the high and low sides of the system. With the valve stem turned to the extreme counterclockwise position (or backseated), the main port is open while the opening to the gage port is closed. The unit should operate with the main port open. By turning the valve stem to the extreme clockwise position, the valve is front-seated. In this position the main port is closed, leaving the gage port open to the compressor. At any intermediate position, both the main port and the gage port are open. The valve stems on all service valves are protected from breakage and leakage by caps, which should be replaced after servicing of the unit is finished.

GAGES

Gages used in servicing are *low-pressure (compound) gages* and *high-pressure gages*. The standard type of low-pressure gage is often called a *compound gage* because its construction permits a reading of both pressure (in pounds per square inch) and vacuum

MAINTENANCE

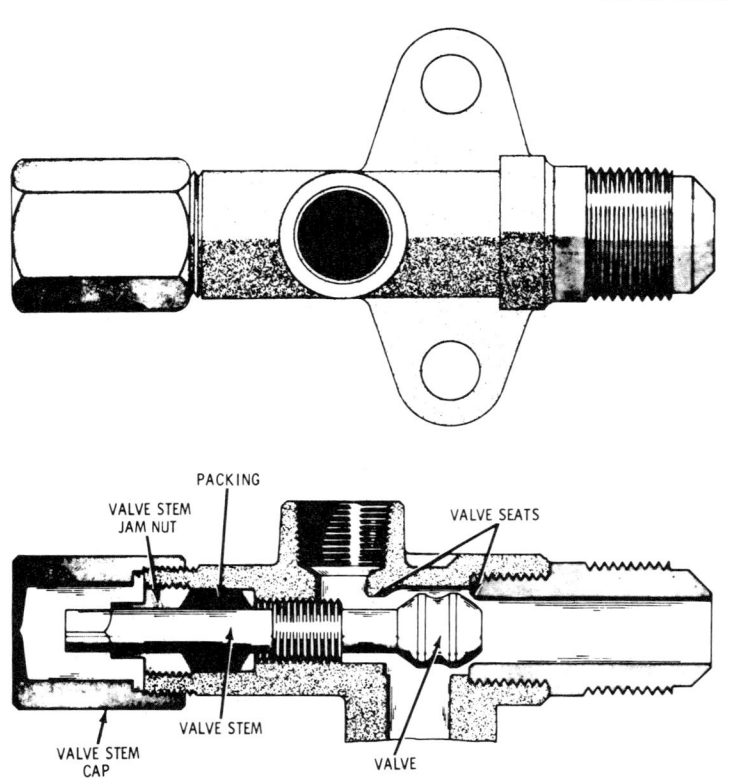

Fig. 16-1. Exterior and interior view of a typical compressor suction or discharge valve.

(in inches of mercury), as shown in Fig. 16-2. A standard compound gage dial is graduated to record a pressure range of from 30 inches of vacuum to 60 psi of pressure. The high-pressure gage is equipped with a dial for measuring pressures of from 0 to 300 psi. Gages used in refrigeration service should be graduated to read approximately double the actual working pressure.

Combination Gages

A test gage set for testing manifolds (Fig. 16-3) consists essentially of two tee valves built into one valve body with stems ex-

AIR CONDITIONING

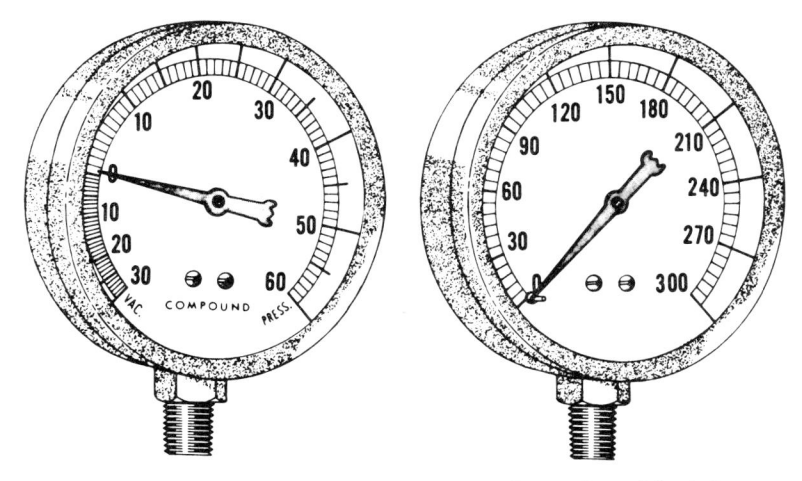

Fig. 16-2. Typical compound pressure and plain pressure gage.

tending out on each end for hand-wheel operations. The tee valves are so constructed that the valve works only on the leg that is attached to the tee. Thus, the opening or closing of the stem only affects this one opening, the other two openings on the valve being open at all times.

ADDING REFRIGERANT

To add refrigerant, use the following procedure:

1. Attach the proper refrigerant drum to the center tap on the charging manifold, as shown in Fig. 16-4.
2. Adjust the manifold valves so that the refrigerant will flow through the suction gage into the suction-gage hose.
3. Attach the suction-gage hose to the charging fitting on the suction service valve. Purge a small amount of refrigerant from the drum to remove any air trapped in hoses before tightening the fittings.
4. Set the refrigerant drum on scales to observe the rate of charging in pounds of refrigerant.
5. Front-seat (close) the suction service valve. Make certain the refrigerant drum is in an upright position, and start the

MAINTENANCE

Fig. 16-3. Typical combination gage set.

Courtesy Mueller Brass

compressor. Open the valve on the refrigerant drum slightly. Throttle the drum valve to keep the suction pressure below 45 psi for Freon-12 and 80 psi for Freon-22 on the suction gage to prevent compressor damage. *Note:* Setting the drum in a pan of warm water will speed up the refrigerant flow into the system.
6. If the system requires a full charge, charge the amount of refrigerant in pounds, shown on the unit nameplate. If only a small amount of refrigerant is to be added, charge the system for a few seconds and then open the suction service valve and let the system stabilize for about 5 minutes. If the system is still undercharged, repeat the previous procedure.

REMOVING REFRIGERANT

If the system is overcharged, purge a small amount of refrigerant from the suction-service-valve gage port. Close the gage port

AIR CONDITIONING

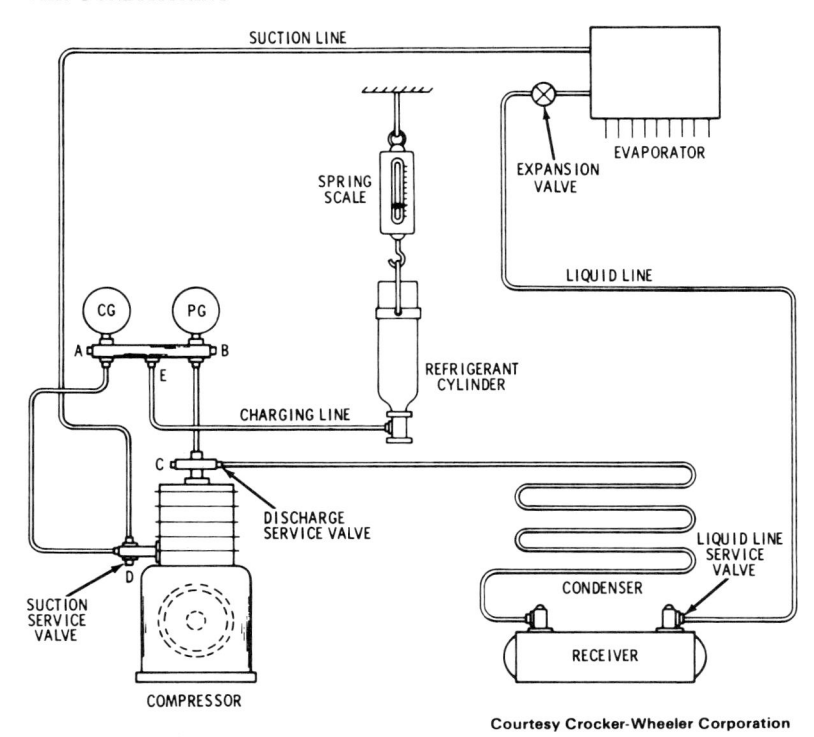

Fig. 16-4. Schematic diagram showing the connections for adding refrigerant through the low side.

and allow the system to stabilize. Continue purging the system until the system becomes undercharged. Use the procedure in the preceding section to correct the refrigerant charge in the system.

ADDING OIL

To add oil to the compressor, use the following procedure:

1. Pump down the unit. (Refer to the section entitled "System Pumpdown.")
2. Remove the suction line from the compressor.
3. Pour the oil into the suction outlet of the compressor until

the oil level is approximately halfway up on the oil-level sight glass. Do not overcharge.
4. Replace the suction line.
5. Operate the compressor for 20 minutes, and then recheck the oil level.

REMOVING OIL

To remove oil from the compressor, proceed as follows:

1. Pump down the unit. (See the following section.)
2. Front-seat (close) the compressor suction, and discharge service valves.
3. Remove both the suction and discharge lines from the compressor.
4. Disconnect capillary tubes and all electrical connections from the compressor.
5. Remove nuts and mounting assemblies that secure the compressor to the unit.
6. Remove the compressor, and pour oil out of the suction opening.

SYSTEM PUMPDOWN

The pumpdown procedure is as follows:

1. Back-seat (open) the suction and discharge service valves.
2. Make sure the suction gage is connected to the suction-service-valve gage port and that the high-pressure gage is connected to the discharge-service-valve gage port.
3. Front-seat service valves slightly to purge the gage lines, and let pressure into the gages.
4. Start the compressor and allow it to run.
5. Remove the cover of the dual-pressure control, and block the low-pressure contacts closed. This prevents the suction pressure from shutting off the unit when the pressure is pumped below the normal cutout setting. To block the con-

AIR CONDITIONING

tacts, force the low-pressure-control contacts into a closed position and insert a nonmetallic object to keep the contacts closed during pumpdown procedure.
6. Close the liquid shutoff valve to retain all of the liquid in the high side of the system.
7. Watch the suction gage as the compressor is operating. When the suction pressure has been reduced to 0 psi, stop the compressor. Normally the pressure will increase after this first pumpdown cycle. Start the compressor again, and repeat this procedure until suction pressure remains at 1 to 2 psi when the compressor is stopped. If the pressure stays at 0 psi after the pumpdown, be sure the center port of the gage manifold is capped; then open both the valves on the gage manifold slightly to allow the pressure from the discharge side of the unit to bleed into the suction side so that a 1 to 2 psi pressure remains in the suction side. Always have a slight positive pressure in the system when it is opened to the atmosphere to prevent air and moisture from being drawn into the system.
8. Front-seat (close) the suction and discharge service valves.
9. The low side of the system can now be repaired.
10. Evacuate and install the new drier when the system is opened.
11. When the unit is repaired, open the compressor suction and discharge service valves.
12. Start the compressor. After determining that the system is operating satisfactorily, remove the nonmetallic object from the dual-pressure control and replace the cover. Back-seat the service valves, then disconnect the gage manifold from the valves. Disconnect the fittings and replace the gage port caps.

PURGING SYSTEM OF NONCONDENSABLES

Air or noncondensable gases may enter the system during original installation or at any time the system is opened for repair. These noncondensables will result in increased head pressure and reduced capacity. To remove air or noncondensable gases, pump

MAINTENANCE

down the system (see the preceding section). Shut down the system for at least 20 minutes. The air or noncondensables will collect in the upper part of the receiver and can be purged from the system by cracking the discharge service valve. After a short period of time, close the discharge valve and start the unit. Check the refrigerant charge since some of the refrigerant may have been lost during the purging operation.

LEAK TEST

Halide-Torch Method

To test for leaks before the initial startup of a system, use the following procedure:

1. Front-seat (close) the suction and discharge service valves if they are not already closed.
2. Attach the charging manifold-suction hose to the process valve.
3. Attach a drum of the proper refrigerant to the center port of the charging manifold.
4. Open the process valve and suction-gage valve and introduce enough refrigerant into the system to build up the pressure as high as possible.
5. Close the process valve and suction-gage valve.
6. Test the system with a halide leak detector (Fig. 16-5), and then a soap solution, if necessary, at all points where connections have been made. Make a complete check of all joints before attempting any repairs since the pressure in the system must be released to make a correction.
7. If leaks exist, disconnect the refrigerant drum hose from the charging manifold.
8. Open the suction valve and suction-gage hose to purge the refrigerant from the system. On refrigerant Freon-12 systems, where pressure is 50 psi or less, and Freon-22 systems, where pressure is 90 psi or less, the expansion valve will open, allowing piping from expansion valve to suction service valve to purge also. If pressure is higher, it is necessary to purge the low side through the suction service valve.

493

AIR CONDITIONING

Fig. 16-5. Gas leak detection is easy. This sensitive halide detector pinpoints leaks in a short time. Used on all noncombustible refrigerant leaks, the flame changes color when the exact location of the leak is determined.

Courtesy Turner Corporation

This can be accomplished by removing the suction-valve gage port plug and carefully opening the suction valve slightly. When the gage reads 0 lb pressure, close the suction service and suction gage valves.
9. Repeat this leak check until all leaks are found and repaired.
10. When repaired, close the suction valve and disconnect the charging manifold from the unit.
11. Back-seat (open) the suction and discharge service valves.

Electronic Leak Detector

Although leak testing with a halide torch is considered satisfactory in most instances, a more reliable test is made by means of the electronic-type tester. This easy-to-use instrument reduces the guesswork in leak testing because it is more sensitive, faster re-

MAINTENANCE

sponding, and capable of detecting a leak even though the surrounding air may be contaminated.

The electronic detector contains an internal pump, which draws air and leaking halogen tracer gas (halogen gases are chlorine, bromine, fluorine, and iodine, or any of their compounds) through a probe and hose as the probe is passed over leaking joints, gaskets, welds, etc. The air/gas mixture is drawn between two electrodes in the control unit — the ion-emitter and collector.

The leak detector contains an automatic balance circuit so that a constant or slowly changing contamination in the atmosphere will not cause the lamp to flash. Refrigerant gas will cause the lamp to flash only when there is an abrupt increase in the amount of gas going into the tip of the probe, such as when the probe tip passes near a leak.

To check for a leak with the electronic detector, proceed as follows:

1. Plug the test detector into a 120-volt ac outlet.
2. Turn the sensitivity knob to the right and allow a 1-minute warm-up.
3. Check the operation by turning the sensitivity knob quickly from one position to another, which should light the probe lamp.
4. Probe for leaks, starting with maximum sensitivity.
5. If the probe lamp lights twice for each leak, reduce the sensitivity.
6. Recheck the suspected leaks for confirmation.

After replacing a component in the refrigerating system, always test all joints for leaks before recharging. The extra time it takes is negligible compared to the loss of a charge due to a faulty connection. Be sure to clean the excess soldering flux (if used) from the new joints before testing since the flux could seal off pinhole leaks that would show up later.

EVACUATING THE SYSTEM

When necessary to remove the entire refrigerant charge from the system, as in the case when an exchange of compressor or any

AIR CONDITIONING

other component of the condensing unit is necessary, proceed as follows:

1. Front-seat (close) the suction and discharge service valves.
2. Attach the charging manifold-suction hose to the suction valve.
3. Connect the vacuum pump to the center port on the charging manifold.
4. Start the vacuum pump and open the suction valve. Adjust the test manifold valves so that the vacuum pump is open to the suction valve. Evacuate the system to obtain the highest vacuum possible with the vacuum pump used.
5. During the early stage of evacuation, it may be desired to shut off the pump at least once to see if there is a rapid loss of vacuum, indicating the presence of leaks in the system.
6. Time required for proper evacuation depends on the system and the vacuum pump. It is recommended that the vacuum pump be allowed to run for 2 hours or even overnight after maximum vacuum is obtained to evacuate the system.
7. Close the charging manifold valve to the vacuum pump and stop the pump. Note the reading on the vacuum gage and allow the system to stand for 1 hour. If the vacuum has not decreased, the system is tight.
8. When the unit is checked and repaired, close the suction valve and disconnect the charging manifold from the valve.
9. To put the system in operation, back-seat (open) the suction and discharge service valves.

CHECKING COMPRESSOR VALVES

To check the compressor valves, proceed as follows:

1. Front-seat (close) the suction service valve.
2. Start the compressor and operate the unit until the compound gage shows 15 inches of vacuum.
3. Stop the compressor.
4. Back-seat (open) the suction service valve slightly to permit

MAINTENANCE

the vapor in the suction line to bring the gage reading back to zero. Close the valve.
5. If the pressure on the gage starts to increase, a discharge valve leak is indicated and the compressor should be replaced.
6. If the compressor will not pull at least 15 inches of vacuum on the compound gage, a suction-valve leak is indicated and the compressor should be replaced.

TROUBLESHOOTING GUIDE

In addition to the foregoing servicing details, the following guide shows the most common operating faults that can occur in residential year-round air conditioning.

System and Possible Cause   *Possible Remedy*

Drafts

(a) Poor air distribution. (a) Readjust air grilles and outlets.

(b) Room temperature too low. (b) Check thermostat setting, return-air damper, and low-pressure cutout.

Shortage of Air Supply at Grilles

(a) Short on cooling capacity. (a) Check for correct change in wet-bulb temperature between entering and leaving air to cooling coil. Check refrigerating unit for capacity.

(b) Incorrect fan speed and incorrect pressure drop across fan. (b) Check pressure drop across the fan with draft gage, and correct fan speed.

(c) Incorrect pressure drop across evaporator coil. (c) Check pressure drop across coil with draft gage. Check filters, baffles, air supply,

497

Air Conditioning

Symptom and Possible Cause	Possible Remedy
	and coil fins for obstructions.
(d) Excessive air turbulence at fan discharge.	(d) Install turning vanes in short radius elbows.
(e) Improper adjustment of splitters or dampers.	(e) Readjust dampers and check air.
(f) High duct resistance.	(f) Check duct sizes and elbows. Inspect duct for internal obstructions. Check drive belt slippage.
(g) Low fan speed.	(g) Check fan motor static control. If multispeed fan is employed, check pulley sizes of drive.
(h) Wrong rotation of fan.	(h) Check rotation with arrow marking on fan housing if supplied, or by throat opening in fan housing. Change rotation by reversing motor direction.
(i) Dirty filters.	(i) Change if throwaway type.

Excessive Air

(a) Motor drive pulley too large.	(a) Check air and change pulley to correct speed.
(b) Improper damper adjustments.	(b) Readjust dampers.
(c) Lower resistance than estimated.	(c) Lower fan speed.

Air Noise

(a) Sharp obstruction in airstream.	(a) Streamline internal surfaces of ducts.
(b) Small slits or openings in duct or plenum.	(b) Smooth out rough edges. Close all holes or openings.
(c) Fan speed too high in rela-	(c) Decrease fan speed, if possi-

MAINTENANCE

Symptom and Possible Cause

tion to noise level and system.

(d) Obstructions in outlets.

(e) Air velocity too high at outlets.

Scraping Noise in Plenum

(a) Fan hitting the fan housing.

Squeaking and Rumble

(a) Loose belts.
(b) Dirt in bearings.
(c) Bearings sticking.

Rattles

(a) Loose filter compartment.
(b) Electrical BX cable loose.
(c) Thermal-bulb tube vibrating.
(d) Housing touching pipe.
(e) Loose screws in housing.

Water Leaking or Dripping from Unit

(a) Improper drain piping.

Possible Remedy

ble, or install sound absorption material in ducts. (Sound insulation should be fireproof material.)

(d) Remove obstruction, if possible, without throwing system out of balance. If removal is not possible, try changing location of angle.

(e) Rebalance air in system.

(a) Adjust fan to turn free on all sides.

(a) Tighten belts.
(b) Clean out bearings.
(c) Remove and replace; examine shaft.

(a) Tighten.
(b) Fasten securely.
(c) Clear vibrating point of tube.
(d) Insulate metal-to-metal location.
(e) Tighten.

(a) Drain should pitch $1/4$ to 1 in/ft. Drain should be properly trapped. Conden-

499

AIR CONDITIONING

Symptom and Possible Cause	Possible Remedy
	sate pan must pitch water toward drain pipe.
(b) Dirt in condensate pan and screen.	(b) Remove screen, clean pan and flush with water.
(c) Loose pipe fittings and valves.	(c) Tighten all connections and packing glands of valves.
(d) Condensate overshooting drain pan.	(d) Air velocity too high through coils. Check resistance pressure and fan speed to determine if they are in accordance with design conditions.

Vibration

(a) Loose motor mounting.	(a) Tighten motor mounting bolts.
(b) Improper weight distribution on hangers or feet.	(b) Level unit on all sides.
(c) Belt too tight causing whip in shaft.	(c) Adjust belt tension.
(d) Pulleys loose or out of line.	(d) Align pulleys and tighten pulley set screws.
(e) Fan shaft sprung.	(e) Remove and replace with new shaft.
(f) Shaft bearings worn.	(f) Replace with new bearings.
(g) Fans out of balance or loose on shaft.	(g) Tighten if shaft is not worn from loose play. Replace if badly damaged.
(h) No canvas breaker strips to duct connection.	(h) Install breaker strips.

Hissing Noise at Expansion Valve

(a) Shortage of refrigerant.	(a) Add refrigerant to proper liquid level.
(b) Clogged strainer.	(b) Remove and clean screen.
(c) Pressure drop in liquid.	(c) Check head pressure.

Symptom and Possible Cause *Possible Remedy*

(d) Heating of liquid line from outside surroundings.
(d) Subcooling necessary.

(e) No liquid at top expansion valve in bank of coils.
(e) Gas head on liquid line about two feet higher than top valve.

Partial Frosting and Sweating of Coil

(a) Shortage of refrigerant.
(a) Add refrigerant.

(b) Restricted or clogged liquid line.
(b) Pump down system and clean out.

(c) Too much outside air being taken on one side of coil.
(c) Change mixing chamber for better distribution.

(d) Expansion valve too small or improperly adjusted.
(d) Check valve capacity and superheat adjustment.

(e) Solenoid valve sticking.
(e) Remove and clean.

Failure to Cool and No Frost

(a) Solenoid valve clogged.
(a) Remove and clean.

(b) Solenoid coil burned out.
(b) Replace coil.

(c) Condensing unit shut off.
(c) Start compressor.

(d) Liquid line valve closed.
(d) Open valve.

(e) Thermostat faulty.
(e) Repair or replace.

Failure to Cool and Complete Frosting

(a) Fan not running.
(a) Start fan.

(b) Insufficient air.
(b) Dirty filters. Dampers stuck shut.

(c) Air outlets closed.
(c) Correct air distribution.

(d) Coil clogged with dirt.
(d) Remove coil and clean thoroughly.

Too Cool

(a) Improper setting of room thermostat, or thermostat faulty.
(a) Correct thermostat.

Symptom and Possible Cause *Possible Remedy*

(b) Wrong outlet distribution.   (b) Balance air distribution correctly.

Frequent Cycling of Unit

(a) Clogged liquid line.   (a) Clean out.
(b) Insufficient refrigerant.   (b) Add refrigerant.
(c) Improper setting of expansion valves.   (c) Adjust valves properly.
(d) Fan stopped running.   (d) Check fan circuit and start fan.
(e) Thermostat differential setting too close.   (e) Reset differential.
(f) Thermostat in direct line of air throw from grille.   (f) Relocate thermostat.
(g) Evaporator coil frosted.   (g) Run fan with compressor shut off to defrost coil.

Motor Overheats

(a) Lack of oil or improper oil.   (a) Oil bearings with recommended oil.
(b) Wrong voltage.   (b) Check with electrician and power company.
(c) Overload.   (c) Reduce fan speed, if possible, without impairing air quantity. Replace with larger motor if necessary.
(d) Belts too tight.   (d) Adjust belt tension.
(e) Ambient room temperature high.   (e) Check motor rating with manufacturer of motor.

Motor Fails to Run

(a) Improper fuse.   (a) Check fuses for proper sizing.
(b) Thermal overload too small.   (b) Check with motor data sheet.
(c) Improper wiring.   (c) Check with wiring diagram.

Maintenance

Symptom and Possible Cause	Possible Remedy
Failure of System to Heat Properly	
(a) Heating coils not properly trapped.	(a) Separate trap should be provided for each coil.
(b) Air trapped in coils.	(b) Coils should be properly vented.
(c) Heating coils improper distance above boiler.	(c) Distance above boiler line should be at least 18 inches.
(d) Supply or return lines too small.	(d) Install proper size piping for coils.
(e) Resistance too great in system.	(e) Install proper size piping.
(f) Too much outside air being introduced.	(f) Use as much recirculated air as possible without impairing the calculated ratio of outside recirculated air.
(g) Heating coils not leveled.	(g) Level coils to obtain proper coil drainage.
(h) Condensate in coils.	(h) Traps may contain scale and sediment preventing operation.
Failure of System to Humidify	
(a) Water supply failure.	(a) Water valve closed. Strainer plugged. Nozzle clogged.
(b) Too much outside air.	(b) Close down on outside damper.
(c) Control valve on humidifier supply closed or clogged.	(c) Remove and clean.
(d) Capacity of humidifier insufficient.	(d) Install larger or additional humidifier and supply line.
(e) Humidistat not functioning.	(e) Inspect humidistat for adjustment and dirt, especially the hair suspension.
(f) Air passing directly over and through humidistat.	(f) Place baffle on upstream side of humidistat to break air around instrument.

SUMMARY

A malfunctioning system may be caused by one portion of the system or a combination of several portions. It is necessary and advisable to check the more obvious causes first. Each part of the system has a certain function to perform, and if any individual part does not function properly and efficiently, the performance of the entire air-conditioning system will be affected.

To simplify servicing a system, the serviceman should remember that the refrigerant, under proper operating conditions, travels through the system in one specific direction. The serviceman can trace the path of refrigerant through any system, beginning at the compressor.

Pressure, temperature, volume, and heat-absorbing characteristics of the refrigerant in the various parts of the system are essential in analyzing air-conditioning problems. When a refrigerant piping system is completed, the system should be checked very carefully for leaks. The most effective way of finding a leak in a Freon system is with a halide leak detector. Testing with soap and water, or with oil at the joints will locate only the larger leaks.

REVIEW QUESTIONS

1. What is the general practice in the diagnosis of a malfunctioning air-conditioning system?
2. Give a step-by-step procedure in checking to determine the source of trouble in an inoperative system.
3. Describe head-pressure checking methods.
4. What may the reason be for a low suction pressure?
5. What are the usual causes of low back pressure?
6. What is the method used in checking refrigerant charge?
7. Describe the types of compressor gages and give pressure ranges.
8. What is the advantage in using a combination gage set?
9. Describe the method of adding and removing refrigerant.
10. What is meant by *system pumpdown*, and how is it accomplished?

Maintenance

11. Why is it necessary to leak test a refrigeration system prior to operation?
12. Describe the method of evacuating a refrigeration system.
13. What is the procedure used in checking compression valves?

Glossary

Absolute Humidity — The weight of water vapor per unit volume; grains per cubic foot or grams per cubic meter.

Absolute Pressure — The sum of gage pressure and atmospheric pressure. Thus, for example, if the gage pressure is 154 psi, the absolute pressure will be 154 + 14.7, or 168.7 psi.

Absolute Zero — A temperature equal to $-459.6°F$ or $-273°C$. At this temperature the volume of an ideal gas maintained at a constant pressure becomes zero.

Absorption — The action of a material in extracting one or more substances present in the atmosphere or a mixture of gases or liquids accompanied by physical change, chemical change, or both.

Acceleration — The time rate of change of velocity. It is the derivative of velocity with respect to time.

Air Conditioning

Accumulator—A shell placed in a suction line for separating the liquid entrained in the suction gas.

Acrolein—A warning agent often used with methyl chloride to call attention to the escape of refrigerant. The material has a compelling, pungent odor and causes irritation of the throat and eyes. Acrolein reacts with sulfur dioxide to form a sludge.

Activated Alumina—A form of aluminum oxide (Al_2O_3) that absorbs moisture readily and is used as a drying agent.

Adiabatic—Referring to a change in gas conditions where no heat is added or removed except in the form of work.

Adiabatic Process—Any thermodynamic process taking place in a closed system without the addition or removal of heat.

Adsorbent—A sorbent that changes physically, chemically, or both during the sorption process.

Aeration—Exposing a substance or area to air circulation.

Agitation—A condition in which a device causes circulation in a tank containing fluid.

Air, Ambient—Generally speaking, the air surrounding an object.

Air Changes—A method of expressing the amount of air leakage into or out of a building or room in terms of the number of building volumes or room volumes exchanged per unit of time.

Air Circulation—Natural or imparted motion of air.

Air Cleaner—A device designed for the purpose of removing airborne impurities such as dust, gases, vapors, fumes, and smoke. An air cleaner includes air washers, air filters, electrostatic precipitors, and charcoal filters.

GLOSSARY

Air Conditioner — An assembly of equipment for the control of at least the first three items enumerated in the definition of *air conditioning*.

Air Conditioner, Room — A factory-made assembly designed as a unit for mounting in a window, through a wall, or as a console. It is designed for free delivery of conditioned air to an enclosed space without ducts.

Air Conditioning — The simultaneous control of all, or at least the first three, of the following factors affecting the physical and chemical conditions of the atmosphere within a structure — temperature, humidity, motion, distribution, dust, bacteria, odors, toxic gases, and ionization — most of which affect human health or comfort.

Air-Conditioning System, Central-Fan — A mechanical indirect system of heating, ventilating, or air conditioning in which the air is treated or handled by equipment located outside the rooms served, usually at a central location and conveyed to and from the rooms by means of a fan and a system of distributing ducts.

Air-Conditioning System, Year-Round — An air-conditioning system that ventilates, heats, and humidifies in winter, and cools and dehumidifies in summer to provide the desired degree of air motion and cleanliness.

Air-Conditioning Unit — A piece of equipment designed as a specific air-treating combination, consisting of a means for ventilation, air circulation, air cleaning, and heat transfer, with a control means for maintaining temperature and humidity within prescribed limits.

Air Cooler — A factory-assembled unit including elements whereby the temperature of air passing through the unit is reduced.

Air Conditioning

Air Cooler, Spray Type — A forced-circulation air cooler wherein the coil surface capacity is augmented by a liquid spray during the period of operation.

Air Cooling — A reduction in air temperature due to the removal of heat as a result of contact with a medium held at a temperature lower than that of the air.

Air Diffuser — A circular, square, or rectangular air distribution outlet, generally located in the ceiling, and comprised of deflecting members discharging supply air in various directions and planes, arranged to promote mixing of primary air with secondary room air.

Air, Dry — In psychrometry, air unmixed with or containing no water vapor.

Air Infiltration — The in-leakage of air through cracks, crevices, doors, windows, or other openings caused by wind pressure or temperature difference.

Air, Recirculated — Return air passed through the conditioner before being again supplied to the conditioned space.

Air, Return — Air returned from conditioned or refrigerated space.

Air, Saturated — Moist air in which the partial pressure of the water vapor is equal to the vapor pressure of water at the existing temperature. This occurs when dry air and saturated water vapor coexist at the same dry-bulb temperature.

Air, Standard — Air with a density of 0.075 lb/cu ft and an absolute viscosity of 1.22×10 lb mass/ft-sec. This is substantially equivalent to dry air at 70°F and 29.92 in. Hg barometer.

Air Washer — An enclosure in which air is forced through a spray of water in order to cleanse, humidify, or precool the air.

GLOSSARY

Ambient Temperature — The temperature of the medium surrounding an object. In a domestic system having an air-cooled condenser, it is the temperature of the air entering the condenser.

Ammonia Machine — An abbreviation for a compression-refrigerating machine using ammonia as a refrigerant. Similarly, Freon machine, sulfur dioxide machine, etc.

Analyzer — A device used in the high side of an absorption system for increasing the concentration of vapor entering the rectifier or condenser.

Anemometer — An instrument for measuring the velocity of air in motion.

Antifreeze, Liquid — A substance added to the refrigerant to prevent formation of ice crystals at the expansion valve. Antifreeze agents in general do not prevent corrosion due to moisture. The use of a liquid should be a temporary measure where large quantities of water are involved, unless a drier is used to reduce the moisture content. Ice crystals may form when moisture is present below the corrosion limits, and in such instances, a suitable noncorrosive antifreeze liquid is often of value. Materials such as alcohol are corrosive and, if used, should be allowed to remain in the machine for a limited time only.

Atmospheric Condenser — A condenser operated with water that is exposed to the atmosphere.

Atmospheric Pressure — The pressure exerted by the atmosphere in all directions as indicated by a barometer. Standard atmospheric pressure is considered to be 14.695 psi (pounds per square inch), which is equivalent to 29.92 in. Hg (inches of mercury).

Atomize — To reduce to a fine spray.

Air Conditioning

Automatic Air Conditioning — An air-conditioning system that regulates itself to maintain a definite set of conditions by means of automatic controls and valves usually responsive to temperature or pressure.

Automatic Expansion Valve — A pressure-actuated device which regulates the flow of refrigerant from the liquid line into the evaporator to maintain a constant evaporator pressure.

Baffle — A partition used to divert the flow of air or a fluid.

Balanced Pressure — The same pressure *in* a system or container that exists *outside* the system or container.

Barometer — An instrument for measuring atmospheric pressure.

Blast Heater — A set of heat-transfer coils or sections used to heat air that is drawn or forced through it by a fan.

Bleeder — A pipe sometimes attached to a condenser to lead off liquid refrigerant parallel to the main flow.

Boiler — A closed vessel in which liquid is heated or vaporized.

Boiler Horsepower — The equivalent evaporation of 34.5 pounds of water per hour from and at 212°F, which is equal to a heat output of 970.3 × 34.5 = 33,475 Btu.

Boiling Point — The temperature at which a liquid is vaporized upon the addition of heat, dependent on the refrigerant and the absolute pressure at the surface of the liquid and vapor.

Bore — The inside diameter of a cylinder.

Brine — Any liquid cooled by a refrigerant and used for transmission of heat without a change in its state.

Brine System — A system whereby brine cooled by a refrigerating system is circulated through pipes to the point where the refrigeration is needed.

British thermal unit (Btu) — The amount of heat required to raise the temperature of one pound of water one degree Fahrenheit. It is also the measure of the amount of heat removed in cooling one pound of water one degree Fahrenheit and is so used as a measure of refrigerating effect.

Butane — A hydrocarbon, flammable refrigerant used to a limited extent in small units.

Calcium Chloride — A chemical having the formula $CaCl_2$, which, in granular form, is used as a drier. This material is soluble in water, and in the presence of large quantities of moisture may dissolve and plug up the drier unit or even pass into the system beyond the drier.

Calcium Sulfate — A solid chemical of the formula $CaSO_4$, which may be used as a drying agent.

Calibration — The process of dividing and numbering the scale of an instrument; also of correcting and determining the error of an existing scale.

Calorie — Heat required to raise the temperature of one gram of water one degree Celsius (actually, from 4° to 5°C). Mean calorie = one-hundredth part of the heat required to raise one gram of water from zero to one-hundred degrees Celsius.

Capacity, Refrigerating — The ability of a refrigerating system, or part thereof, to remove heat. Expressed as a rate of heat removal, it is usually measure in Btu/hr or tons/24 hr.

Capacity Reducer — In a compressor, a device, such as a clearance pocket, movable cylinder head, or suction bypass, by which compressor capacity can be adjusted without otherwise changing the operating conditions.

Capillarity — The action by which the surface of a liquid in contact with a solid (as in a slender tube) is raised or lowered.

Capillary Tube — In refrigeration practice, a tube of small internal diameter used as a liquid refrigerant-flow control or expansion device between high and low sides; also used to transmit pressure from the sensitive bulb of some temperature controls to the operating element.

Carbon Dioxide Ice — Compressed solid CO_2. Dry ice.

Celsius — A thermometric system in which the freezing point of water is called 0° C and its boiling point 100° C at normal pressure.

Centrifugal Compressor — A compressor employing centrifugal force for compression.

Centrifuge — A device for separating liquids of different densities by centrifugal action.

Change of Air — Introduction of new, cleansed, or recirculated air to a conditioned space, measured by the number of complete changes per unit time.

Change of State — A change from one state to another, as from a liquid to a solid, from a liquid to a gas, etc.

Charge — The amount of refrigerant in a system.

Chimney Effect — The tendency of air or gas in a duct or other vertical passage to rise when heated due to its lower density compared with that of the surrounding air or gas. In buildings, the tendency toward displacement, caused by the difference in temperature, of internal heated air by unheated outside air due to the difference in density of outside and inside air.

GLOSSARY

Clearance — Space in a cylinder not occupied by a piston at the end of the compression stroke or volume of gas remaining in a cylinder at the same point, measured in percentage of piston displacement.

Coefficient of Expansion — The fractional increase in length or volume of a material per degree rise in temperature.

Coefficient of Performance (Heat Pump) — Ratio of heating effect produced to the energy supplied, each expressed in the same thermal units.

Coil — Any heating or cooling element made of pipe or tubing connected in series.

Cold Storage — A trade or process of preserving perishables on a large scale by refrigeration.

Comfort Chart — A chart showing effective temperatures with dry-bulb temperatures and humidities (and sometimes air motion) by which the effects of various air conditions on human comfort maybe compared.

Compression System — A refrigerating system in which the pressure-imposing element is mechanically operated.

Compressor — That part of a mechanical refrigerating system which receives the refrigerant vapor at low pressure and compresses it into a smaller volume at higher pressure.

Compressor, Centrifugal — A nonpositive displacement compressor that depends on centrifugal effect, at least in part, for pressure rise.

Compressor, Open-Type — A compressor with a shaft or other moving part, extending through a casing, to be driven by an outside source of power, thus requiring a stuffing box, shaft seals, or equivalent rubbing contact between a fixed and moving part.

Compressor, Reciprocating—A positive displacement compressor with a piston or pistons moving in a straight line but alternately in opposite directions.

Compressor, Rotary—One in which compression is attained in a cylinder by rotation of a positive displacement member.

Compressor Booster—A compressor for very low pressures, usually discharging into the suction line of another compressor.

Condenser—A heat-transfer device that receives high-pressure vapor at temperatures above that of the cooling medium, such as air or water, to which the condenser passes latent heat from the refrigerant, causing the refrigerant vapor to liquefy.

Condensing—The process of giving up latent heat of vaporization in order to liquefy a vapor.

Condensing Unit—A specific refrigerating machine combination for a given refrigerant, consisting of one or more power-driven compressors, condensers, liquid receivers (when required), and the regularly furnished accessories.

Condensing Unit, Sealed—A mechanical condensing unit in which the compressor and compressor motor are enclosed in the same housing, with no external shaft or shaft seal, the compressor motor operating in the refrigerant atmosphere.

Conduction, Thermal—Passage of heat from one point to another by transmission of molecular energy from particle to particle through a conductor.

Conductivity, Thermal—The ability of a material to pass heat from one point to another, generally expressed in terms of Btu per hour per square foot of material per inch of thickness per degree temperature difference.

Conductor, Electrical — A material which will pass an electric current as part of an electrical system.

Connecting Rod — A device connecting the piston to a crank and used to change rotating motion into reciprocating motion, or vice versa, as from a rotating crankshaft to a reciprocating piston.

Constant-Pressure Valve — A valve of the throttling type, responsive to pressure, located in the suction line of an evaporator to maintain a desired constant pressure in the evaporator higher than the main suction line pressure.

Constant-Temperature Valve — A valve of the throttling type, responsive to the temperature of a thermostatic bulb. This valve is located in the suction line of an evaporator to reduce the refrigerating effect on the coil to just maintain a desired minimum temperature.

Control — Any device for regulation of a system or component in normal operation either manual or automatic. If automatic, the implication is that it is responsive to changes of temperature, pressure, or any other property whose magnitude is to be regulated.

Control, High-Pressure — A pressure-responsive device (usually an electric switch) actuated directly by the refrigerant-vapor pressure on the high side of a refrigerating system (usually compressor-head pressure).

Control, Low-Pressure — An electric switch, responsive to pressure, connected into the low-pressure part of a refrigerating system (usually closes at high pressure and opens at low pressure).

Control, Temperature — An electric switch or relay that is responsive to the temperature change of a thermostatic bulb or element.

Convection — The circulatory motion that occurs in a fluid at a nonuniform temperature owing to the variation of its density and the action of gravity.

Convection, Forced — Convection resulting from forced circulation of a fluid as by a fan, jet, or pump.

Cooling Tower, Water — An enclosed device for evaporatively cooling water by contact with air.

Cooling Unit — A specific air-treating combination consisting of a means for air circulation and cooling within prescribed temperature limits.

Cooling Water — Water used for condensation of refrigerant. Condenser water.

Copper Plating — Formation of a film of copper, usually on compressor walls, pistons, or discharge valves caused by moisture in a methyl chloride system.

Corrosive — Having a chemically destructive effect on metals (occasionally on other materials).

Counterflow — In the heat exchange between two fluids, the opposite direction of flow, the coldest portion of one meeting the coldest portion of the other.

Critical Pressure — The vapor pressure corresponding to the critical temperature.

Critical Temperature — The temperature above which a vapor cannot be liquefied, regardless of pressure.

Critical Velocity — The velocity above which fluid flow is turbulent.

Crohydrate — An eutectic brine mixture of water and any salt mixed in proportions to give the lowest freezing temperature.

GLOSSARY

Cycle — A complete course of operation of working fluid back to a starting point measured in thermodynamic terms. Also used in general for any repeated process in any system.

Cycle, Defrosting — The portion of a refrigeration operation that permits the cooling unit to defrost.

Cycle, Refrigeration — A complete course of operation of a refrigerant back to the starting point measured in thermodynamic terms. Also used in general for any repeated process for any system.

Dalton's Law of Partial Pressure — Each constituent of a mixture of gases behaves thermodynamically as if it alone occupied the space. The sum of the individual pressures of the constituents equals the total pressure of the mixture.

Defrosting — Removal of accumulated ice from the cooling unit.

Degree Day — A unit based on temperature difference and time used to specify the nominal heating load in winter. For one day there exist as many degree days as there are degrees Fahrenheit difference in temperature between the average outside air temperature, taken over a 24-hour period, and a temperature of 65°F.

Dehumidifier — An air cooler used for lowering the moisture content of the air passing through it. An absorption or adsorption device for removing moisture from the air.

Dehumidify — To remove water vapor from the atmosphere. To remove water or liquid from stored goods.

Dehydrator — A device used to remove moisture from the refrigerant.

Density — The mass or weight per unit of volume.

Air Conditioning

Dew Point, Air — The temperature at which a specified sample of air, with no moisture added or removed, is completely saturated. The temperature at which the air, on being cooled, gives up moisture, or dew.

Differential (of a Control) — The difference between the cut-in and cut-out temperature or pressure.

Direct Connected — Driver and driven, as motor and compressor, positively connected in line to operate at the same speed.

Direct Expansion — A system in which the evaporator is located in the material or space refrigerated or in the air-circulating passages communicating with such space.

Displacement, Actual — The volume of gas at the compressor inlet actually moved in a given time.

Displacement, Theoretical — The total volume displaced by all the pistons of a compressor for every stroke during a definite interval (usually measured in cubic feet per minute).

Drier — Synonymous with dehydrator.

Dry-Type Evaporator — An evaporator of the continuous-tube type where the refrigerant from a pressure-reducing device is fed into one end and the suction line connected to the outlet end.

Duct — A passageway made of sheet metal or other suitable material, not necessarily leak-tight, used for conveying air or other gas at low pressure.

Dust — An air suspension (aerosol) of solid particles of earthy material, as differentiated from smoke.

Economizer — A reservoir or chamber wherein energy or material from a process is reclaimed for further useful purpose.

Efficiency, Mechanical — The ratio of the output of a machine to the input in equivalent units.

Efficiency, Volumetric — The ratio of the volume of gas actually pumped by a compressor or pump to the theoretical displacement of the compressor.

Ejector — A device that utilizes static pressure to build up a high fluid velocity in a restricted area to obtain a lower static pressure at that point so that fluid from another source maybe drawn in.

Element, Bimetallic — An element formed of two metals having different coefficients of thermal expansion, such as used in temperature-indicating and -controlling devices.

Emulsion — A relatively stable suspension of small, but not colloidal, particles of a substance in a liquid.

Engine — Prime mover; device for transforming fuel or heat energy into mechanical energy.

Entropy — The ratio of the heat added to a substance to the absolute temperature at which it is added.

Equalizer — A piping arrangement to maintain a common liquid level or pressure between two or more chambers.

Eutectic Solution — A solution of such concentration as to have a constant freezing point at the lowest freezing temperature for the solution.

Evaporative Condenser — A refrigerant condenser utilizing the evaporation of water by air at the condenser surface as a means of dissipating heat.

Evaporative Cooling — The process of cooling by means of the evaporation of water in air.

Evaporator — A device in which the refrigerant evaporates while absorbing heat.

Expansion Valve, Automatic — A device that regulates the flow of refrigerant from the liquid line into the evaporator to maintain a constant evaporator pressure.

Expansion Valve, Thermostatic — A device that regulates the flow of refrigerant into an evaporator so as to maintain an evaporation temperature in a definite relationship to the temperature of a thermostatic bulb.

Extended Surface — The evaporator or condenser surface that is not a primary surface. Fins or other surfaces that transmit heat from or to a primary surface, which is part of the refrigerant container.

External Equalizer — In a thermostatic expansion valve, a tube connection from the chamber containing the pressure-actuated element of the valve to the outlet of the evaporator coil. A device to compensate for excessive pressure drop through the coil.

Fahrenheit — A thermometric system in which 32°F denotes the freezing point of water and 212°F the boiling point under normal pressure.

Fan — An air-moving device comprising a wheel, or blade, and housing or orifice plate.

Fan, Centrifugal — A fan rotor or wheel within a scroll-type housing and including driving-mechanism supports for either belt-drive or direct connection.

Fan, Propeller — A propeller or disk-type wheel within a mounting ring or plate and including driving-mechanism supports for either belt-drive or direct connection.

Fan, Tubeaxial — A disk-type wheel within a cylinder, a set of air-guide vanes located either before or after the wheel, and

GLOSSARY

driving-mechanism supports for either belt-drive or direct connection.

Filter — A device to remove solid material from a fluid by a straining action.

Flammability — The ability of a material to burn.

Flare Fitting — A type of connector for soft tubing that involves the flaring of the tube to provide a mechanical seal.

Flash Gas — The gas resulting from the instantaneous evaporation of the refrigerant in a pressure-reducing device to cool the refrigerant to the evaporation temperature obtained at the reduced pressure.

Float Valve — Valve actuated by a float immersed in a liquid container.

Flooded System — A system in which the refrigerant enters into a header from a pressure-reducing valve and the evaporator maintains a liquid level. Opposed to dry evaporator.

Fluid — A gas or liquid.

Foaming — Formation of a foam or froth of oil refrigerant due to rapid boiling out of the refrigerant dissolved in the oil when the pressure is suddenly reduced. This occurs when the compressor operates; and, if large quantities of refrigerant have been dissolved, large quantities of oil may "boil" out and be carried through the refrigerant lines.

Freezeup — Failure of a refrigeration unit to operate normally due to formation of ice at the expansion valve. The valve maybe frozen closed or open, causing improper refrigeration in either case.

Freezing Point — The temperature at which a liquid will solidify upon the removal of heat.

Air Conditioning

Freon-12 — The common name for dichlorodifluoromethane (CCl_2F_2).

Frostback — The flooding of liquid from an evaporator into the suction line, accompanied by frost formation on the suction line in most cases.

Furnace — That part of a boiler or warm-up heating plant in which combustion takes place. Also a complete heating unit for transferring heat from fuel being burned to the air supplied to a heating system.

Fusible Plug — A safety plug used in vessels containing refrigerant. The plug is designed to melt at high temperatures (usually about 165°F) to prevent excessive pressure from bursting the vessel.

Gage — An instrument used for measuring various pressures or liquid levels.

Gas — The vapor state of a material.

Generator — A basic component of any absorption-refrigeration system.

Gravity, Specific — The density of a standard material usually compared to that of water or air.

Grille — A perforated or louvered covering for an air passage, usually installed in a sidewall, ceiling or floor.

Halide Torch — A leak tester generally using alcohol and burning with a blue flame; when the sampling tube draws in halocarbon refrigerant vapor, the color of flame changes to bright green. Gas given off by the burning halocarbon is phosgene, a deadly gas used in World War I in Europe against Allied troops (can be deadly if breathed in a closed or confined area).

GLOSSARY

Heat — Basic form of energy which may be partially converted into other forms and into which all other forms may be entirely converted.

Heat of Fusion — Latent heat involved in changing between the solid and the liquid states.

Heat, Sensible — Heat that is associated with a change in temperature; specific heat exchange of temperature, in contrast to a heat interchange in which a change of state (latent heat) occurs.

Heat, Specific — The ratio of the quantity of heat required to raise the temperature of a given mass of any substance one degree to the quantity required to raise the temperature of an equal mass of a standard substance (usually water at 59°F) one degree.

Heat of Vaporization — Latent heat involved in the change between liquid and vapor states.

Heat Pump — A refrigerating system employed to transfer heat into a space or substance. The condenser provides the heat, while the evaporator is arranged to pick up heat from air, water, etc. By shifting the flow of the refrigerant, a heat-pump system may also be used to cool the space.

Heating System — Any of several heating methods usually termed according to the method used in its generation, such as steam heating, warm-air heating, etc.

Heating System, Electric — Heating produced by the rise of temperature caused by the passage of an electric current through a conductor having a high resistance to the current flow. Residence electric-heating systems generally consist of one or several resistance units installed in a frame or casing, the degree of heating being thermostatically controlled.

Heating System, Steam — A heating system in which heat is transferred from a boiler or other source to the heating units by steam at, above, or below, atmospheric pressure.

Heating System, Vacuum — A two-pipe steam heating system equipped with the necessary accessory apparatus to permit operating the system below atmospheric pressure.

Heating System, Warm-Air — A warm-air heating plant consisting of a heating unit (fuel burning furnace) enclosed in a casing from which the heated air is distributed to various rooms of the building through ducts.

Hermetically Sealed Unit — A refrigerating unit containing the motor and compressor in a sealed container.

High-Pressure Cutout — A control device connected into the high-pressure part of a refrigerating system to stop the machine when the pressure becomes excessive.

High Side — That part of the refrigerating system containing the high-pressure refrigerant. Also the term used to refer to the condensing unit, consisting of the motor, compressor, condenser, and receiver mounted on a single base.

High-Side Float Valve — A float valve that floats in high-pressure liquid. Opens on an increase in liquid level.

Hold Over — In an evaporator, the ability to stay cold after heat removal from the evaporator stops.

Horsepower — A unit of power. Work done at the rate of 33,000 foot-pounds per minute, or 550 foot-pounds per second.

Humidifier — A device to add moisture to the air.

Humidify — To add water vapor to the atmosphere; to add water vapor or moisture to any material.

GLOSSARY

Humidistat — A control device actuated by changes in humidity and used for automatic control of relative humidity.

Humidity, Absolute — The definite amount of water contained in a definite quantity of air (usually measured in grains of water per pound or per cubic foot of air).

Humidity, Relative — The ratio of the water-vapor pressure of air compared to the vapor pressure it would have if saturated at its dry-bulb temperature. Very nearly the ratio of the amount of moisture contained in air compared to what it could hold at the existing temperature.

Humidity, Specific — The weight of vapor associated with one pound of dry air; also termed *humidity ratio*.

Hydrocarbons — A series of chemicals of similar chemical nature, ranging from methane (the main constituent of natural gas) through butane, octane, etc., to heavy lubricating oils. All are more or less flammable. Butane and isobutane have been used to a limited extent as refrigerants.

Hydrolysis — Reaction of a material, such as Freon-12 or methyl chloride, with water. Acid materials in general are formed.

Hydrostatic Pressure — The pressure due to liquid in a container that contains no gas space.

Hygroscope — *See* Humidistat.

Ice-Melting Equivalent — The amount of heat (144 Btu) absorbed by one pound of ice at 32°F in liquefying to water at 32°F.

Indirect Cooling System — *See* Brine System.

Infiltration — The leakage of air into a building or space.

Insulation — A material of low heat conductivity.

AIR CONDITIONING

Irritant Refrigerant — Any refrigerant that has an irritating effect on the eyes, nose, throat, or lungs.

Isobutane — A hydrocarbon refrigerant used to a limited extent. It is flammable.

Kilowatt — Unit of electrical power equal to 1000 watts, or 1.34 horsepower, approximately.

Lag of Temperature Control — The delay in action of a temperature-responsive element due to the time required for the temperature of the element to reach the surrounding temperature.

Latent Heat — The quantity of heat that may be added to a substance during a change of state without causing a temperature change.

Latent Heat of Evaporation — The quantity of heat required to change one pound of liquid into a vapor with no change in temperature. Reversible.

Leak Detector — A device used to detect refrigerant leaks in a refrigerating system.

Liquid — The state of a material in which its top surface in a vessel will become horizontal. Distinguished from solid or vapor forms.

Liquid Line — The tube or pipe that carries the refrigerant liquid from the condenser or receiver of a refrigerating system to a pressure-reducing device.

Liquid Receiver — That part of the condensing unit that stores the liquid refrigerant.

Load — The required rate of heat removal.

Glossary

Low-Pressure Control — An electric switch and pressure-responsive element connected into the suction side of a refrigerating unit to control the operation of the system.

Low Side — That part of a refrigerating system which normally operates under low pressure, as opposed to the high side. Also used to refer to the evaporator.

Low-Side Float — A valve operated by the low-pressure liquid, which opens at a low level and closes at a high level.

Main — A pipe or duct for distributing to or collecting conditioned air from various branches.

Manometer — A U-shaped liquid-filled tube for measuring pressure differences.

Mechanical Efficiency — The ratio of work done by a machine to the work done on it or energy used by it.

Mechanical Equivalent of Heat — An energy-conversion ratio of 778.18 foot-pounds = 1 Btu.

Methyl Chloride — A refrigerant having the chemical formula CH_3Cl.

Micron (μ) — A unit of length; the thousandth part of one millimeter or the millionth part of a meter.

Mollier Chart — A graphical representation of thermal properties of fluids, with total heat and entropy as coordinates.

Motor — A device for transforming electrical energy into mechanical energy.

Motor Capacitor — A device designed to improve the starting ability of single-phase induction motors.

Air Conditioning

Noncondensables — Foreign gases mixed with a refrigerant which cannot be condensed into liquid form at the temperatures and pressures at which the refrigerant condenses.

Oil Trap — A device to separate oil from the high-pressure vapor from the compressor. Usually contains a float valve to return the oil to the compressor crankcase.

Output — Net refrigeration produced by the system.

Ozone — The O_3 form of oxygen, sometimes used in air conditioning or cold-storage rooms to eliminate odors; can be toxic in concentrations of 0.5 ppm and over.

Packing — The stuffing around a shaft to prevent fluid leakage between the shaft and parts around the shaft.

Packless Valve — A valve that does not use packing to prevent leaks around the valve stem. Flexible material is usually used to seal against leaks and still permit valve movement.

Performance Factor — The ratio of the heat moved by a refrigerating system to heat equivalent of the energy used. Varies with conditions.

Phosphorous Pentoxide — An efficient drier material that becomes gummy reacting with moisture and hence is not used alone as a drying agent.

Pour Point, Oil — The temperature below which the oil surface will not change when the oil container is tilted.

Power — The rate of doing work measured in horsepower, watts, kilowatts, etc.

Power Factor, Electrical Devices — The ratio of watts to volt-amperes in an alternating current circuit.

GLOSSARY

Pressure — The force exerted per unit of area.

Pressure Drop — Loss in pressure, as from one end of a refrigerant line to the other, due to friction, static head, etc.

Pressure Gage — *See* Gage.

Pressure-Relief Valve — A valve or rupture member designed to relieve excessive pressure automatically.

Psychrometric Chart — A chart used to determine the specific volume, heat content, dew point, relative humidity, absolute humidity, and wet- and dry-bulb temperatures, knowing any two independent items of those mentioned.

Purging — The act of blowing out refrigerant gas from a refrigerant containing vessel usually for the purpose of removing noncondensables.

Pyrometer — An instrument for the measurement of high temperatures.

Radiation — The passage of heat from one object to another without warming the space between. The heat is passed by wave motion similar to light.

Refrigerant — The medium of heat transfer in a refrigerating system that picks up heat by evaporating at a low temperature and gives up heat by condensing at a higher temperature.

Refrigerating System — A combination of parts in which a refrigerant is circulated for the purpose of extracting heat.

Relative Humidity — The ratio of the water-vapor pressure of air compared to the vapor pressure it would have if saturated at its dry-bulb temperature. Very nearly the ratio of the amount of moisture contained in air compared to what it could hold at the existing temperature.

Relief Valve — A valve designed to open at excessively high pressures to allow the refrigerant to escape.

Resistance, Electrical — The opposition to electric-current flow, measured in ohms.

Resistance, Thermal — The reciprocal of thermal conductivity.

Room Cooler — A cooling element for a room. In air conditioning, a device for conditioning small volumes of air for comfort.

Rotary Compressor — A compressor in which compression is attained in a cylinder by rotation of a semiradial member.

Running Time — Usually indicates percent of time a refrigerant compressor operates.

Saturated Vapor — Vapor not superheated but of 100 percent quality, i.e., containing no unvaporized liquid.

Seal, Shaft — A mechanical system of parts for preventing gas leakage between a rotating shaft and a stationary crankcase.

Sealed Unit — *See* Hermetically Sealed Unit.

Shell and Tube — Pertaining to heat exchangers in which a coil of tubing or pipe is contained in a shell or container. The pipe is provided with openings to allow the passage of a fluid through it, while the shell is also provided with an inlet and outlet for a fluid flow.

Silica Gel — A drier material having the formula SiO_2.

Sludge — A decomposition product formed in a refrigerant due to impurities in the oil or due to moisture. Sludges may be gummy or hard.

GLOSSARY

Soda Lime — A material used for removing moisture. Not recommended for refrigeration use.

Solenoid Valve — A valve opened by a magnetic effect of an electric current through a solenoid coil.

Solid — The state of matter in which a force can be exerted in a downward direction only when not confined. As distinguished from fluids.

Solubility — The ability of one material to enter into solution with another.

Solution — The homogeneous mixture of two or more materials.

Specific Gravity — The weight of a volume of a material compared to the weight of the same volume of water.

Specific Heat — The quantity of heat required to raise the temperature of a definite mass of a material a definite amount compared to that required to raise the temperature of the same mass of water the same amount. May be expressed as Btu/per pound per degrees Fahrenheit.

Specific Volume — The volume of a definite weight of a material. Usually expressed in cubic feet per pound. The reciprocal of density.

Spray Pond — An arrangement for lowering the temperature of water by evaporative cooling of the water in contact with outside air; the water to be cooled is sprayed by nozzles into the space above a body of previously cooled water and allowed to fall by gravity into it.

Steam — Water in the vapor phase.

Steam Trap — A device for allowing the passage of condensate, or air and condensate, and preventing the passage of steam.

Subcooled — Cooled below the condensing temperature corresponding to the existing pressure.

Sublimation — The change from a solid to a vapor state without an intermediate liquid state.

Suction Line — The tube or pipe that carries refrigerant vapor from the evaporator to the compressor inlet.

Suction Pressure — Pressure on the suction side of the compressor.

Superheater — A heat exchanger used on flooded evaporators, wherein hot liquid on its way to enter the evaporator is cooled by supplying heat to dry and superheat the wet vapor leaving the evaporator.

Sweating — Condensation of moisture from the air on surfaces below the dew-point temperature.

System — A heating or refrigerating scheme or machine, usually confined to those parts in contact with the heating or refrigerating medium.

Temperature — Heat level or pressure. The thermal state of a body with respect to its ability to pick up heat from or pass heat to another body.

Thermal Conductivity — The ability of a material to conduct heat from one point to another. Indicated in terms of Btu/per hour per square foot per inches of thickness per degrees Fahrenheit.

Thermocouple — A device consisting of two electrical conductors having two junctions — one at a point whose temperature is to be measured, and the other at a known temperature. The temperature between the two junctions is determined by the material characteristics and the electrical potential setup.

Thermodynamics — The science of the mechanics of heat.

GLOSSARY

Thermometer — A device for indicating temperature.

Thermostat — A temperature-actuated switch.

Thermostatic Expansion Valve — A device to regulate the flow of refrigerant into an evaporator so as to maintain an evaporation temperature in a definite relationship to the temperature of a thermostatic bulb.

Ton of Refrigeration — Refrigeration equivalent to the melting of one ton of ice per twenty-four hours. 288,000 Btu per day, 12,000 Btu per hour, or 200 Btu per minute.

Total Heat — The total heat added to a refrigerant above an arbitrary starting point to bring it to a given set of conditions (usually expressed in Btu/per pound). For instance, in a super-heated gas, the combined heat added to the liquid necessary to raise its temperature from an arbitrary starting point to the evaporation temperature to complete evaporation, and to raise the temperature to the final temperature where the gas is superheated.

Total Pressure — In fluid flow, the sum of static pressure and velocity pressure.

Turbulent Flow — Fluid flow in which the fluid moves transversely as well as in the direction of the tube or pipe axis, as opposed to streamline or viscous flow.

Unit Heater — A direct-heating, factory-made, encased assembly including a heating element, fan, motor, and directional outlet.

Unit System — A system that can be removed from the user's premises without disconnecting refrigerant containing parts, water connection, or fixed electrical connections.

Unloader — A device in a compressor for equalizing high- and low-side pressures when the compressor stops and for a brief period after it starts so as to decrease the starting load on the motor.

Vacuum — A pressure below atmospheric, usually measured in inches of mercury below atmospheric pressure.

Valve — In refrigeration, a device for regulation of a liquid, air, or gas.

Vapor — A gas, particularly one near to equilibrium with the liquid phase of the substance, which does not follow the gas laws. Frequently used instead of gas for a refrigerant and, in general, for any gas below the critical temperature.

Viscosity — The property of a fluid to resist flow or change of shape.

Water Cooler — Evaporator for cooling water in an indirect refrigerating system.

Wax — A material that may separate when oil/refrigerant mixtures are cooled. Wax may plug the expansion valve and reduce heat transfer of the coil.

Wet-Bulb Depression — Different between dry- and wet-bulb temperatures.

Wet Compression — A system of refrigeration in which some liquid refrigerant is mixed with vapor entering the compressor so as to cause discharge vapors from the compressor to tend to be saturated rather than superheated.

Xylene — A flammable solvent, similar to kerosene, used for dissolving or loosening sludges and for cleaning compressors and lines.

Zero, Absolute, of Pressure — The pressure existing in a vessel that is entirely empty. The lowest possible pressure. Perfect vacuum.

GLOSSARY

Zero, Absolute, of Temperature — The temperature at which a body has no heat in it ($-459.6°F$, or $-273.1°C$).

Zone, Comfort (Average) — The range of effective temperature during which the majority of adults feel comfortable.

Index

A

Absorption-type air conditioning, 380
  cycle of operation, 381
Air
  compressor and motor, 154
    amperage, 160
    automatic expansion valve, 160
    evaporator temperature, 159
    filters, 158
    leaks, 157
    oiling, 157
    refrigerant
      leaks, 157
    troubleshooting, 161
    winter care, 158
  properties of, 10
    carbon monoxide, 10
    neon, 10
    ozone, 10
    water vapor, 25
Air circulation, 11
Air cleaning and filtering, 11
Air conditioner(s)
  capacity requirements, 120
  cooling capacity, 120
  dismantling, 149
  electrical, 140
    components, 142
      compressors, 143, 155
      fan motor, 147
      motors, 155
      overload protector, 144
      running capacitor, 146
      starting
        capacitors, 146
        relay, 144
      thermostats, 147

Air conditioner(s) (cont.)
  electrical components (cont.)
    unit-control switch, 148
    testing, 150
  filters, 158
  interior cleaning, 158
    compressor-motor
      overload, 154
      relay, 154
    current supply, 150
    fan motors, 147, 150
    relay check, 154
    starting capacitor, 153
  installation, 121
    casement window, 128
    console type, 133
    double-hung window, 123
    ducts, 138
    outside support bracket, 133
    wall method, 131
  service operations, 148
  troubleshooting chart, 163
  winter care, 158
Air conditioning
  absorption type, 380
  central (*see* Central air conditioning)
  control methods, 325
    automatic, 326
    basic control types, 327
    dampers, 336, 337
    duct thermostats, 332
    hygrostats, 333
    immersion thermostats, 332
    limit controls, 335
    manual control, 326
    pressure regulators, 335
    refrigerant control devices, 338
    relays, 336, 337
    room thermostats, 328

INDEX

Air conditioning (cont.)
  control methods (cont.)
    semiautomatic, 327
    twin-type thermostats, 331
    valves, 336
  properties of
    dew point, 28
    effective temperature, 28
    moisture, 27
    pressure, 21
    relative humidity, 26
Air duct(s)
  calculations, 96
  heat gains, 94
  resistance losses, 100
  sizes, 97
  systems, 92
  velocities, 75
Air filters
  classification of, 75
  installation, 81
  types of
    dry, 75
      plate, 76
    wet
      automatic viscous, 77
      flushing, 78
      immersion, 78
      manually cleaned, 77
      replacement, 77
Air filtration, 73
Air leakage, 65
Air volume, 106
Ammonia, 175
Atmospheric water vapor, 25
Attic-fan installation, 111
Automatic expansion valve, 338
Automobile air conditioners, 411
  component description, 418
    air distribution, 428, 431
    compressor, 419
    condenser, 419
    evaporators, 423, 433
    magnetic-clutch control, 425
    receiver-dehydrator, 419
    solenoid control, 424
    suction-throttle control, 426
    thermostatic expansion valve, 420, 432
  control system, 423–426
  installation procedure, 434
  service and maintenance, 428

Automobile air conditioners (cont.)
  service and maintenance (cont.)
    air
      distribution, 431
      output normal, 429
      output not normal, 429
    checking
      control systems, 430
      suction pressures, 430
      compressor clutch, 432
      condenser, 432
      defective compressor, 431
      evaporator, 423, 433
      refrigerant lines, 433
      suction-throttle valve, 433
      thermostatic expansion valve, 420
      use of receiver sight glass for diagnosis, 434
    system operation, 416
    troubleshooting guides, 438

B

Back pressure, 484
Barometer, 18
Boyle's law, 15
British thermal unit, 20

C

Calcium chloride, 177
Capillary tubes, 344
Carbon dioxide, 176
Carbon monoxide, 10
Celsius thermometric scale, 17
Central air conditioning
  absorption-type, 380
  features, 372
  air supply
    fans, 375
    system, 372
    air washers, 376
    cooling coils, 375
    control method, 377
    filters, 376
    heating coils, 374
    mixing chamber, 373
    preheaters, 374
    system selection, 379
    zone control, 379
  nonportable, 398
  purpose, 371

540

INDEX

Central air conditioning (cont.)
  reverse-cycle (heat pumps), 384
  system types, 372
Centrifugal compressors, 216
  construction features, 219
  lubrication system, 220
Chart
  comfort, 36
  psychometric, 30
Check valves, 298
Climate conditions, 354
Combined law of Boyle and Charles, 16
Combined wind and temperature
  forces, 68
Comfort chart, 36
Components of air, 10
Compound pressure gage, 21
Compressed refrigerant systems, 197
Compressor(s), 197, 203, 431
  connecting rods, 103, 209
  crankshaft, 210
  drive methods, 204
  installation, 221
  lubrication and cooling, 213
  maintenance, 221
    flowback protection, 224
    foundation, 221
    knocks, 226
    motor and belts, 221
    removal, 227
    replacement, 227
    shaft-seal leaks, 211, 225
    startup, 228
    stuck or tight compressor, 227
    valve maintenance, 225
    winter shutdown, 229
  methods of compression, 203
  pistons and piston rings, 204
  reciprocating, 198, 204
  service guide, 231
  shaft seals, 211
  troubleshooting, 232
  types of
    centrifugal, 216
    rotary, 213
      rotating blade, 213
      stationary blade, 213
  valves, 204
Condensers, 199
Condensing equipment, 243
  classification of, 243
    air-cooled, 244, 258

Condensing equipment (cont.)
  classification of (cont.)
    water-cooled, 244, 246
      double-tube type, 247
      shell-and-coil, 251
      shell-and-tube, 244
  condenser maintenance, 259
    charging the system, 262
    cleaning
      evaporative condensers, 263
      precautions, 260
      time, 263
    flushing the system, 263
    gravity and forced circulation, 265
    removing scale formation, 264
  cooling-water circuits, 246
  evaporative condenser, 251, 256
  head-pressure-control methods, 255
    air-cooled condensers, 244
    evaporative condensers, 251
    water-cooled condensers, 244,
      255
Conduction, 12
Control
  dampers, 337
  relays, 337
  valves, 338
Convection, 12
Cooling ponds, 310
Cooling towers, 306
Crankcase-pressure-regulating valve,
  344
Critical pressure, 16
Critical temperature, 16, 184

D

Dampers, 337
Dehumidifiers, 88
  adsorption-type, 91
  control of, 89
  electric, 89
Dehumidifying, 26
Dehydrators, 295
Design conditions, 353
  winter, 353
  summer, 353
  weather-oriented, 366
Dew point temperature, 28
Dismantling window air conditioners,
  149
Discharge lines, 317

541

INDEX

Drive methods, fan, 107
Dry air filter
  dry plate, 76
  packed type, 77
Dry-bulb
  temperature, 28
  thermometer, 20
Duct-system resistance, 71
Duct velocities, 95
Dust
  effect on health, 73
  sources, 74

E

Effective temperature, 28
English scale, 17
Ethyl chloride, 174
Evaporator(s), 271
  accessory equipment, 286
    check valves, 298
    dehydrator, 295
    heat exchangers, 286
    mufflers, 297
    oil separators, 291
    refrigerant driers, 294
    strainers, 296
    subcoolers, 297
    surge drums, 290
  calculations, 280
    exterior conditions, 282
    surface area determination, 282
  capacity requirements, 280
  classification, use of, 273
  coil construction, 271
  finned, 270
  maintenance, 284
    time clocks, 285
  pressure-regulating valves, 355
  refrigerant-control methods, 273
  types of
    finned coil, 270
    plain tubing, 271
Expansion valves, 200, 420

F

Fahrenheit thermometric scale, 17
Fan, 102
  air volume, 103
  applications, 109
    attic, 109
      installation, 111

Fan (cont.)
  applications (cont.)
    circulating, 109
    exhaust, 109
    kitchen, 109
  drive methods, 107
  horsepower requirements, 106
  motor, 147
    oiling, 153
  operation, 110, 113
  selection, 108
  types of
    centrifugal, 102, 106, 375
    propeller, 102, 106
    tubeaxial, 102, 106
    vaneaxial, 102, 106
Filter
  care, 154
  installation, 81
First aid, 192
Flue effect, 67
Freon refrigerants
  physical properties of, 179, 180
    amount of refrigerant circulated, 181
    flammability, 181
    volume displacement, 182
  types of
    Freon-11, 179
    Freon-12, 179
    Freon-13, 179
    Freon-21, 179
    Freon-22, 179
    Freon-113, 179
    Freon-114, 180
    Freon-115, 180
    Freon-502, 180
Freezol (isobutane), 177
Frequency of occurrence, weather conditions, 366
Fresh air requirements, 69
Fundamentals, 9–24

G

Gage pressure, 19
Grains of moisture, 27

H

Head pressure, 225, 482
Health, effect of dust on, 73

INDEX

Heat
  sensible, 13
  specific, 13
  types of
    latent heat, 13
    evaporation, 14
    fusion, 14
Heat exchangers, 287
Heat gains in air ducts, 94
Heat leakage, 41–64
Heat pumps, 384
  control equipment, 388
  cooling cycle, 384
  defrost cycle operation, 389
  heating cycle, 384
  indoor arrangement, 390
  installation and maintenance, 389
  refrigerant tubing, 392
  reversing-valve operation, 385
  system troubles, 393
  troubleshooting, 395
High-side float, 342
Horsepower requirements, 107
Humidifiers,
  air-operated, 87
  air-washer, 84
  electrically operated, 86
  pan-type, 84
Humidifying, 26
Humidity, 26
  control methods, 83
Hygrostats, 333

I

Interpolation between weather stations, 365
  adjustment for
    air-mass modification, 366
    elevation, 365
    vegetation, 366
Isobutane, 177

L

Latent heat, 13
  of evaporation, 14, 186
  of fusion, 14
Law(s)
  Boyle's, 15
  of Boyle and Charles, combined, 16
  of Charles, 15

Law(s) (cont.)
  of conservation of energy, 15
  of conservation of matter, 15
Leakage
  air, 65
  condensing unit, 157
  heat, 41–64
Limit controls, 335
Liquid lines, 315
Low-side float valve, 342

M

Magnetic clutch, 425
Maintenance, 481
  adding oil, 490
  adding refrigerant, 488
  back pressure, 484
  checking compressor valves, 496
  compressor service valves, 486
  evacuating system, 495
  gages, 486
    combination gages, 487
  general procedure, 481
  head pressures, 482
  leak test, 493
    electronic leak detector, 494
    halide torch, 494
  purging system of
    noncondensables, 492
  refrigerant charge, 485
  removing oil, 491
  removing refrigerant, 489
  suction pressure, 484
  system pumpdown, 491
  troubleshooting guide, 497
Measurements and measuring devices, 17
Mechanical ventilation, 71
Methyl chloride, 177
Motor(s), 445
  ac motor control, 461, 467
    squirrel cage, 461
    synchronous, 451, 466
    wound rotor, 463
  belt drives, 471
  compressor controls, 469
    condensing water, 471
    pressure, 470
    solenoid valves, 470
    temperature, 469
  maintenance, 472
    horsepower calculations, 473

543

INDEX

Motor(s) (cont.)
  maintenance (cont.)
    lubrication, 472
    power conversion factors, 475
  polyphase, 446
    squirrel cage, 446
      class A, 447
      class B, 447
      class C, 447
      class D, 447
    synchronous, 451
    wound rotor, 448
  single-phase, 452
    capacitors, 458
    -start, 454
    overload protectors, 467
    permanent split-capacitor, 456
    repulsion induction, 456
    -start, 457
    split phase, 453
    starting relays, 454, 455, 468
    two-valve connection, 455
  troubleshooting, 475
  voltage, 446
Mufflers, 297

N

Natural ventilation, 66
Neon, 10
Nonportable air conditioner(s), 398
  construction principles, 399
  heating coils, 374
  plumbing connections, 400
  suspended-type, 403
  troubleshooting guide, 406
  water-cooling tower, 400

O

Oil separators, 291
Operating pressures
  condenser, 183
  evaporator, 183
Oxygen, 10
Ozone, 10

P

Pressure gages, 21
Pressure regulators, 338
Pressure-temperature chart, 185

Pressure-temperature relationship, 15
Propeller fan, 102, 106
Psychrometric charts, 30
  instructions, 30
Psychometry, 25–40

R

Radiation, 12
Reamur thermometric scale, 17
Receivers, 199
Reciprocating compressors, 204
Refrigerant(s), 202
  capacity measurements, 21
  characteristics, 184
    critical pressure, 16
    critical temperature, 184
    latent heat of evaporation, 186
    power consumption, 187
    specific heat, 187
    volume of liquid circulated, 188
  classifications, 170
  control devices, 200
    automatic expansion valve, 338
    capillary tubes, 344
    crankcase-pressure-regulating valve, 344
    evaporator-pressure-regulating valve, 344
    high-side floats, 342
    low-side floats, 342
    reversing valves, 345
    sizing of thermostatic expansion valves, 341
    thermostatic expansion valve, 341
  cycle, 201
    processes
      compression, 183
      condensing, 183
      pressure reduction, 183
      vaporizing, 183
  cylinder capacity, 192
  definition of, 180, 202
  desirable properties of, 180
  driers, 294
  first aid, 192
    ammonia, 175
    Freons, 171, 178
    methyl chloride, 171, 173–180
    sulfur dioxide, 171, 173
  handling, 190, 191
  operating pressures, 183

544

INDEX

Refrigerants (cont.)
  piping arrangement, 315, 318
  pressure temperature chart, 172
    insulation, 320
    supports, 319
  properties of, 173
  simple system, 188
  storing and handling cylinders of, 191
  system service, 159
    automatic expansion valves, 160
    compressor amperage, 160
    evaporator temperature, 159
    refrigerant leaks, 160
  types of
    ammonia, 175, 193
    calcium chloride, 177
    carbon dioxide, 176
    ethyl chloride, 174
    Freons (see Freon refrigerants), 171, 178, 193
    Freezol, 177
    methyl chloride, 171, 173–180
    sulfur dioxide, 171, 173, 193
Relative humidity, 26
  measurements, 26
Resistance losses in duct system, 100
Reversing valves, 345
Roof ventilators, 68
Room air conditioner, 117–168
  capacity requirements, 120
  cooling capacity, 120
  installation methods, 121
  operation, 118
Rotary compressors, 213
Running capacitor, 146, 459

S

Sensible heat, 13
Simple refrigeration system, 189
Sling psychrometer, 19, 26
Specific heat, 13, 187
Specific volume, 28
Spray-cooling ponds, 312
Standard atmospheric pressure, 19
Starting capacitor, 146
Starting relay, 144
Strainers, 296
Subcoolers, 297
Sulfur dioxide, 171, 173, 193
Suction lines, 316

Suction lines (cont.)
  design of, 316
  oil circulation, 316
Suction pressure, 484
Superheat, 14, 275
Surge drums, 290

T

Tables
  air-duct calculations, 96
  air velocities, 95
  atmospheric pressure for various barometer readings, 22
  characteristics of typical refrigerants, 171
  climatic conditions for the United States, 351–369
  fresh-air requirements, 69
  heat leakage
    brick walls (4 in. cut stone), 54
    brick veneer (frame walls), 63
    brick veneer on hollow tile wall, 50
    cinder and concrete block wall, 48
      (brick veneer), 49
    concrete floors and ceilings, 61
    concrete floors on ground, 60
    concrete wall
      brick veneer, 46
      exterior stucco finish, 45
      4 in. cut stone, 47
      no exterior finish, 44
    frame floors and ceilings (no fill), 58
    frame floors and ceilings (with fill), 62
    hollow tile wall (4 in. cut stone veneer), 51
    hollow tile wall (stucco exterior), 52
    interior walls and plastered partitions (no fill), 57
    limestone or sandstone wall, 53
    masonry partitions, 59
    solid brick wall (no exterior finish), 55
    wood siding clapboard frame or shingle walls, 56
  quantities of refrigerant
    circulated per minute under standard ton conditions, 182

545

INDEX

Tables (cont.)
  recommended air velocities in ducts, 37
  refrigerant pressure vs. temperature, 172
  room-cooling requirements in Btu, 121
  volume of air required, 106
Temperature-humidity index, 29
Thermometer
  operating principle of, 17
  scales on, 17
  types of, 17
Thermostatic expansion valve, 338, 420
Thermostats, 147
  duct, 332
  immersion, 332
  room, 328
  twin-type, 331
Ton of refrigeration, 21
Total heat content, 29
Troubleshooting guides
  automobile air conditioner, 438
  compressor, 232
  electrical system, 475
  maintenance, 483
  reversing valve, 385
  room air conditioner, 117
  self-contained air conditioner, 406

U

Unit-control switch, 148

V

Ventilation requirements, 65–116
Ventilator capacity, 69
Volume of air required, 71

W

Water-cooling systems, 305
  discharge lines, 317
  liquid lines, 315
  new developments, 312
  piping, 314, 315
    methods, 318
    principles, 318

Water-cooling systems (cont.)
  purpose of, 305
  suction lines, 316
  towers, 306
  types of
    cooling ponds, 310
    spray-cooling ponds, 312
    water towers, 306
  vapor, 25
Weather data and design conditions
  factors, 366
  indoor, 351
  interpolation between stations, 365
  adjustment for
    air mass modification, 366
    elevation, 365
    vegetation, 366
  outdoor, 353
    summer, 353
    winter, 353
Wet-air filter
  types of
    automatic viscous, 78
    flushing, 79
    immersion, 78
    manually cleaned, 77
    replacement, 77
Wet-bulb
  depression, 29
  temperature, 29
  thermometer, 20
Wind forces, 66
Window installation
  casement, 128
    alternative method, 131
    concrete block or brick construction, 132
    frame construction, 131
    wall installations, 131
  double-hung, 123

Y

Year round air conditioning, 371

Z

Zone control, 379

546

AUDEL®

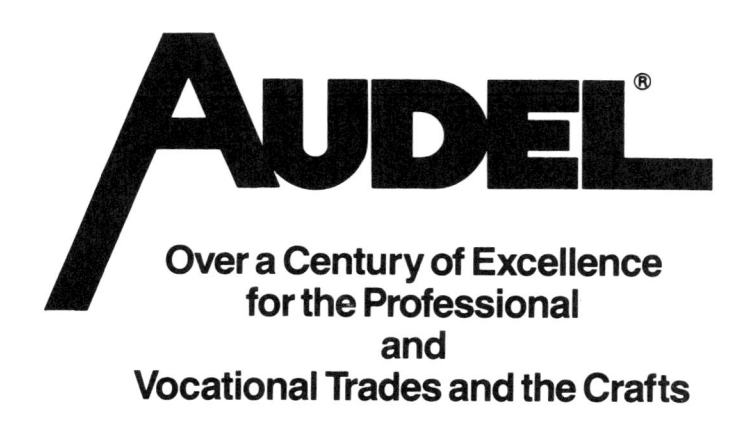

**Over a Century of Excellence
for the Professional
and
Vocational Trades and the Crafts**

Order now from your local bookstore
or use the convenient order form
at the back of this book.

AUDEL

These fully illustrated, up-to-date guides and manuals mean a better job done for mechanics, engineers, electricians, plumbers, carpenters, and all skilled workers.

CONTENTS

Electrical II	Carpentry and Construction VI
Machine Shop and	Woodworking VI
Mechanical Trades III	Maintenance and Repair VII
Plumbing IV	Automotive and Engines VII
HVAC V	Drafting VIII
Pneumatics and Hydraulics V	Hobbies VIII

ELECTRICAL

House Wiring (Sixth Edition)
ROLAND E. PALMQUIST
5 1/2 x 8 1/4   *Hardcover*   150 Illus.
ISBN: *[NEW EDITION FOR 1991]*
The ... National
Elect... apply to residential
wiring ... with examples and illustrations.

Practical Electricity
(Fifth Edition)
ROBERT G. MIDDLETON;
revised by L. DONALD MEYERS
5 1/2 x 8 1/4   Hardcover   512 pp.   335 Illus.
ISBN: 0-02-584561-6 $19.95
The fundamentals of electricity for electrical workers, apprentices, and others requiring concise information about electric principles and their practical applications.

Guide to the 1990 National Electrical Code
ROLAND E. PALMQUIST;
revised by PAUL ROSENBERG
5 1/2 x 8 1/4   Hardcover   664 pp.   230 Illus.
ISBN: 0-02-594565-3 $24.95
The most authoritative guide available to interpreting the National Electrical Code for electricians, contractors, electrical inspectors, and homeowners. Examples and illustrations.

Mathematics for Electricians and Electronics Technicians
REX MILLER
5 1/2 x 8 1/4   Hardcover   312 pp.   115 Illus.
ISBN: 0-8161-1700-4 $14.95
Mathematical concepts, formulas, and problem-solving techniques utilized on-the-job by electricians and those in electronics and related fields.

Fractional-Horsepower Electric Motors
REX MILLER and
MARK RICHARD MILLER
5 1/2 x 8 1/4   Hardcover   436 pp.   285 Illus.
ISBN: 0-672-23410-6 $15.95
The installation, operation, maintenance, repair, and replacement of the small-to-moderate-size electric motors that power home appliances and industrial equipment.

Electric Motors (Fourth Edition)
EDWIN P. ANDERSON;
revised by REX MILLER
5 1/2 x 8 1/4   *[NEW EDITION FOR 1991]*   405 Illus.
ISBN: ...
Insta... ..., and repair of all
types ... ic motors.

II

Home Appliance Servicing
(Fourth Edition)
EDWIN P. ANDERSON;
revised by REX MILLER
5 1/2 x 8 1/4   Hardcover   640 pp.   345 Illus.
ISBN: 0-672-23379-7 $22.50

The essentials of testing, maintaining, and repairing all types of home appliances.

Television Service Manual
(Fifth Edition)
ROBERT G. MIDDLETON;
revised by JOSEPH G. BARRILE
5 1/2 x 8 1/4   Hardcover   512 pp.   395 Illus.
ISBN: 0-672-23395-9 $16.95

A guide to all aspects of television transmission and reception, including the operating principles of black and white and color receivers. Step-by-step maintenance and repair procedures.

Electrical Course for Apprentices and Journeymen
(Third Edition)
ROLAND E. PALMQUIST
5 1/2 x 8 1/4   Hardcover   478 pp.   290 Illus.
ISBN: 0-02-594550-5 $19.95

This practical course in electricity for those in formal training programs or learning on their own provides a thorough understanding of operational theory and its applications on the job.

Questions and Answers for Electricians Examinations
(Tenth Edition)
Revised by PAUL ROSENBERG
5 1/2 x 8 1/4   Hardcover   316 pp.   110 Illus.
ISBN: 0-02-604955-4 $22.95

Based on the 1990 National Electrical Code, this book reviews the subjects included in the various electricians examinations—apprentice, journeyman, and master. Question and Answer format.

MACHINE SHOP AND MECHANICAL TRADES

Machinists Library
(Fourth Edition, 3 Vols.)
REX MILLER
5 1/2 x 8 1/4   Hardcover   1352 pp.   1120 Illus.
ISBN: 0-672-23380-0 $52.95

An indispensable three-volume reference set for machinists, tool and die makers, machine operators, metal workers, and those with home workshops. The principles and methods of the entire field are covered in an up-to-date text, photographs, diagrams, and tables.

Volume I: Basic Machine Shop
REX MILLER
5 1/2 x 8 1/4   Hardcover   392 pp.   375 Illus.
ISBN: 0-672-23381-9 $17.95

Volume II: Machine Shop
REX MILLER
5 1/2 x 8 1/4   Hardcover   528 pp.   445 Illus.
ISBN: 0-672-23382-7 $19.95

Volume III: Toolmakers Handy Book
REX MILLER
5 1/2 x 8 1/4   Hardcover   432 pp.   300 Illus.
ISBN: 0-672-23383-5 $14.95

Mathematics for Mechanical Technicians and Technologists
JOHN D. BIES
5 1/2 x 8 1/4   Hardcover   342 pp.   190 Illus.
ISBN: 0-02-510620-1 $17.95

The mathematical concepts, formulas, and problem-solving techniques utilized on the job by engineers, technicians, and other workers in industrial and mechanical technology and related fields.

Millwrights and Mechanics Guide (Fourth Edition)
CARL A. NELSON
5 1/2 x 8 1/4   Hardcover   1,040 pp.   880 Illus.
ISBN: 0-02-588591-x $29.95

The most comprehensive and authoritative guide available for millwrights, mechanics, maintenance workers, riggers, shop workers, foremen, inspectors, and superintendents on plant installation, operation, and maintenance.

Welders Guide (Third Edition)
JAMES E. BRUMBAUGH
5 1/2 x 8 1/4   Hardcover   960 pp.   615 Illus.
ISBN: 0-672-23374-6 $23.95

The theory, operation, and maintenance of all welding machines. Covers gas welding equipment, supplies, and process; arc welding equipment, supplies, and process; TIG and MIG welding; and much more.

Welders/Fitters Guide
HARRY L. STEWART
8 1/2 x 11   Paperback   160 pp.   195 Illus.
ISBN: 0-672-23325-8 $7.95
Step-by-step instruction for those training to become welders/fitters who have some knowledge of welding and the ability to read blueprints.

Sheet Metal Work
JOHN D. BIES
5 1/2 x 8 1/4   Hardcover   456 pp.   215 Illus.
ISBN: 0-8161-1706-3 $19.95
An on-the-job guide for workers in the manufacturing and construction industries and for those with home workshops. All facets of sheet metal work detailed and illustrated by drawings, photographs, and tables.

Power Plant Engineers Guide
(Third Edition)
FRANK D. GRAHAM;
revised by CHARLIE BUFFINGTON
5 1/2 x 8 1/4   Hardcover   960 pp.   530 Illus.
ISBN: 0-672-23329-0 $27.50
This all-inclusive, one-volume guide is perfect for engineers, firemen, water tenders, oilers, operators of steam and diesel-power engines, and those applying for engineer's and firemen's licenses.

Mechanical Trades Pocket Manual (Third Edition)
CARL A. NELSON
4 x 6   Paperback   364 pp.   255 Illus.
ISBN: 0-02-588665-7 $14.95
A handbook for workers in the industrial and mechanical trades on methods, tools, equipment, and procedures. Pocket-sized for easy reference and fully illustrated.

PLUMBING

Plumbers and Pipe Fitters Library (Fourth Edition, 3 Vols.)
CHARLES N. McCONNELL
5 1/2 x 8 1/4   Hardcover   952 pp.   560 Illus.
ISBN: 0-02-582914-9 $68.45
This comprehensive three-volume set contains the most up-to-date information available for master plumbers, journeymen, apprentices, engineers, and those in the building trades. A detailed text and clear diagrams, photographs, and charts and tables treat all aspects of the plumbing, heating, and air conditioning trades.

Volume I: Materials, Tools, Roughing-In
CHARLES N. McCONNELL;
revised by TOM PHILBIN
5 1/2 x 8 1/4   Hardcover   304 pp.   240 Illus.
ISBN: 0-02-582911-4 $20.95

Volume II: Welding, Heating, Air Conditioning
CHARLES N. McCONNELL;
revised by TOM PHILBIN
5 1/2 x 8 1/4   Hardcover   384 pp.   220 Illus.
ISBN: 0-02-582912-2 $22.95

Volume III: Water Supply, Drainage, Calculations
CHARLES N. McCONNELL;
revised by TOM PHILBIN
5 1/2 x 8 1/4   Hardcover   264 pp.   100 Illus.
ISBN: 0-02-582913-0 $20.95

Home Plumbing Handbook
(Third Edition)
CHARLES N. McCONNELL
8 1/2 x 11   Paperback   200 pp.   100 Illus.
ISBN: 0-672-23413-0 $14.95
An up-to-date guide to home plumbing installation and repair.

The Plumbers Handbook
(Seventh Edition)
JOSEPH P. ALMOND, SR.
4 x 6   Paperback   352 pp.   170 Illus.
ISBN: 0-672-23416-5
NEW EDITION FOR 1991
A handy sized book for plumbers and pipe fitters, and apprentices. It has a rugged binding for use on the job, and fits in the tool box or conveniently in the pocket.

Questions and Answers for Plumbers Examinations (Second Edition)
JULES ORAVETZ
5 1/2 x 8 1/4   Paperback   256 pp.   145 Illus.
ISBN: 0-8161-...
NEW EDITION FOR 1991
A study guide for those preparing to take a licensing examination for master, apprentice, journeyman, or plumber. Question and Answer format.

IV

HVAC

Air Conditioning: Home and Commercial (Fourth Edition)
EDWIN P. ANDERSON;
revised by REX MILLER
5 1/2 x 8 1/4 Hardcover 528 pp. 180 Illus.
ISBN: 0-02-584885-2 $29.95
A guide to the construction, installation, operation, maintenance, and repair of home, commercial, and industrial air conditioning systems.

Heating, Ventilating, and Air Conditioning Library
(Second Edition, 3 Vols.)
JAMES E. BRUMBAUGH
5 1/2 x 8 1/4 Hardcover 1,840 pp. 1,275 Illus.
ISBN: 0-672-23388-6 $53.85
An authoritative three-volume reference library for those who install, operate, maintain, and repair HVAC equipment commercially, industrially, or at home.

Volume I: Heating Fundamentals, Furnaces, Boilers, Boiler Conversions
JAMES E. BRUMBAUGH
5 1/2 x 8 1/4 Hardcover 656 pp. 405 Illus.
ISBN: 0-672-23389-4 $17.95

Volume II: Oil, Gas and Coal Burners, Controls, Ducts, Piping, Valves
JAMES E. BRUMBAUGH
5 1/2 x 8 1/4 Hardcover 592 pp. 455 Illus.
ISBN: 0-672-23390-8 $17.95

Volume III: Radiant Heating, Water Heaters, Ventilation, Air Conditioning, Heat Pumps, Air Cleaners
JAMES E. BRUMBAUGH
5 1/2 x 8 1/4 Hardcover 592 pp. 415 Illus.
ISBN: 0-672-23391-6 $17.95

Oil Burners (Fifth Edition)
EDWIN M. FIELD
5 1/2 x 8 1/4 Hardcover 360 pp. 170 Illus.
ISBN: 0-02-537745-0 $29.95
An up-to-date sourcebook on the construction, installation, operation, testing, servicing, and repair of all types of oil burners, both industrial and domestic.

Refrigeration: Home and Commercial (Fourth Edition)
EDWIN P. ANDERSON;
revised by REX MILLER
5 1/2 x 8 1/4 Hardcover 768 pp. 285 Illus.
ISBN: 0-02-584875-5 $34.95
A reference for technicians, plant engineers, and the homeowner on the installation, operation, servicing, and repair of everything from single refrigeration units to commercial and industrial systems.

PNEUMATICS AND HYDRAULICS

Hydraulics for Off-the-Road Equipment (Second Edition)
HARRY L. STEWART;
revised by TOM PHILBIN
5 1/2 x 8 1/4 Hardcover 256 pp. 175 Illus.
ISBN: 0-8161-1701-2 $13.95
This complete reference manual on heavy equipment covers hydraulic pumps, accumulators, and motors; force components; hydraulic control components; filters and filtration, lines and fittings, and fluids; hydrostatic transmissions; maintenance; and troubleshooting.

Pneumatics and Hydraulics
(Fourth Edition)
HARRY L. STEWART;
revised by TOM STEWART
5 1/2 x 8 1/4 Hardcover 512 pp. 315 Illus.
ISBN: 0-672-23412-2 $19.95
The principles and applications of fluid power. Covers pressure, work, and power; general features of machines; hydraulic and pneumatic symbols; pressure boosters; air compressors and accessories; and much more.

Pumps (Fourth Edition)
HARRY STEWART;
revised by TOM *NEW EDITION FOR 1991*
5 1/2 x 8 1/4 Hardcover ?60 Illus.
ISBN: ...
The principles of day-to-day operation of pumps ... controls, and hydraulics are thoroughly detailed and illustrated.

v

CARPENTRY AND CONSTRUCTION

Carpenters and Builders Library
(Fifth Edition, 4 Vols.)
JOHN E. BALL;
revised by TOM PHILBIN
5 1/2 x 8 1/4 Hardcover 1,224 pp. 1,010 Illus.
ISBN: 0-02-506450-9 $43.95

This comprehensive four-volume library has set the professional standard for decades for carpenters, joiners, and woodworkers.

Volume I: Tools, Steel Square, Joinery
JOHN E. BALL;
revised by TOM PHILBIN
5 1/2 x 8 1/4 Hardcover 384 pp. 345 Illus.
ISBN: 0-672-23365-7 $10.95

Volume II: Builders Math, Plans, Specifications
JOHN E. BALL;
revised by TOM PHILBIN
5 1/2 x 8 1/4 Hardcover 304 pp. 205 Illus.
ISBN: 0-672-23366-5 $10.95

Volume III: Layouts, Foundations, Framing
JOHN E. BALL;
revised by TOM PHILBIN
5 1/2 x 8 1/4 Hardcover 272 pp. 215 Illus.
ISBN: 0-672-23367-3 $10.95

Volume IV: Millwork, Power Tools, Painting
JOHN E. BALL;
revised by TOM PHILBIN
5 1/2 x 8 1/4 Hardcover 344 pp. 245 Illus.
ISBN: 0-672-23368-1 $10.95

Complete Building Construction
(Second Edition)
JOHN PHELPS;
revised by TOM PHILBIN
5 1/2 x 8 1/4 Hardcover 744 pp. 645 Illus.
ISBN: 0-672-23377-0 $22.50

Constructing a frame or brick building from the footings to the ridge. Whether the building project is a tool shed, garage, or a complete home, this single fully illustrated volume provides all the necessary information.

Complete Roofing Handbook
JAMES E. BRUMBAUGH
5 1/2 x 8 1/4 Hardcover 536 pp. 510 Illus.
ISBN: 0-02-517850-4 $29.95

Covers types of roofs; roofing and reroofing; roof and attic insulation and ventilation; skylights and roof openings; dormer construction; roof flashing details; and much more.

Complete Siding Handbook
JAMES E. BRUMBAUGH
5 1/2 x 8 1/4 Hardcover 512 pp. 450 Illus.
ISBN: 0-02-517880-6 $24.95

This companion volume to the *Complete Roofing Handbook* includes comprehensive step-by-step instructions and accompanying line drawings on every aspect of siding a building.

Masons and Builders Library
(Second Edition, 2 Vols.)
LOUIS M. DEZETTEL;
revised by TOM PHILBIN
5 1/2 x 8 1/4 Hardcover 688 pp. 500 Illus.
ISBN: 0-672-23401-7 $27.95

This two-volume set provides practical instruction in bricklaying and masonry. Covers brick; mortar; tools; bonding; corners, openings, and arches; chimneys and fireplaces; structural clay tile and glass block; brick walls; and much more.

Volume 1: Concrete, Block, Tile, Terrazzo
LOUIS M. DEZETTEL;
revised by TOM PHILBIN
5 1/2 x 8 1/4 Hardcover 304 pp. 190 Illus.
ISBN: 0-672-23402-5 $14.95

Volume 2: Bricklaying, Plastering, Rock Masonry, Clay Tile
LOUIS M. DEZETTEL;
revised by TOM PHILBIN
5 1/2 x 8 1/4 Hardcover 384 pp. 310 Illus.
ISBN: 0-672-23403-3 $14.95

WOODWORKING

Wood Furniture: Finishing, Refinishing, Repairing
(Second Edition)
JAMES E. BRUMBAUGH
5 1/2 x 8 1/4 Hardcover 352 pp. 185 Illus.
ISBN: 0-672-23409-2 $12.95

A fully illustrated guide to repairing furniture and finishing and refinishing wood surfaces. Covers tools and supplies; types of wood; veneering; inlaying; repairing, restoring, and stripping; wood preparation; and much more.

Woodworking and Cabinetmaking
F. RICHARD BOLLER
5 1/2 x 8 1/4 Hardcover 360 pp. 455 Illus.
ISBN: 0-02-512800-0 $18.95

Essential information on all aspects of working with wood. Step-by-step procedures for woodworking projects are accompanied by detailed drawings and photographs.

MAINTENANCE AND REPAIR

Building Maintenance
(Second Edition)
JULES ORAVETZ
*5 1/2 x 8 1/4   Paperback   384 pp.   210 Illus.
ISBN: 0-672-23278-2 $11.95*
Professional maintenance procedures used in office, educational, and commercial buildings. Covers painting and decorating; plumbing and pipe fitting; concrete and masonry; and much more.

Gardening, Landscaping and Grounds Maintenance
(Third Edition)
JULES ORAVETZ
*5 1/2 x 8 1/4   Hardcover   424 pp.   340 Illus.
ISBN: 0-672-23417-3 $15.95*
Maintaining lawns and gardens as well as industrial, municipal, and estate grounds.

Home Maintenance and Repair: Walls, Ceilings and Floors
GARY D. BRANSON
*8 1/2 x 11   Paperback   80 pp.   80 Illus.
ISBN: 0-672-23281-2 $6.95*
The do-it-yourselfer's guide to interior remodeling with professional results.

Painting and Decorating
REX MILLER and GLEN E. BAKER
*5 1/2 x 8 1/4   Hardcover   464 pp.   325 Illus.
ISBN: 0-672-23405-x $18.95*
A practical guide for painters, decorators, and homeowners to the most up-to-date materials and techniques in the field.

Tree Care (Second Edition)
JOHN M. HALLER
*8 1/2 x 11   Paperback   224 pp.   305 Illus.
ISBN: 0-02-062870-6 $16.95*
The standard in the field. A comprehensive guide for growers, nursery owners, foresters, landscapers, and homeowners to planting, nurturing and protecting trees.

Upholstering (Updated)
JAMES E. BRUMBAUGH
*5 1/2 x 8 1/4   Hardcover   400 pp.   380 Illus.
ISBN: 0-672-23372-x $15.95*
The esentials of upholstering fully explained and illustrated for the professional, the apprentice, and the hobbyist.

AUTOMOTIVE AND ENGINES

Diesel Engine Manual
(Fourth Edition)
PERRY O. BLACK;
revised by WILLIAM E. SCAHILL
*5 1/2 x 8 1/4   Hardcover   512 pp.   255 Illus.
ISBN: 0-672-23371-1 $15.95*
The principles, design, operation, and maintenance of today's diesel engines. All aspects of typical two- and four-cycle engines are thoroughly explained and illustrated by photographs, line drawings, and charts and tables.

Gas Engine Manual
(Third Edition)
EDWIN P. ANDERSON;
revised by CHARLES G. FACKLAM
*5 1/2 x 8 1/4   Hardcover   424 pp.   225 Illus.
ISBN: 0-8161-1707-1 $12.95*
How to operate, maintain, and repair gas engines of all types and sizes. All engine parts and step-by-step procedures are illustrated by photographs, diagrams, and troubleshooting charts.

Small Gasoline Engines
REX MILLER and
MARK RICHARD MILLER
*5 1/2 x 8 1/4   Hardcover   640 pp.   525 Illus.
ISBN: 0-672-23414-9 $16.95*
Practical information for those who repair, maintain, and overhaul two- and four-cycle engines—including lawn mowers, edgers, grass sweepers, snowblowers, emergency electrical generators, outboard motors, and other equipment with engines of up to ten horsepower.

Truck Guide Library (3 Vols.)
JAMES E. BRUMBAUGH
*5 1/2 x 8 1/4   2,144 pp.   1,715 Illus.
ISBN: 0-672-23392-4 $50.95*

This three-volume set provides the most comprehensive, profusely illustrated collection of information available on truck operation and maintenance.

Volume 1: Engines
JAMES E. BRUMBAUGH
5 1/2 x 8 1/4   Hardcover   416 pp.   290 Illus.
ISBN: 0-672-23356-8 $16.95

Volume 2: Engine Auxiliary Systems
JAMES E. BRUMBAUGH
5 1/2 x 8 1/4   Hardcover   704 pp.   520 Illus.
ISBN: 0-672-23357-6 $16.95

Volume 3: Transmissions, Steering, and Brakes
JAMES E. BRUMBAUGH
5 1/2 x 8 1/4   Hardcover   1,024 pp.   905 Illus.
ISBN: 0-672-23406-8 $16.95

DRAFTING

Industrial Drafting
JOHN D. BIES
5 1/2 x 8 1/4   Hardcover   544 pp.   Illus.
ISBN: 0-02-510610-4 $24.95

Professional-level introductory guide for practicing drafters, engineers, managers, and technical workers in all industries who use or prepare working drawings.

Answers on Blueprint Reading
(Fourth Edition)
ROLAND PALMQUIST; revised by THOMAS J. MORRISEY
5 1/2 x 8 1/4   Hardcover   320 pp.   275 Illus.
ISBN: 0-8161-1704-7 $12.95

Understanding blueprints of machines and tools, electrical systems, and architecture. Question and answer format.

HOBBIES

Complete Course in Stained Glass
PEPE MENDEZ
8 1/2 x 11   Paperback   80 pp.   50 Illus.
ISBN: 0-672-23287-1 $8.95

The tools, materials, and techniques of the art of working with stained glass.